Concrete Construction

Akhtar Surahyo

Concrete Construction

Practical Problems and Solutions

 Springer

Akhtar Surahyo
IBI Group
Toronto, ON, Canada

ISBN 978-3-030-10509-9   ISBN 978-3-030-10510-5   (eBook)
https://doi.org/10.1007/978-3-030-10510-5

Library of Congress Control Number: 2018966808

This Springer imprint is published by the registered company Springer Nature Switzerland AG
The registered company address is: Gewerbestrasse 11, 6330 Cham, Switzerland

Dedicated To
My Father
Mohammad Ahsan Surahyo
(Deceased 1971)

Foreword to the First Edition

Cement concrete is the most abundantly used material in all types of construction, particularly in developing countries, primarily because of the cheap availability of its ingredients, and its versatility of being formed into any practically desired shape to give the intended structure the size, strength, and properties required for its satisfactory performance. Through proper selection of materials and adoption of the correct construction method appropriate for each individual construction project, concrete can be prepared to satisfy all the requirements of strength and performance in any kind of location and environmental conditions. Good concrete is, no doubt, required to achieve this purpose for which standard specifications and procedures are available in many books and manuals of practices.

Many engineers and technical staff working on construction sites do not have access to this large volume of literature, besides which they are often unaware of the results of research carried out on this subject in the recent past. Quite often, they are ignorant of the specific problems encountered on construction sites. If not properly handled, these problems can lead to poor quality of construction, and sometimes even failures.

Based on his practical experience, Mr. Surahyo has tried to put in one volume, and in a concise form, both the aspects: (a) the properties required of good quality concrete-making materials and, (b) the factors and mistakes that create construction problems on the sites. He has also quoted the relevant British and American standards/specifications with a view to creating awareness among readers about the importance of adherence to these standards, besides improving their knowledge of both the quality of concrete and the correct construction practices.

This book is written in simple and easy-to-understand language for the benefit of engineers and technicians employed on construction sites. It will also be useful for engineering and technical students in their subjects of Concrete Technology and Construction Engineering.

Professor Emeritus, Mehran University
of Engineering and Technology,   Dr. A. F. Abbasi,
Jamshoro, Sindh, Pakistan

Preface to the Second Edition

The aim of this thorough and comprehensive update of 2002 edition by the author is twofold: addresses concrete technology as well as concrete construction highlighting the principal causes of concrete deterioration along with protective measures in the light of Canadian, American, and British standards. Apart from minor changes in the text, many new topics are included explaining various topics in depth with examples and pictures, so that the reader looking for any information about concrete construction should find at one place.

The book is designed to provide the reader with advanced, in-depth knowledge of both the theory and practical application of concrete technology to prepare the civil engineering students for a variety of senior roles. The book is equally useful for all those civil engineering professionals who are involved in construction process. Providing references from the highly adopted standards within the construction industry, the book allows the reader to embrace knowledge enabling him to work on site with advanced understanding of concrete construction. The book would also serve as a basic tool of information and knowledge for immigrant professionals.

In practice, concrete construction is based on various codes and standards. The book will enable the reader to understand different nomenclature used for the same materials in different parts of the world. For example, in the United States, ASTM-C150 refers various types of cement as Type-I, Type-II, Type-III, Type-IV, and Type-V, whereas in Canada CSA refers differently same type of cements as GU, HE, MS, HS, MH, and LH, and in the United Kingdom, the categories of cements are divided differently in five types based on their composition as per European standard BS EN 197-1-2000, which has replaced most of the British standards. Similarly, types, grades, and sizes of steel reinforcement are also referred differently by ASTM, CAN/CSA, and BS-4449. Additionally, grading requirements for concrete aggregates are also different in the United States, the United Kingdom, and Canada. Hence, this book will provide the reader practical information with respect to concrete construction in the light of various standard codes being used in many parts of the world.

The book covers the basic information about normal concrete in its first part, about its grades, and about 14 different kinds of modified concretes including polymer concretes, fiber-reinforced concrete, sulfur concrete, and high-performance and ultrahigh-performance concretes. It provides complete information on constituent materials of concrete such as Portland cement, its chemical composition, types of cement used in various countries, blended cements, and special cements including physical properties of Portland cement, such as fineness, setting time, soundness, hydration, and strength with references to ACI, CSA, and BS standards. About 13 admixtures and additives are explained in the book including requirements for the use of curing and mixing water and coarse and fine aggregates including recycled concrete aggregates. The book further highlights the physical properties of concrete: workability and its methods of measurement, segregation, bleeding, plastic shrinkage, and air entrainment with various methods of measurement. The book also describes the structural properties of the concrete: strength, durability, permeability and porosity, shrinkage, modulus of elasticity, Poisson's ratio, creep, and thermal properties of concrete. The information on concrete British and ACI mix design methods with examples is also provided.

On the other hand, the second part of the book highlights the principal causes of concrete deterioration along with protective measures, resulting from incorrect selection of constituent materials and poor construction methods such as the use of excess water content, segregation, poor compaction, and inadequate cover to reinforcement, incorrect placement of steel, inadequate formwork, poor curing, inadequate mixing, and incorrect placement of construction joints. The book further addresses external factors affecting concrete production such as freezing and thawing, wetting and drying, leaching, abrasion, overloading, settlement, fire resistance, and the type of joints including causes of joint failures. Additionally, the book provides information on chemical attacks such as chloride attack, sulfate attack, carbonation, alkali-aggregate reaction, and acid attack. The book further explains hot and cold weather effects, corrosion of embedded metals in concrete, and various errors in designing and detailing. A chapter discussing the methods for achieving quality control and ensuring quality assurance in concrete construction is also added.

Many new standards have been introduced in this second edition with updating of some old standards. However, still some old standards and references are retained because as per A. M. Neville (*Properties of Concrete*—Fifth edition), they contribute to knowledge of what is desirable in the understanding of a relevant property. Secondly, the old references contain the development of our knowledge. Additionally, the British standards are being replaced by European standards denoted by BS EN, and this process will be completed up to 2020. Hence, wherever possible, revised BS standards to BS EN are used; otherwise the old BS standards are retained. However, the author has provided information about old BS standard used in this book and its related revised standard at the end of the book under Annexture. Given continuing evolution of standards, for specific use, the reader can refer to the new or revised standards specified in "Bibliography" section provided at the end of each chapter and Annexture, if required.

Mississauga, ON   Akhtar Surahyo, P. Eng

Acknowledgments

I would like to recognize with much appreciation my friends, colleagues, and seniors in the Ministry of Housing (Technical Affairs Directorate), State of Bahrain, and particularly my daughter, whose brilliant technical contribution to several figures of this book (continued from the first edition) has served to ensure its quality and integrity.

I would also like to record my gratitude to Dr. A. F. Abbasi, Professor Emeritus, Mehran University of Engineering and Technology, Jamshoro, Sindh Pakistan, for kindly writing the foreword to the first edition of this book and for his guidance and encouragement.

I would also like to thank Gour Pado Saha, P.Eng., Quality Control Manager, from the Coco Paving Inc., Toronto, Ontario, who generously donated his time and provided technical review of the complete book including with additional input.

I gratefully acknowledge the courtesy extended by the following and others named in the text, who have kindly given permission for the use of the copyright material.

ACI—American Concrete Institute
CSA Group—Canadian Standards Association Group
ASTM International—American Society for Testing and Materials
BSI—British Standards Institute
PCA—Portland Cement Association
CIRIA—Construction Industry Research and Information Association
HMSO—Her Majesty's Stationery Office, UK
Tata McGraw-Hill
R.S. Means Company Inc.
BRE: Building Research Establishment, UK
ICC—International Code Council
OBC—Ontario Building Code
NRC Canada—National Research Council Canada
OPSS—Ontario Provincial Standard Specification

I would like to acknowledge the reproduction of material from Ontario Provincial Standards for Roads and Public Works and Queen's Printer for Ontario 2008 (last amended January 2016). The material referred is an unofficial version of the Government of Ontario Legal materials.

Extracts from "BS Number & Year" are reproduced with the permission of BSl under license number 2000SK/0472; British Standards can be obtained by post from BS1 Customer Services at the following address:

BS1 Customer Services
389 Chiswick High Road, London W4 4AL, United Kingdom (Tel UK 44 20 8996 7553)

Figures-XX and Tables-YY are reproduced from *Design of Normal Concrete Mixes,* second edition (1997), by permission of Building Research Establishment Ltd, UK. Copies are available from CRC, 151 Rosebery Avenue, London, ECIR 4GB, UK: Tel + 4420 75056622, Fax + 4420 75056606, email crc@construct.emap. co.uk.

Every effort has been made to trace all the copyright holders, but if any have been inadvertently overlooked, the author/publisher will be pleased to make the necessary arrangement at the first opportunity.

Contents

About the Author

Akhtar Surahyo is a Graduate Civil Engineer with a Postgraduate Diploma in Structural Engineering. He is a registered Professional Engineer with Professional Engineers Ontario, Canada, life Member Pakistan Engineering Council, Member Project Management Institute (PMI), Member American Society of Testing and Materials (ASTM), and a Member (ex) American Concrete Institute (ACI). He has extensive practical construction experience of more than 30 plus years in executing and managing multidisciplinary projects in many countries and has successfully completed many projects from inception through to completion. Mr. Surahyo is also author of the book *Understanding Construction Contracts: Canadian and International Conventions* published by Springer International Publishing in 2018. Currently, Mr. Surahyo is working as a Contract Administrator with IBI Group Canada.

Part I
The Concrete

Chapter 1
Concrete

1.1 Introduction

The word "concrete" has its origin from the Latin word "concretus", which means to grow together. It is a very strong and versatile construction material as it can be moulded to take up the shapes required for the various structural forms. It is produced by mixing cement, sand, and aggregate (e.g. gravel or crushed rock) with water. The oldest concrete was discovered around 7000 BC. It was found in 1985 when a concrete floor was uncovered during the construction of a road at Yiftah El in Galilee, Israel. It consisted of a lime concrete, made from burning limestone to produce quicklime, which when mixed with water and stone hardened to form concrete [1].

Throughout history, cementing materials have played a vital role and were used widely in the ancient world. A cementing material was used between the stone blocks in the construction of the Great Pyramid at Giza in ancient Egypt around 2500 BC. Greeks and Romans used lime made by heating limestone and added sand to make mortar, with coarser stones for concrete. Examples of early Roman concrete have been found dating back to 300 BC [1]. The Romans found that a cement could be made which set under water and this was used for the construction of harbours. This cement was made by adding crushed volcanic ash to lime and was later called a "pozzolanic" cement, named after the village of Pozzuoli near Vesuvius. British engineer John Smeaton made the first modern concrete in the year 1756. He added pebbles as the course aggregate with brick powder. It is believed that the Romans used concrete that is similar to the ones that are in use today.

Portland cement was developed from natural cements made in Britain beginning in the middle of the eighteenth century. In 1824, Joseph Aspdin, a British stonemason, produced a cement in his kitchen. He prepared that cement by heating a mixture of finely divided clay and hard limestone in a furnace until CO_2 had been driven off. The type of modern cement was made in 1845 by Isaac Johnson, who burnt a mixture of clay and chalk by clinkering. The name "Portland cement" was given originally due to the resemblance of the colour and quality of the hardened cement to Portland stone, a type of building stone (limestone) quarried in Dorset.

© Springer Nature Switzerland AG 2019
A. Surahyo, *Concrete Construction*,
https://doi.org/10.1007/978-3-030-10510-5_1

There are a number of benefits to use concrete as a material for construction:

- It is durable and economical.
- It can be formed into a variety of shapes and sizes.
- It is an easily available material for construction.
- Its properties can be altered according to the requirements of the job.
- It provides resistance to high and low temperatures, sulphates, and chlorides when proper specifications are being implemented.
- It is non-combustible which makes it an improved fire-resistant material.
- Used concrete is recyclable and serves as aggregate in roadbeds or as granular material in new concrete.
- Reinforced concrete has a high compressive strength compared to other building materials.
- Due to the provided reinforcement, it can also withstand a good amount of tensile stress.

1.2 Normal Concrete

Concrete can be defined as a mixture of cement, water, aggregate (fine and coarse), and admixture, which is sometimes added to modify certain of its properties. It is a temporarily plastic material but is later converted to a rock-like material by chemical reaction with a high compressive strength.

Normal concrete has comparatively low tensile strength and for structural applications it is normal practice to incorporate steel bars to resist tensile forces. The strength, durability, and other characteristics of concrete depend upon the properties of its ingredients, the proportions of mix, and the method of transporting, placing, compacting, and curing. Good concrete has to satisfy performance requirements in the plastic and hardened states. In the plastic state, the concrete should be workable and free from segregation and bleeding. In the hardened state, concrete should be strong, durable, and impermeable, and it should have minimum dimensional changes.

The popularity of the concrete is due to the fact that from the common ingredients, it is possible to tailor the properties of concrete to meet the demands of any particular situation. Among the various properties of concrete, its compressive strength is considered to be the most important and is taken as a measure of its overall quality.

1.2.1 Concrete Grades

The grade of concrete is defined as that number, which indicates the characteristic compressive strength of concrete in N/mm^2, determined by cube or cylinder tests made at twenty-eight days. Thus a grade 30 concrete has a characteristic strength of 30 N/mm^2. According to BS-5337 [2] the grade 25(C25) is the lowest grade that may be used as reinforced concrete. CSA-A23.3-04 [3] provides that specified

Table 1.1 Concrete
compressive strengths

Concrete grade	Characteristic strength (N/mm²)
C20	20
C25	25
C30	30
C35	35
C40	40
C45	45
C50	50

concrete compressive strengths used in design shall not be less than 20 MPa. Higher strength concretes may be used for special applications and in prestressed concrete applications. However, in practice, a C30 mix is invariably necessary because of durability considerations.

Table 1.1 shows the characteristic strengths of various grades of concrete normally specified in reinforced concrete design.

1.3   Modified Concretes

Modifications have been made from time to time to improve the particular properties of concrete. There are many types of concrete, which have been developed for special purposes.

1.3.1   Polymer Concretes

Polymers are combined with concrete in several ways, with the aim of modifying properties of the concrete, particularly increasing its strength, reducing its permeability, and increasing resistance to chemicals and abrasion resistance, thereby improving durability. Basically a polymer consists of numerous monomers, which are linked together in a chain-like structure and the chemical process which causes these linkages is called polymerisation. Generally, polymerisation is achieved by gamma radiation or by thermal-catalytic means. Polymers are used to produce three types of polymer-concrete composites:

1.3.1.1   Polymer Portland Cement Concrete (PPCC)

It is prepared by adding a polymer or monomer to Portland cement concrete during mixing. Polymer Portland cement concrete (PPCC) is also known as polymer-modified concrete (PMC) or latex-modified concrete (LMC). Like Portland cement concrete, the primary curing mechanism for polymer-modified concrete is hydration of the cement binder.

The properties of PPCC include increased bond strength, flexural and tensile strengths, split strength, reduced elastic modulus, and reduced water permeability. PPCC can also improve corrosion resistance and resistance to chemical attack and severe environment (such as sulphuric acid attack, penetration by water and dissolved salts, and freezing-and-thawing resistance), and it reduces the need for sustained moist curing. Reduced elastic modulus might be particularly helpful when LMC is applied as a bridge deck overlay or repair surface. The reduced elastic modulus results in a reduction of the stresses developed due to differential shrinkage and thermal strains that would reduce the tendency of the material to crack [4].

The main application of latex-containing polymer cement concrete is in floor surfacing, as it is non-dusting and relatively cheap. Because of lower shrinkage, good resistance to permeation by various liquids such as water and salt solutions, and good bonding properties to old concrete, it is particularly suitable for thin (25 mm) floor toppings, concrete bridge deck overlays, anti-corrosive overlays, concrete repairs, and patching.

Common polymers used to modify hydraulic cementitious material are natural rubber and synthetic latexes, epoxy resin, acrylic polymers, thermoplastics, and bituminous materials like asphalt and coal tar.

Modification of concrete with a polymer latex (colloidal dispersion of polymer particles in water) results in greatly improved properties, at a reasonable cost. Therefore, a great variety of latexes are now available for use in polymer cement concrete products and mortars. The most common latexes are based on poly (methyl methacrylate) also called acrylic latex, poly(vinyl acetate), vinyl chloride copolymers, poly(vinylidene chloride), (styrene–butadiene) copolymer, nitrile rubber, and natural rubber. Each polymer produces characteristic physical properties. The acrylic latex provides a very good water-resistant bond between the modifying polymer and the concrete components, whereas use of latexes of styrene-based polymers results in a high compressive strength [5].

1.3.1.2 Polymer-Impregnated Concrete (PIC)

It is prepared by monomer or polymer impregnation of hardened cement concrete, followed by inside concrete polymerisation. Impregnation results in markedly improved strength and durability (e.g. resistance to freeze/thaw damage and corrosion) in comparison with conventional concrete.

Impregnation of concrete results in a remarkable improvement in tensile, compressive, and impact strength; enhanced durability; and reduced permeability to water and aqueous salt solutions such as sulphates and chlorides. The compressive strength can be increased from 35 MPa to 140 MPa, the water absorption can be reduced significantly, and the freeze/thaw resistance is considerably enhanced. The greatest strength can be achieved by impregnation of autoclaved concrete. This material can have a compressive strength-to-density ratio nearly three times that of steel. Although its modulus of elasticity is only moderately greater than that of non-autoclaved polymer-impregnated concrete, the maximum strain at break is significantly higher.

The monomers most widely used in the impregnation of concrete are the vinyl type, such as methyl methacrylate (MMA), styrene, acrylonitrile, *t*-butyl styrene, and vinyl acetate.

1.3.1.3 Polymer Concrete (PC)

It is prepared by mixing synthetic resin or monomer as a binder with an aggregate (filler), followed by polymerisation. When sand is used as a filler, the composite is referred to as a polymer mortar. Other fillers include chalk, gravel, limestone, crushed stone, condensed silica fume (silica flour, silica dust), quartz, clay, granite, expanded glass, and metallic fillers. Generally, any dry, non-absorbent, solid material can be used as a filler.

Polymer concretes are costly and are used for precast chemical-resistant pipes, lightweight drainage channels, and quick repair and strengthening applications, particularly in structures that are subjected to severe deleterious effects of corrosion and erosive elements. Generally they are not suitable for use where fire-resistant properties are required. The polymer composites also undergo considerable creep at high temperatures [6]. Polyethylene (PE), polystyrene (PC), polyvinyl chloride (PVC), and epoxy resins are well-known examples of polymer materials.

1.3.2 Fibre-Reinforced Concretes

Fibre-reinforced concretes are manufactured to increase the tensile strength, impact strength, fatigue strength, and resistance to corrosion.

According to ACI 544.1R [7], fibre-reinforced concrete is defined as concrete made primarily with hydraulic cements, aggregates, and discrete reinforcing fibres. Many different properties of the fresh and hardened concrete can be effectively influenced by adding fibres. There are numerous different types of fibres with different material characteristics and shapes. Correct selection for different uses is important. As well as the actual material, the shape of the fibres is also a critical factor. Fibres include steel fibres, polypropylene fibres, polyvinyl alcohol fibres, carbon fibres, vegetable fibres, glass fibres, polyester fibres, and so on, each of which lends varying properties to the concrete.

Fibre reinforcement reduces plastic shrinkage, cracking, and crazing. It has been used in road pavements where both flexural and impact strengths are important. Generally, the length and diameter of the fibres used for FRC do not exceed 3 in. (76 mm) and 0.04 in. (1 mm), respectively.

Fibre-reinforced concrete is generally made with a high cement content and low water/cement ratio. When well compacted and cured, concretes containing steel fibres seem to possess excellent durability as long as fibres remain protected by cement paste. Fibres are usually added to conventional 10–20 mm aggregate concrete mixes at the rate of 0.6–0.9 kg/m^3. Fibres may be added either at a conventional batching/mixing plant or by hand to the ready-mix truck on site. Fibre-dosed concrete mixes may be hand tamped or vibrated by conventional means to provide the necessary compaction. Placed concrete mixes containing fibres may be floated and troweled using all normal hand or power tools, to provide a smooth, fibre-free surface appearance.

Glass-fibre composites in the form of I-beams and reinforcing bars (Fig. 1.1) can reinforce concrete slabs and beams instead of conventional steel reinforcement. Glass-fibre reinforcement is further described in Chap. 10. Fibre-reinforced plastics (FRP) offer a revolutionary approach to solving the problems of steel corrosion in concrete structures [8].

1.3.3 Sulphur Concretes (SC)

SC is extremely corrosion resistant. It is unaffected by salt, acids, and mild alkalis. When made with acid-resistant aggregates such as granite and other siliceous materials, it is unaffected by continuous exposure to hydrochloric acid and to

Fig. 1.1 Fibre-reinforced polymer (FRP) reinforcing bars and I-Beam

I Beam Source: ePlastics San Diego, CA 92123 USA

sulphuric acid up to 98% concentration [9]. Owing to such properties, it is normally used in the fertiliser and metal refining industries.

SC gain strength very rapidly and attain about 90% of its ultimate strength in 6–8 h under normal temperatures. This property makes SC suitable for precast units. One of its disadvantages is that it does not support combustion. Sulphur present in the surface will slowly burn when exposed to direct flame but extinguishes when the flame is removed. It has a low melting point of 119 °C (240 °F) [10] and a poor resistance to freezing and thawing. It also becomes brittle and has a high creep. Owing to all these disadvantages, SC becomes unsuitable for most structural applications. However, in order to modify the properties of sulphur, polymer modifiers are used to stabilise the sulphur such that the crystal structure will not change.

SC mix usually consists of sulphur (2% by weight), fine aggregate (32% by weight), coarse aggregate (48% by weight), and a mineral filler such as a crusher dust or silica flour (5% by weight), but contains no water or cement [10]. The quantity of sulphur can be increased to 10% based on product requirements including with 1.2% of sulphur polymer modifier. To produce sulphur concrete, sulphur is heated to approximately 130 °C. The sulphur is in a melted state at this temperature. The other ingredients (as mentioned above) for sulphur concrete are then added to the melted material. These ingredients will also have been preheated. Once all the ingredients have been added, the material is poured and cooled in a controlled setting. The solidified product is sulphur concrete. In fact, the technology of the sulphur concrete is very similar to the technology of the asphalt concrete.

Many products are being manufactured by using modified sulphur concrete, such as [11]:

- Tanks for various substances, cesspits, drains.
- Sewerage pipes, drainages, sewerage channels, weights for electric traction lines.
- Telecommunications drains, elements reinforcing the wharf and harbour constructions.
- Surfaces of landing strips, roadwork elements.

1.3.4 Roller-Compacted Concrete (RCC)

RCC is usually used for the construction of dams, heavy-duty pavements, and parking areas. It is proportioned in such a way that it has the ability to support a roller and spreading equipment. It contains less cement than normal concrete to avoid heat of hydration, and has little to no slump. The amount of water used in the mix is very low and the consolidation is achieved by vibratory rollers. The admixture usually added is fly ash. Typically, RCC is constructed without joints. It needs neither forms nor finishing, nor does it contain dowels or steel.

RCC has the same basic ingredient as conventional concrete: cement, water, and aggregates. Generally, RCC mixture is mixed in a twin-shaft pug-mill mixer. The mixing action should be vigorous to ensure that the small amount of water added is properly distributed through the entire mix and the constituents are homogeneously spread.

Although made of similar materials, RCC pavements are unlike conventional concrete pavements in many ways, especially during production and placement, such as the following [12]:

- RCC requires placement with asphalt-type pavers.
- RCC mixtures require compaction with the use of vibratory, tamper bar screeds, and subsequently with vibratory rollers to achieve a target density. RCC does not require the internal vibration necessary to consolidate conventional concrete.
- RCC has little to no slump.
- Due to its dry nature, RCC pavements do not allow for or require finishing operations like conventional concrete pavement.
- The surface of an RCC pavement does not resemble the surface of a conventional concrete. Rather an RCC surface is more inconsistent and typically includes minor surface tearing, pitting, or pockmarks. It more closely resembles an asphalt pavement surface.
- RCC pavement, unlike conventional concrete pavement, cannot be reinforced with steel nor include dowel bars to provide joint load transfer. Joint load transfer must rely on aggregate interlock and base support for long-term joint performance.

High-density pavers (modified pavers) are also used for placing and compaction operations, thereby eliminating the need for rolling, as well as providing a surface suitable for high-speed traffic. While placing RCC with a modified asphalt paver, it is suggested that pavement thicker than 10 in. (250 mm) should be placed in two or more lifts. When placing multi-lift construction, a good bond must be achieved between the lifts so as to have monolithic pavement. Compaction is the most important stage of construction: it provides density, strength, smoothness, and surface texture. Compaction begins immediately after placement and continues until the pavement meets density requirements.

The RCC surface dries quickly even under normal temperatures and therefore a combination of moist curing and membrane curing is used in order to prevent drying and scaling. The high strength of RCC pavements eliminates many problems traditionally associated with asphalt pavements. They provide better resistance to rutting; do not deform under heavy, concentrated loads; do not deteriorate from spills of fuels; and are resistant to high temperatures.

The Cement Association of Canada has developed design and quality control manuals on RCC and can be consulted for additional information on RCC. Information on RCC can also be found in ACI 207.5R, ACI 325.1R, ACI 325.10R, and ACI 309.5R.

1.3.5   High-Performance Concrete (HPC)

High-performance concrete (HPC) exceeds the properties and constructability of normal concrete. Normal and special materials are used to make these specially designed concretes that must meet a combination of performance requirements. ACI has defined HPC as concrete that meets special performance and uniformity requirements that cannot always be achieved using conventional materials and normal mixing, placing, and curing practices. High-performance concretes are currently used in tunnels, bridges, and tall buildings for its strength, durability, and high modulus of elasticity. It has also been used in shotcrete repair, poles, parking garages, and agricultural applications.

The strength of HPC varies from 60 to 120 MPa. The requirements may involve enhancements of placement and compaction without segregation. Normally a slump of 100 mm (4 in.) will provide proper workability to fulfil these requirements. HPC usually consists water/cement ratio ranging from 0.27 to 0.45; super-plasticisers are, therefore, used to achieve a normal workable mix. Common cement contents in HPC range from 311 to 513 kg/m^3. Silica fume and fly ash are generally used to achieve required durability.

The high-performance concrete is normally designed for one or more parameters falling under the strength criteria, i.e. compressive strength, modulus of elasticity, shrinkage, and creep, and durability criteria, i.e. freeze and thaw, abrasion resistance, scaling, and chloride permeability. Since HPC has been used primarily in high-strength applications, it can also be called as high-strength concrete (HSC). However, as a fact, HPC is specially designed to achieve certain particular characteristics such as high abrasion assistance and compaction without segregation. HPC is defined in terms of strength and durability. However, HSC is defined purely on the basis of its compressive strength, even though admixtures and additives are also added in designing HSC mix.

The advantages of HPC are that it can be used in bridges, longer spans, increased girder spacing and shallower sections. Figure 1.2 shows variation in depth of prestressed concrete girders achieved by using various compressive strengths of

Fig. 1.2 Girder depths for 100 ft. span

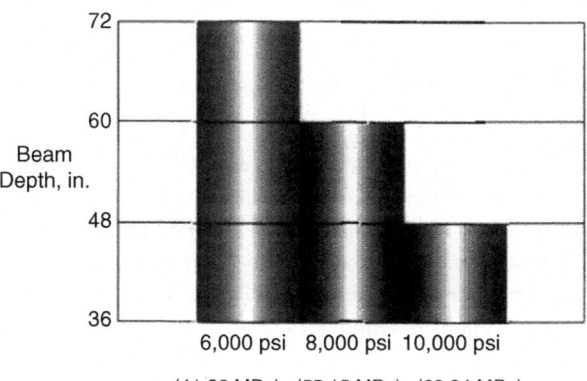

HPC [13]. In buildings, column sections can be reduced in size or for same cross section the amount of steel reinforcement can be reduced. HPC increases durability, enhances mechanical properties, and reduces permeability. However, the disadvantages are that it has a low shear strength, increased creep, and shrinkage due to a low aggregate content.

1.3.6 High-Strength Concrete

The definition of high-strength concrete has changed over the years, and should not be considered static. In 1984, ACI defined high-strength concrete as concrete with a compressive strength greater than 6000 psi (41 MPa). In the draft update of ACI 363R-92, high-strength concrete has been defined as concrete having a specified compressive strength for design of 8000 psi (55 MPa), or greater. Upper limit on the strength of concrete has not been specified by ACI. The Concrete Society, UK, provides that HSC may be defined as concrete with a specified characteristic cube strength between 60 and 100 N/mm^2, although higher strengths have been achieved and used. Canadian Standards Association (CSA) defines high-strength concrete having a specified compressive strength of at least 70 MPa at a specified age not exceeding 91 days.

High-strength concrete is specified where reduced weight is important or where smaller load-carrying elements are required. The most common use of high-strength concrete is for construction of high-rise buildings. The main applications for HSC in situ concrete construction are in offshore structures, columns for tall buildings, long-span bridges, and other highway structures.

For producing HSC, low w/c ratio is considered in the range of 0.30–0.35 or even lower with addition of super-plasticisers to achieve workability. HSC mixture will have high cementitious material content typically around 415 kg/m^3. Most mixtures contain admixtures such as silica fume, fly ash, or pozzolanic materials. For resistance to freeze/thaw air entrainment is required, even though it reduces the strength.

1.3.7 Ultra-High-Performance Concrete

Ultra-high-performance concrete (UHPC) is one of the major breakthroughs in concrete technology in the recent past. It is a fibre-reinforced, cementitious material that offers exceptionally high mechanical and durability performance, including compressive strengths varying from 120 to 150 MPa (17,000 to 22,000 psi) and flexural strengths 15 to 25 MPa (2200 to 3600 psi) [14] and excellent resistance against environmental degradation.

Ultra-high-performance concrete (UHPC) is also known as reactive powder concrete (RPC), and is formulated by combining Portland cement, fine sand, silica

Table 1.2 Typical
composition of UHPC [15]

Constituent	Amount (lb/yd^3)	% by weight
Portland cement	1200	28.5
Silica fume	390	9.3
Fine sand	1720	41.0
Ground quartz	355	8.5
Super-plasticiser	51	1.2
Water	218	5.2
Steel fibres	263	6.3

fume, high-range water-reducing admixture (HRWR), fibres (usually steel), and water. Small aggregates are sometimes used, as well as a variety of chemical admixtures. Different combinations of these materials may be used, depending on the application and supplier. Table 1.2 shows typical composition of UHPC constituent materials.

Ultra-high-performance concrete (UHPC) in its present form became commercially available in the United States in about 2000. In Canada, the first UHPC bridge was constructed in 1997. This pedestrian bridge consists of a precast, post-tensioned space truss. At least 26 bridges have been built in Canada using UHPC in one or more components.

1.3.8 Flowing Concrete/Self-Consolidating Concrete (SCC)

Flowing or SCC concretes have been used successfully on many projects with areas of congested reinforcement. Particularly in heavy reinforced walls, columns, and congested mat foundations, concrete of conventional workability is difficult to be placed and consolidated; hence flowing concrete has been introduced. It can be placed and consolidated/compacted without the use of vibrators.

In order to get high slump and to keep water content low, super-plasticisers are used. Adjustments to traditional mix designs and use of super-plasticisers create flowing concrete that meets tough performance requirements. If needed, low dosages of viscosity modifier can eliminate unwanted bleeding and segregation. This concrete offers advantages particularly where there is congested concrete-reinforcing steel or restricting formwork that prevents traditional consolidation/compaction. It is widely used for improved architectural appearance of the concrete surface due to its extreme flow characteristics, and resistance to concrete segregation under adverse site conditions. However, it needs to be ensured that the formwork must have a high quality of surface finish since the concrete will mirror any imperfections that exist. Additionally, it must also be ensured that formwork is tight to prevent leakage at form panel edges and should be able to withstand any pressure initially generated by the concrete.

1.3.9 Ferro-Cement

It is a kind of reinforced concrete and consists of one or more layers of thin wire mesh (steel or fibre) and a skeletal reinforcement of either mild steel bars or welded mesh, completely surrounded by cement mortar. The skeletal steel frame is made conforming exactly to the geometry and shape of the structure, and is used for holding the wire meshes in position to conform to the shape of the structure.

Ferro-cement is much thinner than reinforced concrete; usually 10–60 mm in thickness, the mesh can be formed to any shape and then plastered or mortared by hand or by shotcreting. Shotcreting is the process of applying concrete at a high velocity onto a surface. Proper placement is the most important element in achieving adequate compaction and good encasement of the reinforcement with no entrapped rebound or overspray.

The cement-mortar matrix usually consists of Portland cement or Portland-pozzolana cement and fine aggregate in the ratio of 1:1½ or 1:2 with water/cement ratio of 0.40–0.50. Super-plasticisers may be used in wet-mix shotcrete mixes to provide the necessary workability for pumping and shooting.

Ferro-cement structures provide very high resistance to cracking, and properties such as toughness, fatigue resistance, and impermeability are superior to ordinary reinforced concrete. It is usually used in boats, mobile homes, swimming pools, silos and bins, water tanks, folded plates, shell roofs, and kiosks.

1.3.10 Lightweight Concrete

Normal concrete is a heavy material with a density of about 2400 kg/m^3. In order to reduce the dead load of concrete structures and to improve thermal insulation and fire resistance, lightweight concrete is mostly used. The lightweight concrete can be classified into the following types:

1.3.10.1 Lightweight Aggregate Concrete (LWAC)

This type of concrete is produced by using porous lightweight aggregates with fines which can be natural sand. The density depends on the type of grading of the aggregates used, the moisture content, and the mix proportions. Similarly, LWAC compressive strength depends on the properties of the particular aggregates being used and on other conditions. Accordingly strengths from 3000 to 5000 psi (21 to 35 MPa) can be achieved normally. The lightweight aggregate concrete can provide the same level of structural performance as normal-weight concrete so it can be used in exactly the same way.

The lightweight aggregates consist of pumice, diatomite, scoria, volcanic cinders, expanded shale or clay or perlite or vermiculite, etc. Structural grade lightweight aggregates are produced in manufacturing plants from raw materials, including suitable shales, clays, slates, fly ashes, or blast-furnace slags. Naturally occurring lightweight aggregates are mined from volcanic deposits that include pumice and scoria.

1.3.10.2 Aerated or Cellular or Foamed or Gas Concrete

This concrete is known by many names. In fact, it is produced without adding coarse aggregates, and by introducing gas bubbles or air bubbles into the plastic mortar mix (cement and fine aggregate with or without fly ash), which result in the formation of a material having a cellular structure containing voids between 0.1 and 1 mm (0.004 to 0.04 in.) in size. For introducing gas bubbles, aluminium powder is most commonly used, whereas for air bubble foaming agents like resin soap are added to the mortar mix.

Foamed concrete is self-compacting concrete and requires no compaction and may be poured or pumped into moulds, or directly into structural elements. It can be pumped successfully over significant height and distances. After its initial set, the blocks or panels cast are sometimes baked (autoclaved). The heat helps the material to cure faster so that blocks and panels maintain their dimensions. Reinforcement is placed within panels prior to curing.

The 28-day compressive strength of the concrete varies according to its composition depending primarily on the density produced and usually they range from 0.7 to 18 MPa with density between 300 and 1600 kg/m^3. This concrete carries superior thermal insulation and fire-resisting properties.

1.3.10.3 No-Fines Concrete

This type of concrete is produced by omitting fine aggregate from the mix, due to which large number of voids are formed into the body of concrete. Both lightweight and normal-weight aggregates (10/20 mm) can be used to obtain no-fines concrete depending upon the requirement for density.

Strength is between 5 and 15 MPa at 28 days. No-fines concrete has large uniformly distributed interconnected voids (up to 30%) between the aggregate particles. This structure makes it ideal as a free-draining medium. The ratio of aggregate to cement by volume is generally 8:1 or 10:1 by mass. No-fines concrete, when conventional aggregates are used, may show a density of about 1600–1900 kg/m^3, but when no-fines concrete is made by using lightweight aggregate the density may come to about 360 kg/m^3 [16].

For preparing no-fines concrete, the water/cement ratio must be chosen carefully. With too low water/cement ratio the paste will become too dry and the aggregates will not get properly coated with paste resulting in insufficient adhesion between the particles. On the other hand, if the water/cement ratio is too high, the cement paste will flow to the bottom of the concrete resulting in poor adhesion between the overall aggregates.

Since it does not contain any fine aggregate the mix cannot segregate and consequently it can be dropped from a height. No-fines concrete is mainly used in load-bearing wall of domestic buildings. It is also used as a fill behind retaining walls to dissipate water pressure. In case of framed structure, they can be used in infill walls. Normally reinforcement is not included in no-fines concrete because the bond strength of no-fines concretes is very low. Additionally, having the porous structure, no-fines concrete has high absorption quality. It is therefore not advisable to use no-fines concrete for foundations and in places that are continuously in touch with moisture.

1.3.11 Shrinkage-Compensating Concrete

It is manufactured by adding expansive cements (described in Chap. 2) or expansive compounds, and is used to reduce cracking caused by drying shrinkage in concrete. Drying shrinkage is the contraction caused by moisture loss from the hardened concrete. The amount of drying shrinkage in concrete depends on the characteristics of the materials, mix proportions, placing methods, curing, and restraint. When properly restrained by reinforcement or other means, these concretes expand an equal amount to or slightly greater than the anticipated drying shrinkage. When concrete is dry and shrinkage takes place, it will try to reduce or relieve the expansive strains already created by the initial expansion, as a result of which there will be no reduction/shrinkage in concrete volume. Hence the cracks, which usually result on drying shrinkage, are eliminated.

These concretes are mostly used in ground-floor slabs. Restraint by reinforcing steel is essential to the effective performance of shrinkage-compensating concrete. The elasticity of the reinforcing steel restores the concrete to almost its original length as shrinkage progresses and cracking is avoided. The expansive/shrinkage compensative cements used mostly consist of 75–90% constituents of conventional Portland cement, added with materials like aluminates and calcium sulphates. The tensile, flexural, and compressive strength development of shrinkage-compensating concrete after expansion will be completed similar to that of Portland cement concrete under both moist- and steam-curing conditions [17]. When properly designed and adequately cured, shrinkage-compensating concrete is equally resistant to freezing and thawing as Portland cement concrete of the same water/cement ratio. The effects of air content and aggregate are essentially the same, but the water requirement of shrinkage-compensating concrete is greater than that of Portland cement concrete for a given consistency.

Proper attention should be given to the admixture's effect on slump, restrained expansion, drying shrinkage, and temperature gain control. The use of certain admixtures with some shrinkage-compensating cements has resulted in excessive shrinkage. These effects may be experienced for both normal-range and high-range admixtures. CSA A23.1-14 [18] provides that shrinkage-reducing admixtures may affect the stability of the air void system of the concrete. Some adjustments may be required to compensate. Recommended air-entraining admixtures may be used for shrinkage-compensating concrete.

With respect to curing of shrinkage-compensating concrete, the usually accepted methods of curing can be considered; however, those that provide additional moisture to the concrete such as ponding, continuous sprinkling, and wet coverings are preferred. Other methods such as moisture-proof covers and sprayed-on membranes have been successfully utilised, provided that coverage is complete so that it prevents loss of moisture from the entire concrete surface. ACI 223-98 [17] recommends that the curing of shrinkage-compensating concrete shall be continued for a minimum of 7 days.

1.3.12   Microsilica Concrete

It is produced by adding microsilica to ordinary Portland cement. Normally, microsilica concrete ranges from 5 to 15% of Portland cement content. Microsilica also known as silica dust and silica fume is a by-product from the operation of electric arc furnaces, resulting from the reduction of quartz with coal and wood chips in the production of ferrosilicon alloys. The fume, which has a high content of amorphous silicon dioxide and consists of very fine spherical particles, is collected from the gases escaping from the furnaces.

Silica fume consists of much smaller particles than cement particles having the specific surface area of 20,000 m^2/kg compared to 275 m^2/kg of that of cement [19]. The particle/size distribution of a typical silica fume shows most particles to be smaller than 1 μm with an average diameter of about 0.1 μm, which is approximately 100 times smaller than the average cement particle [20].

Due to higher fineness, microsilica fume fills the interstices, provides a finer pore structure within the concrete, and accelerates the hydration process. These properties produce a concrete of high strength and durability with reduced permeability.

Microsilica concrete is used in industrial flooring, warehouses, etc. in situations where exceptional strength, wear resistance, durability, and impact resistance are necessary. As an example of recent advances in flooring technology, microsilica concrete was used along with steel-fibre reinforcement to provide 42,530 m^2 internal joint-less flooring for a food retailing chain, "Somerfield" in the United Kingdom [19]. Microsilica concrete has also been used in producing high-strength concrete by using super-plasticiser as an admixture.

Microsilica concrete has a tendency to develop plastic shrinkage cracks; hence it is necessary to quickly cover the surface of freshly placed concrete to prevent rapid water evaporation. A light fog spray of water should be continuously used to keep the concrete surface moist between finishing operations. Full curing must be started right after the concrete finishing is completed. Use of a proprietary evaporation retarder to prevent moisture loss between initial and final finishing will also be helpful.

1.3.13   High-Early-Strength Concrete (HES)

This type of concrete provides a strength gain at earlier times than normal-set concrete. As such the properties of concrete depend closely on the degree of cement hydration; hence the hydration period is reduced for quick turnaround times by adding accelerators. These chemical admixtures speed up the process of hydration and hardening of the concrete while still achieving the high strength requirements. The time period in which a specified strength should be achieved may range from a few hours (or even minutes) to several days. High-early strength can be attained by using traditional concrete ingredients and concreting practices, although sometimes special materials or techniques are needed.

For HES concrete, a minimum compressive strength of 5000 psi (35 MPa) is mostly recommended in 24 h. Some of the advantages of HES concrete are the following:

- Ideal solution where there is limited time to carry out renovation or repair of structures, slabs, or pavements in a wide variety of applications
- Faster construction time
- Early stripping of formwork for reuse
- Development of most cost-efficient mix designs
- Rapid rotation of moulds for precast concretes and finishing of concrete surfaces
- Early prestressing at low temperatures
- Precast concrete for rapid production of elements

When HES concrete is intended for highway application, exposure to frost must be expected; it is essential that the concrete be frost resistant with a low w/c ratio around 0.35–0.40.

Depending on the age at which the specified strength must be achieved, high-early strength can be obtained by using one or a combination of type III or high-early-strength cement content ranging from 400 to 600 kg/m³; low w/c ratio ranging from 0.20 to 0.45; and chemical admixtures and silica fume or other supplementary cementing materials. When designing early-strength mixtures, strength development is not the only criteria that should be evaluated; durability, early stiffening, autogenous shrinkage, drying shrinkage, temperature rise, and other properties also should be evaluated for compatible with the project. Special curing procedures, such as fogging, may be needed to control plastic shrinkage cracking.

1.3.14   Unshrinkable Fill Material

Unshrinkable fill is a mixture of aggregates, cementing material, and water, with or without chemical admixtures. It is a self-compacting low-strength material with a flowable consistency. It is used as an alternative to compacted granular fill resulting a saving in labour costs, time, and equipment. ACI Committee 229 uses the term

"Controlled Low Strength Material" instead of unshrinkable fill. Since it contains constituent materials of those of concrete, it is described in this section.

OPSS.MUNI 1359 [21] specifies that coarse aggregate shall be crushed stone, crushed gravel, natural gravel, recycled concrete material (RCM), or a combination thereof. The typical maximum nominal size shall be 25 mm, or as specified in the contract documents. Fine aggregates shall be natural sand, crushed sand, recycled concrete aggregates, or a combination thereof. Slag aggregate, glass, and ceramics shall not be used. It further specifies that the mix requirements shall be according to the following:

(a) The unshrinkable backfill shall contain 25 kg/m^3 of type GU Portland cement according to CAN/CSA A3001 and may contain additional supplementary cementing materials.
(b) Slump at the point of discharge shall be a minimum of 150 mm and the unshrinkable backfill shall be uniformly mixed throughout.
(c) The material shall be designed so that it can flow into the excavation and fill the entire space without vibration and segregation.
(d) The 28-day compressive strength shall be a maximum of 0.40 MPa.
(e) The mixture may contain foaming agents.

OPSS.MUNI 1359 recommends 28-day compressive strength as a maximum of 0.40 MPa; however, on many road projects unshrinkable fill with 28-day compressive strength of 0.70 MPa is also being used. Unshrinkable fill material is widely used in sewer and utility trenches, bridge abutments, any road cuttings, retaining walls, and pile excavations. The unshrinkable fill surface should be screeded while it is still sufficiently flowable to achieve the desired grades and elevation. The surface should be uniformed and free from undulations and projections.

Bibliography

1. Portland Cement Association- PCA, Design and Control of Concrete Mixtures-EB001- Introduction to Concrete.
2. BS 5337-76 (Amended-82), Code of Practice for the Structural Use of Concrete for Retaining Aqueous Liquids.
3. Cement Association of Canada-CSA-A23.3-04-Concrete Design Handbook.
4. ACI 548.3R-09- Report on Polymer-Modified Concrete-American Concrete Institute April 2009.
5. A. Blaga, J.J. Beaudoin, CBD-241. Polymer Modified Concrete-National Research Council Canada, October 1985.
6. Peter H. Emmons, Alexander M. Vaysburd & Jay Thomas, Strengthening Concrete Structures (part 1), ACI Journal International, Vol. 20, No. 3, March 1998.
7. ACI 544.IR-96 (Reapproved-2009), Report on Fiber Reinforced Concrete.
8. Ahmad A. Hamid, Improving Structural Concrete Durability in the Arabian Gulf, ACI Journal, Concrete International, Vol. 17, No. 7, July 1995.
9. Sean M. Crick & David W. Whitmore, Using Sulphur Concrete on a Commercial Scale; Alan H. Vroom, Sulphur Concrete for Precast Products ACI Journal, Concrete International, Vol. 20, No. 2, Feb 1998.

10. A.M. Neville and J.J. Brooks, Concrete Technology, E.L.B.S. Longman, Singapore, 1993.
11. Natalia Ciak, Jolanta Harasymiuk, University of Warmia and Mazury in Olsztyn. SULPHUR CONCRETE'S TECHNOLOGY AND ITS APPLICATION TO THE BUILDING INDUSTRY, Technical Sciences 16(4), 2013, 323–33.
12. American Concrete Pavement Association (ACPA) Guide Specifications- Roller-Compacted Concrete Pavements as Exposed Wearing Surface- Version 1.2 – September 4, 2014.
13. Henry G. Russell, P.E-Technical talk-An article on "Why use High Performance Concrete".
14. Portland Cement Association USA- Brief on Ultra High Performance Concrete.
15. Zach Haber, Ph.D. Bridge Engineering Researcher-Ultra-High Performance Concrete (UHPC)-2016 FDOT Design Training Expo; Daytona Beach, FL, June 13-15, 2016-US Department of Transportation; Federal Highway Administration.
16. The Constructor-Civil Engineering Home-No-Fines Concrete & its Mix Proportions.
17. ACI 223-90/98 (Revised 2010), Standard Practice for the Use of Shrinkage-compensating Concrete.
18. Canadian Standards Association (CSA) A23.1-14—Concrete materials and methods of concrete construction/Test methods and standard practices for concrete.
19. David Martin- Microsilica Steel -Fiber Reinforced Concrete Flooring. Concrete Society Journal, CONCRETE, Vol. 34, No. 4, April 2000.
20. ACI 363R-92 (Revised 2010), State-of-the-Art Report on High-strength Concrete.
21. Ontario Provincial Standard Specifications—OPSS.MUNI 1359, Material Specifications for Unshrinkable Backfill—NOVEMBER 2016.

Chapter 2
Constituent Materials

2.1 Introduction

The properties of constituent materials and variation of the mix ingredients mostly affect the quality of concrete produced. The constituent materials of concrete can be classified into two groups: active and inactive. The active group consists of cement and water, whereas the inactive group comprises fine and coarse aggregates. Cement is the chemically active constituent, but its reactivity is only brought into effect on mixing with water. The aggregate plays no part in chemical reactions, but it serves as a filler material with good resistance to volume changes that take place within the concrete after mixing, and it improves the durability of the concrete.

The proportions of the constituent materials encountered in most concrete mixes are 25–40% for the active group, and 60–75% for the inactive group. In properly proportioned and compacted concrete, the voids are usually less than 2%. Fine aggregates consisting of natural or manufactured sand and coarse aggregate consisting of crushed stone, gravel, or air-cooled iron blast-furnace slag are used to produce "normal-density concrete" with a mass density of about 2400 kg/m³. "Low-density concrete" is produced by using low-density aggregates with a mass density of maximum 1850 kg/m³. When the constituent materials are mixed, cement in the presence of water hydrates and hardens, binding the aggregate particles to form a solid rock-like mass. The quality of cement paste mainly depends on water/cement ratio. All these constituent materials are further discussed in the following pages.

2.2 Cement

There are many types of cement which are artificially manufactured, and each cement is used under certain conditions due to its special properties. However, ordinary Portland cement is the most common type in use.

© Springer Nature Switzerland AG 2019
A. Surahyo, *Concrete Construction*,
https://doi.org/10.1007/978-3-030-10510-5_2

2.2.1 Portland Cements

Portland cements are hydraulic cements as they set and harden by reacting chemically with water and are composed primarily of hydraulic calcium silicates. When water is added to cement and the constituents are mixed to form cement paste, chemical reaction (hydration) takes place and the mix becomes stiffer with time and sets.

The Portland cement is primarily manufactured from limestone and chalk or marl combined with shale, clay, slate, blast-furnace slag, silica sand, and iron ore. The process of manufacture of cement consists essentially of grinding and mixing the raw materials in certain proportions and burning in a huge cylindrical steel rotary kilns lined with special firebrick at a temperature of up to about 1425–1650 °C when the material converts into balls known as clinker. This clinker comes out of the kiln as grey balls, about the size of marbles. After the clinker is cooled, cement plants grind it and mix it with small amounts of gypsum to control setting time, and the cement is then ready to use. The mixing and grinding of the raw materials can be done either in water or in dry condition; however, the most common way to manufacture Portland cement is through a dry method.

Cement must be sound; that is, it must not contain excessive quantities of certain substances such as lime, magnesia, and calcium sulphate that may expand on hydration or react with other substances in the aggregate and cause the concrete to disintegrate. Cements of different types should not be used together. Cement immediately absorbs moisture, and for this reason it is essential to keep cement away from any contact with water and from the effects of moisture at all stages, i.e. during transportation and storage, before its actual use in construction. Set cement is not suitable for use in construction.

2.2.1.1 Chemical Composition of Portland Cement

Tables 2.1 and 2.2 indicate main chemical compounds and oxide composition of Portland cement, respectively. The four compounds shown in Table 2.1 are treated as the major constituents of cement. The silicates C_3S and C_2S, which form about 70–80% of the cement, are responsible for the strength of cement. C_3S hydrate rapidly, releasing most of its heat in the 1st week, and develops early strength. C_2S hydrates slowly and contributes to development in strength after about 7 days. Generally, cements rich in C_2S result in a greater resistance to chemical attack.

The compound C_3A hydrates very quickly, releasing most of its heat in the 1st day, producing some increased strength within about 24 h, after which its contribution to strength is almost zero. C_3A is least stable, and cements containing more than 10% of this compound produce concretes which are prone to sulphate attack. Compound C_4AF is of less importance than the other three compounds, as it does not affect the behaviour significantly. However, it may increase the rate of hydration of the silicates.

Table 2.1 Main chemical compounds of Portland cement

Name of compounds	Chemical composition (oxide composition)	Usual abbreviation	Percentage by weight in cement
Tricalcium silicate	$3CaO \cdot SiO_2$	C_3S	25–50
Dicalcium silicate	$2CaO \cdot SiO_2$	C_2S	20–45
Tricalcium aluminate	$3CaO \cdot Al_2O_3$	C_3A	5–11
Tetracalcium aluminoferrite	$4CaOAl_2O_3 \cdot Fe_2O_3$	C_4AF	8–14

Table 2.2 Approximate chemical composition of Portland cement

Oxide	Content percentage
Lime, CaO	60–66
Silica, SiO_2	19–25
Alumina, Al_2O_3	3–8
Iron oxide, Fe_2O_3	1.0–5.0
Magnesia, MgO	0.1–4.0
Sulphur trioxide, SO_3	1–3
Alkalis ($Na_2O + K_2O$)	0.5–1.3

2.2.1.2 Types of Portland Cement

By changing the chemical composition of the cement by varying the percentage of the above four basic compounds shown in Table 2.1, various types of cement with different characteristics can be obtained. Different types of Portland cement are manufactured to meet various normal physical and chemical requirements for specific purposes.

ASTM-C150-17 Standard Specification for Portland cement [1] provides five basic types of Portland cement. Table 2.3 specifies these various types of Portland cements along with their uses. Additionally, types I to III are also available with air-entraining agent and are then denoted by letter "A":

Type IA: Normal, air entraining
Type IIA: Moderate sulphate resistance, air entraining
Type II (MH): When moderate heat of hydration and moderate sulphate resistance are desired
Type IIIA: High-early strength, air entraining

In Canada, the Canadian Standards Association (CSA) recognises six types of Portland hydraulic cements under Standard A3001 [2], as shown in Table 2.4. However, with an interest in the industry for performance-based specifications, ASTM C1157 [3] also describes same type of cements as CSA by their performance attributes. Comparing ASTM C1157 performance cements to ASTM C150 Portland cements, they have similar strength gain characteristics, can be used under identical environmental conditions, are indistinguishable during mixing and placement, and have the same durability characteristics.

Table 2.3 Types of Portland cement as per ASTM-C150-17

Types of cement		Uses
Type I	Ordinary Portland	For general purpose
Type II	Modified cement	For moderate sulphate resistance
Type II (MH)		Especially when moderate heat of hydration and moderate sulphate resistance are desired
Type III	High-early strength	For high-early strength
Type IV	Low-heat Portland	For low heat of hydration
Type V	Sulphate-resisting Portland	For high sulphate resistance

Table 2.4 Types of cement as per CSA and ASTM

Type	Name	Standard CSA/ASTM
GU	General use	A3001/C1157
HE	High-early strength	A3001/C1157
MS	Moderate sulphate resistance	A3001/C1157
HS	High sulphate resistance	A3001/C1157
MH	Moderate heat of hydration	A3001/C1157
LH	Low heat of hydration	A3001/C1157

The only major difference between the two cements is that the energy and carbon dioxide footprint are decreased 10% or more for ASTM C1157 performance cement compared with ordinary Portland cement and that is something extraordinary.

In the United Kingdom, the categories of cements are divided into the following five types based on their composition as per European standard BS EN 197-1-2000 [4] which has replaced most of the British standards:

CEM I: Portland cement
CEM II: Portland-composite cement
CEM III: Blast-furnace cement
CEM IV: Pozzolanic cement
CEM V: Composite cement

Based on the above main types of cement and combination, the composition of constituents added to Portland clinker of each commonly used cements is provided in Table 2.5.

2.2.1.3   Uses of Portland Cements

Ordinary Portland (Type I) Cement

This is the most commonly used cement in general concrete structures when the special properties specified for the other four types of cement are not required. The rate of strength development and heat evolution of ordinary Portland cement is medium. It is used in the same manner as GU cement.

Table 2.5 Types of cement and combination as per European Standard BS EN 197-1-2000 [4]

Cement notation		Composition of constituents added to Portland clinker
Designation	Cement type	Composition of constituents added to Portland clinker
Portland	CEM I	Clinker 95–100%[a]
Portland pozzolana	CEM II/A-P	Pozzolana 6–20%
	CEM II/B-P	Pozzolana 21–35%
Portland fly ash	CEM II/A-V	Fly ash 6–20% (pfa)
	CEM II/B-V	Fly ash 21–35% (pfa)
Portland slag	CEM II/A-S	Blast-furnace slag 6–20% (ggbs)
	CEM II/B-S	Blast-furnace slag 21–35% (ggbs)
Portland limestone	CEM II/A-L or LL	Lime 6–20%
Portland limestone	CEM II/B-L or LL	Lime 21–35%
Portland silica fume	CEM II/A-D	Silica fume 6–10%
Blast furnace	CEM III/A	Blast-furnace slag 36–65%
	CEM III/B	Blast-furnace slag 66–80%
Pozzolanic	CEM IV/A	Silica fume, pozzolana, and fly ash 11–35%
	CEM IV/B	Silica fume, pozzolana, and fly ash 36–55%
Composite cements	CEM V/A-M	Pozzolana and fly ash 18–30%
	CEM V/B-M	Pozzolana and fly ash 31–50%

Notes: Proportion of silica fume is limited to 10%. Minor additional constituents used are 0–5% *ggbs* ground granulated blast-furnace slag. A and B refer to high and medium clinker levels
Other letters: *S* blast-furnace slag, *D* silica fume, *P* natural pozzolana, *V* siliceous fly ash (e.g. *pfa* pulverised fuel ash). *M* a composite cement. The -L or -LL suffixes identify the total organic carbon (TOC) content of the limestone used: LL cements use limestone with a maximum TOC of 0.2% by mass, while L cements are made with limestones with a TOC of up to 0.5%
[a]Clinker for all other added ingredients varies from 20 to 94%

Modified (Type II) Cement

It is mostly used where a moderate exposure to sulphate attack or a moderate heat of hydration is required. These properties are attained by placing limitations on the C_3A and C_3S content of the cement. The type II cement gains strength a little more slowly than type I but ultimately reaches equal strength.

It is mostly used in normal structures or elements exposed to soil or groundwaters where sulphate concentrations are higher than normal but not severe. Type II cement in concrete must be used with a low w/c ratio and low permeability to control sulphate attack. Concrete exposed to seawater is mostly made with type II cement. Having the properties of moderate heat of hydration, it can also be used in structures having mass concrete. Its use reduces temperature-related cracking, which is especially important when concrete is placed in warm weather. It is used in the same manner as type MS and MH cement.

High-Early-Strength Portland (Type III) Cement

Type III Portland cement is used where high-early strength is required. The rate of strength development and heat evolution is high. Hence it is not advisable to use this cement for mass concreting. Concrete made with type III cement develops in 7 days the same strength that type I or II cements develop in 28 days. This high-early strength is achieved by increasing the C_3S and C_3A content of the cement and finer grinding.

It is used when early stripping of forms is required for reuse or when the structure must be put into service quickly. In cold weather its use permits a reduction in the length of the curing period. Although type I cement can also be used to gain high-early strength using higher cement content mixes, type III may provide it easier and more economically. Type III cement is used in the same way as HE cement.

Low-Heat Portland (Type IV) Cement

This cement is useful in mass concreting. The rate of strength development and heat evolution is low, but the ultimate strength is unaffected. Low heat of hydration in type IV cement is achieved by reducing the contents of C_3S and C_3A.

Type IV cement is particularly used in massive concrete structures, such as large gravity dams, where the temperature rise resulting from heat generated during hardening must be minimised. It is used in the same manner as LH cement.

Sulphate-Resisting (Type V) Cement

This cement is used when there is extensive exposure to sulphates. The rate of strength development and heat evolution is low to medium. The sulphate resistance of type V cement is achieved by reducing the C_3A content of cement to a maximum of 5%.

Use of a low w/c ratio and low permeability is critical to the performance of any concrete with type V cement exposed to sulphates. Type V cement concrete will not be fully effective in a severe sulphate exposure if the concrete has a high w/c ratio. Type V cement is not resistant to acids and other highly corrosive substances. It is used in the same way as type HS cement.

Air-Entraining Portland Cements

As mentioned above, ASTM C-150 [1] specifies three types of air-entraining Portland cements: type IA normal, air-entraining; type IIA moderate sulphate resistance, air-entraining; and type IIIA high-early strength, air entraining.

All these three cements correspond in composition to ASTM types I, II, and III, respectively, except that small quantities of air-entraining material are inter-ground with the clinker during manufacture.

These cements produce concrete with improved resistance to freezing and thawing and to scaling caused by chemicals applied for severe frost and ice removal. Concrete made with this cement contains millions of tiny, well-distributed, and completely separated air bubbles. However, mostly air entrainment for the required concrete is achieved through the use of an air-entraining admixture, rather than through the use of air-entraining cements.

White Portland Cements

White Portland cement differs from grey cement mainly in colour. It is made to conform ASTM C 150 specifications with respect to type I or type III, and CSA specifications, usually general-use (GU) or high-early-strength hydraulic cement (type HE). The manufacturing process is controlled so that the finished product is to be white.

White Portland cement is made from limestone or chalk and white china clay, which contains low iron oxide content and magnesium oxide. White Portland cement produces mortars of brilliant white colour and is, therefore, used for architectural applications. Its use is recommended wherever white or coloured concrete, grout, or mortar is desired. Figure 2.1 shows concrete prepared using white cement.

2.2.2 Blended Cements

Blended hydraulic cements are obtained by mixing/grinding Portland cement clinker with calcium sulphate (gypsum) and supplementary cementitious materials like fly ash, blast-furnace slag, silica fumes, and pozzolans. Blended cements are now

Fig. 2.1 Concrete prepared with white cement

being considered superior as compared to conventional Portland cement category of cements.

Blended cements are used in concrete construction in the same manner as Portland cements. Blended cements can be used directly to a concrete mix or they can be used in combination with other supplementary cementitious materials added at the concrete plant.

ASTM C 595 [5] recognises five primary classes of blended cements as follows:

Type IS: Portland blast-furnace slag cement
Type IP and Type P: Portland-pozzolan cement
Type I (PM): Pozzolan-modified Portland cement
Type S: Slag cement
Type I (SM): Slag-modified Portland cement

However, in Canada, CSA A3001 [2] recognises same Portland cements as mentioned in Table 2.4 with suffix "b" as blended hydraulic cement, such as the following:

GUb (general-use Portland blended cement)
MSb (moderate sulphate-resistant Portland blended cement)
MHb (moderate Heat of hydration Portland blended cement)
HEb (high-early-strength Portland blended cement)
LHb (low heat of hydration Portland blended cement)
HSb (high sulphate-resistant Portland blended cement)

Additionally, CSA A3001 also recognises Portland-limestone cements with suffix "L" as Portland-limestone cement such as:

GUL (general-use Portland-limestone cement)
MHL (moderate heat of hydration Portland-limestone cement)
HEL (high-early-strength Portland-limestone cement)
LHL (low heat of hydration Portland-limestone cement)

Initially, CSA-A23.1-14 [6] specified that Portland-limestone cements (PLC) should not be used in an environment subjected to sulphate exposure. CSA does not recognise sulphate-resistant Portland-limestone cement or blended Portland-limestone cement. However, with recent changes in the CSA-A23.1-14 Standard, concrete containing Portland-limestone cement is now permitted for use in sulphate-exposure environments. PLC-based concrete used in such exposure environments must first meet the mix design requirements of the CSA-A23.1-14 Standard and pass sulphate quality assurance tests as well [7].

2.2.2.1 Slag [Type IS, S, and I (SM)] Cements

Different types of slag cement are produced by inter-grinding varying proportions of granulated blast-furnace slag with ordinary Portland (type I) cement and gypsum or by blending finely ground granulated blast-furnace slag and ordinary

Portland (type I) cement. Blast-furnace slag is a waste product in the manufacture of pig iron.

1. Portland blast-furnace (type IS) cement

   The rate of strength development and heat evaluation is medium. It is suitable for mass concreting. Due to lower C_3A content, it is also used for sulphate-resisting purposes. It is produced by mixing up to 25–70% granulated blast-furnace slag with ordinary Portland (type I) cement [8]. Type IS may be used with air-entrainment (A), moderate sulphate resistance (MS), or moderate heat of hydration (MH) cements.

2. Slag-modified Portland [type I (SM)] cement

   The rate of strength development and heat evolution is moderate. It contains less than 25% slag and its properties are close to those of ordinary Portland (type I) cement. It can be used where moderate sulphate resistance is required or for general concrete construction. Type I (SM) may be used with air-entrained (A), moderate sulphate resistance (MS), or moderate heat of hydration (MH) cements.

3. Slag (type S) cement

   It contains up to 85% slag, 10–15% gypsum, and a small percentage of ordinary Portland (type I) cement [6]. Due to low heat evolution, it is useful for mass concrete work. It has high resistance to sulphate attack, as well as to peaty acids and oils. Its rate of strength development is strongly affected at low and high temperatures; hence it needs proper attention and should not be used alone in structural concrete. Type S cement can be used with air-entrained cements.

2.2.2.2 Pozzolanic [Type IP, P, and I (PM)] Cements

These cements are produced by grinding a pozzolanic material with ordinary Portland (type I) cement clinker. Pozzolans occur naturally as volcanic ash and are also obtained in the form of pulverised fuel ash (pfa), which is also known as fly ash. It is used in the manufacture of following Portland-pozzolan cements:

1. Portland-pozzolan (type IP and P) Cements

   These cements contain 15–40% pfa content [8]. Type P is similar to type IP, but early strength requirements are lower as it is generally made with a little higher pozzolan content.

   These cements are used to reduce the water demand of a concrete mix. The rate of strength development and evolution of heat is lower than the ordinary Portland (type I) cement; due to this property, they are useful for mass concreting. Like sulphate-resisting Portland (type V) cement, the pozzolanic cements have a high resistance to chemical attack.

2. Pozzolan-Modified Portland [Type I (PM)] Cement

   It contains less than 15% pozzolan content. The rate of strength development and heat evolution is moderate. Its properties are close to those of ordinary Portland (type I) cement.

2.2.3 Other Cements or Special Cements

2.2.3.1 High-Alumina Cement

High-alumina cement also known as calcium aluminate cement (CAC) or alumi-nous cement is composed of calcium aluminates, whereas Portland cement is com-posed of calcium silicates.

This cement is manufactured by melting a mixture of limestone or chalk and bauxite (aluminium ore) and finely grinding the cold mass. The chemical composi-tion of CAC may vary over a wide range, with Al_2O_3 contents ranging between 40 and 80%. The rate of strength development and heat evolution is very high. It devel-ops about 80% of the ultimate strength at the age of 24 h.

CAC is a special hydraulic cement, which is distinguished from ordinary Portland cement by its high-performance characteristics such as slow setting but very rapid hardening, high chemical resistance, high corrosion resistance, high resistance to acids, and high refractory properties. These superiorities of CAC enable it to be used within a wide spectrum in the construction industry as well as in other indus-tries such as the refractory industry.

Portland cement and calcium aluminate cement combinations have been used to make rapid setting concretes and mortars. After blended with one or more inorganic materials, e.g. Pozzolanic cement, ground-granulated blast-furnace slag (GGBFS), lime, and gypsum, CAC can be useful in such applications as repair mortars, self-levelling compounds, and tile adhesives.

Due to early strength property, it is mostly used in urgent repair works. Its use is not recommended for structural concrete as its strength is adversely affected by rise in temperature.

2.2.3.2 Coloured Portland Cement

It is made by the addition of mineral pigments to ordinary rapid-hardening or white Portland cements. A wide range of colours are available and many have been used successfully in concrete construction. Its properties are similar to those of ordinary Portland (type 1) cement. It is also used for architectural applications.

2.2.3.3 Waterproof or Water-Repellent Cement

This cement is manufactured by adding water-repelling agents or finely divided powders to ordinary Portland cement during mixing. It is used for waterproofing purposes. Water-repellent cements reduce capillary water transmission when there is little to no pressure but do not stop water vapour transmission. They are used in tile grouts, paint, and stucco finish coats.

2.2.3.4 Hydrophobic Cement

This cement is made by adding substances like stearic acid, boric acid, and oleic acid to ordinary Portland (type I) cement during grinding of cement clinker. A water-repellent film is formed by these acids around the cement particles that prevent the entry of atmosphere moisture. When concrete is mixed, this film breaks down and hydration takes place. Due to the water-repellent film, this type of cement can be stored under unfavourable conditions of humidity for long periods of time without any significant deterioration.

2.2.3.5 Low-Alkali Cement

This cement is produced by adding 0.60% of alkalis (Na_2O and K_2O) with Portland cement. It is used in concrete containing deleteriously reactive aggregates.

2.2.3.6 Shrinkage-Compensating or Expansive Cements

Expansive cement is a hydraulic cement that expands slightly during the early hardening period after setting. They are used to reduce cracking resulting from drying shrinkage. They expand a little during the first few days of hydration so as to cover the effects of shrinkage resulting after the concrete dries. As per ACI-223 [9], three types of expanding cements, type M, type K, and type S, are produced by adding different expansive materials (like aluminates and calcium sulphates) with Portland cement by various methods.

1. Type K expansive cement: contains Portland cement, tetracalcium trialuminate sulphate (C_4A_3S), calcium sulphate ($CaSO_4$), and lime (CaO)
2. Type M expansive cement: Contains Portland cement, calcium aluminate cement, and calcium sulphate
3. Type S expansive cement: Contains Portland cement with high tricalcium aluminate (C_3A) and an amount of calcium sulphate above the usual amount found in Portland cement

The aluminate compounds are added before or after the Portland cement clinker is made; hence approximately 90% of the constituents in each are identical to regular Portland cement. For this reason, expansive cements have almost identical properties, except for higher percentages of sulphates and aluminates, as those of Portland cement.

2.2.3.7 Masonry and Mortar Cements

Masonry and mortar cements are designed for use in mortar for masonry construction. They consist of a mixture of Portland cement or blended hydraulic cement (such as limestone or hydrated or hydraulic lime) and plasticising materials, together with other materials introduced to enhance one or more properties such as setting time, workability, water retention, and durability. An adequate amount of air entrainment is also added sometimes which enhances the important hardened property of mortar for durability, which is very important in the freeze/thaw climate areas.

ASTM C 91 [10], classifies masonry cements as type N, type S, and type M. However, CSA A179 [11] classifies only two types of masonry cements: type N and type S.

Types N, S, and M generally have increasing levels of Portland cement and higher strength, with type M having the highest strength. Type N is used most commonly. Mortar cement is similar to masonry cement in that it is a factory-prepared cement primarily used to produce masonry mortar. However, ASTM C 1329 places lower maximum air content limits on mortar cement than permitted for masonry cements; also, ASTM C 1329 is the only ASTM masonry material specification that includes bond strength performance criteria [12].

CSA A179 [11] Guide for the selection of mortars recommends mortar type S for load-bearing walls requiring high compressive strength on exterior and above-grade locations, whereas for same locations recommends type N for load-bearing walls requiring low-compressive-strength, non-load-bearing, and parapet walls.

For exterior at or below-grade level, CSA A179 recommends type S mortar to be used for foundation walls, retaining walls, manholes, sewers, pavements, sidewalks, and patios. For interior load-bearing and non-load-bearing walls CSA A179 [11] recommends type N mortar.

The workability, strength, and colour of masonry cements and mortar cements stay at a high level of uniformity because of manufacturing controls. In addition to mortar for masonry construction, masonry cements are also used in Portland cement-based plaster or stucco construction. Masonry cements and mortar cements are used for parging. Parging can be used as a base coat for damp proofing concrete block and poured concrete walls, or can even be used to provide a more aesthetically pleasing and smooth appearance.

Parging, which consists of a cementitious material, sand, and water, can be applied as a one-, two-, or three-coat system. A common parging mix would consist of one part type S masonry cement and three parts damp loose sand by volume, and enough water for the desired workability [13]. Masonry cement and mortar cement should not be used for making concrete.

2.2.3.8 Plastic Cements

Plastic cement is a hydraulic cement that meets the requirements of ASTM C 1328 [14]. It is formulated to have high degree of plasticity or workability so as to be used in plaster or stucco. Plastic cements consist of a mixture of Portland and blended hydraulic cement and plasticising materials (such as limestone, hydrated or hydraulic lime), together with materials introduced to enhance one or more properties such as setting time, workability, water retention, and durability.

ASTM C 1328 defines separate requirements for a type M and a type S plastic cement with type M having higher strength requirements. Plaster made from this cement must remain workable for a long enough time for it to be reworked to obtain the desired densification and texture. Plastic cement should not be used to make concrete.

2.2.3.9 Finely Ground Cements (Ultrafine Cements)

Finely ground cements, also called ultrafine cements, are hydraulic cements that are ground very fine for use in rock and soil injection for stabilising week soils and also for filling and sealing of cracks in concrete.

The cement particles are less than 10 µm in diameter with 50% of particles less than 5 µm. Blaine fineness often exceeds 800 m^2/kg [12]. These very fine cements consist of Portland cement, ground granulated blast-furnace slag, and other mineral additives.

2.2.3.10 Oil-Well Cements

Oil-well cements are usually made from Portland cement clinker or from blended hydraulic cements. They are specially used in sealing spaces between oil-well casings and linings and the surrounding rock. Generally they must be slow setting and resistant to high temperatures and pressures. The American Petroleum Institute (API) [15] includes requirements for eight classes of well cements (classes A through H) and three grades (grades O—ordinary, MSR—moderate sulphate resistant, and HSR—high sulphate resistant). Each class is applicable for use at a certain range of well depths, temperatures, pressures, and sulphate environments. The petroleum industry also uses conventional types of Portland cement with suitable cement modifiers. Expansive cements have also performed adequately as well as cements.

2.2.3.11 Magnesium Phosphate Cements

Magnesium phosphate cements are rapid-hardening, early-strength-gaining non-Portland cements. Magnesium phosphate cements are formed by the reaction of magnesium oxide with a soluble phosphate, such as ammonium phosphate, either the mono- or dibasic salt or an agricultural fertiliser solution known as 10-34-0 (NPK designation) can also be used [16]. This cement system has good water and freeze/thaw resistance. Commercial magnesium phosphate cements typically reach a compressive strength of about 2900 psi (20 MPa) after 1 h, with an ultimate strength of 8000 psi (55 MPa) [16].

They are usually used for special applications, such as repair of highway and airport pavements and concrete structures, or for resistance to certain aggressive chemicals. They may be available in two-part cements: consisting of a dry powder and a phosphoric acid liquid with which the powder is mixed, or they are available in one component product to which only water is added.

2.2.3.12 Sulphur Cements

Sulphur cements have excellent resistance to high concentrations of non-oxidising acids such as sulphuric, hydrochloric, and phosphoric acids. They are also resistant to moderate concentrations of oxidising acids such as nitric acid. They are available in different formulations such as sulphur regular cement (silica filled) and sulphur carbon cement (carbon filled). Sulphur regular and sulphur carbon cements are used to bond acid brick to protect floors, trenches, pickling tanks, process vessels, towers, manholes, and sewage systems. Sulphur carbon cement is used specifically in areas involving hydrofluoric acid and fluoride salts where sulphur regular cement would not be suitable.

Sulphur cement is also used with conventional aggregates to make sulphur cement concrete for repairs and chemically resistant applications. Sulphur cement melts at temperatures between 113 °C and 121 °C (235 °F and 250 °F). Sulphur concrete is maintained at temperatures around 130 °C (270 °F) during mixing and placing. The material gains strength quickly as it cools and is resistant to acids and aggressive chemicals. Sulphur cement does not contain Portland or hydraulic cement.

2.2.4 Physical Properties of Portland Cements

Portland cements are commonly characterised by their physical properties for quality control purposes. Owing to their different chemical compositions, each type of cement exhibits different properties. Their physical properties can be used to classify and compare Portland cements. Proper selection of cement to provide specific properties or to meet special service requirements can only be made if the influence of cement upon individual properties of concrete is understood. The principal physical properties of more common types of Portland cement are summarised and defined as below:

2.2.4.1 Fineness

Fineness or particle size is an important property of Portland cement, as it affects hydration rate and is thus used to determine the rate of gain of strength. It increases the workability of concrete mix, as due to fineness the mix becomes more cohesive.

The reaction between water and cement starts on the surface of the cement particles, hence the greater the surface area (means smaller the particle size) of a given volume of cement, the greater the hydration. For a given weight of cement, the surface area is more for finer cement in comparison to coarser cement. Thus finer cements, due to their greater surface area, will have a higher rate of hydration and an early strength gain.

In order to find the property of "fineness", the specific surface of cement is determined. The specific surface of cement can be determined by the air permeability method by BS-4550. The other method of ASTM C-115 or AASHTO T98 using Wagner turbid metre can also be used which is an older method for determining the size of cement particles.

The most common method for characterising the surface area of a cement is the Blaine air permeability test, which is described by ASTM-C 204 or AASHTO-T153. This test is based on the fact that the rate at which air can pass through a porous bed of particles under a given pressure gradient is a function of the surface area of the powder. A chamber of known cross-sectional area and volume is filled with a known mass of cement, and then the time required to pass a known volume of air through the powder is measured. While the surface area can in theory be calculated explicitly from this data, in practice the surface area is determined through an empirical equation developed by measuring powders of a known surface area using the same instrument. The resulting value, called the Blaine fineness, is expressed in units of m^2/kg, although in previous times it was expressed in cm^2/g. The Blaine fineness of OPC usually ranges from 300 to 500 m^2/kg (3000–5000 cm^2/g). ASTM requires a minimum of 280 m^2/kg for type I cement. Cements with a fineness below 280 m^2/kg may produce concrete with poor workability and excessive bleeding. The US Department of the Interior Concrete Manual indicates that the Blaine method correlates to about 1.8 times the Wagner method.

2.2.4.2 Setting Time

The time taken by cement paste to change from a fluid state to a rigid state is termed as setting time. The time at which the cement paste loses its plasticity and begins to stiffen considerably is known as initial setting time and when the paste becomes a hard mass and can sustain some load it is termed as the final setting time. The setting time is usually affected by the amount of mixing water, atmospheric temperature, cement composition, and fineness.

Setting tests are used to characterise how a particular cement paste sets. Most common methods used to evaluate setting time of cement paste are Vicat test method and Gillmore test method. As per Vicat method, initial setting occurs when a 1 mm needle penetrates 25 mm into cement paste. Final set occurs when there is no visible penetration. Gillmore method is less common than Vicat needle test; however as per this test, initial set occurs when a 113.4 g Gillmore needle (2.12 mm in diameter) fails to penetrate. Final set occurs when a 453.6 g Gillmore needle (1.06 mm in diameter) fails to penetrate. Gillmore times tend to be longer than Vicat times. ASTM C150 specifies set times by these both test methods as follows:

Vicat method: Initial set time ≥ 45 min and final set time ≤ 375 min

Gillmore method: Initial set time ≥ 60 min and final set time ≤ 600 min

Early stiffening of cement paste: Early stiffening is the early development of stiffness in the working characteristics or plasticity of cement paste, mortar, or concrete. This includes both false set and flash set.

False set: rapid stiffening without much heat generation. In this case, paste can be remixed without addition of more water to restore plasticity of the paste until it sets under normal procedure and without a loss of strength.

Flash set: rapid stiffening with considerable heat generation, and plasticity cannot be regained. When water is added to cement, the pure C_3A compound reacts very rapidly with water causing the cement to stiffen within a few minutes by generating a considerable amount of heat. This setting is termed as flash set. Such setting is avoided by the addition of appropriate amount of gypsum to the cement clinker.

2.2.4.3 Soundness

Soundness of Portland cement refers to the ability of a hardened cement paste to retain its volume after setting without delayed expansion. Normal cement hydration reactions occurring due to setting and hardening tend to produce small changes in the volume of the hydrating paste. If this change in volume of a cement paste is in excess, then such cement is termed as unsound. This expansion is caused by excessive amounts of free lime (CaO) or magnesia (MgO). The cement paste should not undergo large changes in volume after it has set. However, when excessive amounts of free CaO or MgO are present in the cement, these oxides can slowly hydrate and cause expansion of the hardened cement paste. Soundness is usually considered as the volume stability of the cement paste.

As per ACI, unsoundness was a serious problem for concrete in former times. In more recent times, better manufacturing, testing, and controls have almost completely eliminated unsound cement. The presence of such is detected by the autoclave expansion test ASTM C 151.

Unsound cements are not suitable for the manufacture of concrete and result in cracking and disintegration of surface of the concrete. Le-Chatelier's test and autoclave test specified by BS-4550 [17] (replaced by BS EN 196-3-1995) and ASTM-C151 [18], respectively, are generally performed to find the unsoundness. Accordingly, such expansion for various Portland cements should not be more than 10 mm as per Le-Chatelier's test and 0.8% in autoclave test.

The unsoundness may be reduced by thorough mixing and fine grinding, allowing cement to be aerated for several days and limiting the MgO content to less than 0.5%.

2.2.4.4 Hydration

When Portland cement and water are mixed, a series of chemical reactions take place resulting in setting, hardening, and evolution of heat of hydration and strength development. The overall process is known as cement hydration, usually expressed as joules or calories per gram. In fact, when water is added to cement, the silicates and aluminates of Portland cement form the products of hydration or hydrates (water-containing compound); hence, the reaction is referred to as hydration.

Two methods of determining the heat of hydration are commonly used. The most common method of determining the heat of hydration is by measuring the heats of solution of unhydrated and hydrated cement in a mixture of nitric and hydrofluoric acids and the difference between the two values represents the heat of hydration. The other common method is known as conduction calorimetry. In this procedure, a sample of cement is placed in a conductive container at a specific temperature. Water is added and the energy required to maintain the sample temperature is continuously recorded. By integration, the heat of hydration at any time can be obtained. ASTM C186 [19] proposes that the length of the test should be limited to 3 days as the rate of heat evolution becomes too low to measure beyond that time period.

Based on various tests conducted between 1992 and 1996, ASTM C186 specifies average heat of hydration for type I cement as 83.5 cal/g for 7 days and 95.6 cal/g for 28 days. Type III cement has higher heat of hydration than other cement types (average = 88.5 cal/g at 7 days) and type IV has the lowest (average = 55.7 cal/g at 7 days) [20]. Usually, the greatest rate of heat liberation occurs within the first 24 h and a large amount of heat evolves within the first 3 days.

The heat generated by the cement's hydration raises the temperature of concrete. Since concrete is a poor conductor of heat, the heat generated during hydration can have undesirable effects on the properties of the hardened concrete as a result of microcracking of the set cement paste. The rate of hydration depends on the properties of silicate and aluminate compounds of cement, the fineness of cement, and the temperature. The heat of hydration of cement can be reduced by reducing the proportions of C_3A and C_3S or by decreasing the fineness of cement. Other factors influencing heat development in concrete include the cement content, water/cement ratio, placing and curing temperature, presence of mineral and chemical admixtures, and dimensions of the structural element. In general, higher cement contents result in more heat development.

The water used for preparing concrete needs to be pure in order to prevent side reactions which may weaken the concrete or otherwise interfere with the hydration process. The role of water is important because w/c ratio is the most critical factor in the production of good concrete. Too much water reduces concrete strength, while too little will make the concrete unworkable.

2.2.4.5 Strength

The strength of hardened cement paste is its most important property. Fineness and chemical composition are the major characteristics of cement that influence the strength development of concrete. Generally, finer cements develop strength more early due to higher rate of hydration. For producing strength, components C_3A, C_3S, and C_2S are more responsible. Changing the properties of these components can vary the early strength. Some of the minor components of Portland cement also affect the strength. In particular, the quantity of calcium sulphate is normally chosen to optimise strength and other properties under the most common conditions of curing and use. Additionally, strength of cement decreases with increasing ignition loss. The loss on ignition of a cement is generally an indicator of the amount of water or carbon dioxide, or both, chemically combined with the cement.

Strength of cement is defined in three ways: compression, tensile, and flexural. Primarily, compression test is considered. Usually, neat cement paste is not tested directly for strength purpose due to large variations obtained in results. However, cement-sand mortar or concrete is used to determine the compressive strength of cement. Strength of cement sand paste can be affected by the number of items including w/c ratio, cement-sand ratio, type and grading of sand used, mixing and curing conditions, etc.

Compressive strength test is usually carried out on a 50 mm (2 in.) cement mortar test specimen. The test specimen is subjected to a compressive load until failure. This loading sequence must take no less than 20 s and no more than 80 s. As per ASTM C 150-07 compressive strength for type 1 Portland cement mortar is suggested as:

- 3 days 12.4 MPa (1800 psi)
- 7 days 19.3 MPa (2800 psi)
- 28 days 28.0 MPa (4060 psi)

However, minimum requirements of compressive strengths of masonry cement types N, S, and M specified by ASTM C109 [21] requirements, are produced in Table 2.6. Types of masonry cements are described earlier in this chapter under Sect. 2.2.3.7.

Table 2.6 Compressive strength of masonry cements

Compressive strength—minimum (average three cubes)	Type N	Type S	Type M
7 days psi (MPa)	500 (3.4)	1300 (9.0)	1800 (12.4)
28 days psi (MPa)	900 (6.2)	2100 (14.5)	2900 (20.0)

Table 2.7 Aggregate sieve analysis

Sieve size		Percentage retained
English	SI	
3/8″	9.5 mm	0
No. 4	4.75 mm	2
No. 8	2.36 mm	16
No. 16	1.18 mm	27
No. 30	600 μm	62
No. 50	300 μm	88
No. 100	150 μm	96

2.3 Aggregate

Aggregates comprise about 60–75% by volume of concrete mix. Normal-density aggregates are classified mainly into the following two sizes:

2.3.1 Coarse Aggregate

Particles larger than 5.00 mm (3/16 in.).

2.3.2 Fine Aggregate

Particles in between 0.075 mm and 5 mm (3/16 in.).

Coarse aggregates are generally gravels and crushed stone. Naturally occurring aggregates are called gravels. Natural gravel and sand are mostly dug or dredged from pit, lake, river, or seabed. Crushed aggregate is produced by crushing quarry rock, boulders, cobbles, or large-size gravels. Fine aggregate is basically sand.

There are two types of sand like river sand and manufactured sand. Manufactured sand is crushed from coarse aggregates through crushing machines. OPSS.PROV-1002 [22] specifies that manufactured sand made with carbonate rock (e.g., limestone and dolostone) shall not be accepted for use in a hydraulic cement concrete surface exposed to vehicular traffic due to the risk of polishing, unless the acid-insoluble residue as determined by MTO (Ministry of Transportation Ontario) test LS-613 is greater than 50%.

2.3.3 Classification of Aggregates

Aggregates normally can be classified as follows:

2.3.3.1 Natural Mineral Aggregate

The aggregate types that fall in this category are sand, gravel, and crushed rock derived from natural source, for example granite, basalt, limestone, sandstone, marble, and slate.

Sand and gravel have a bulk density of 95–105 lb/ft.^{3} (1520–1680 kg/m^3) and produce normal-weight concrete (NWC). NWC have unit weight: 150 lb/ft.^{3} (2400 kg/m^3).

Aggregates with bulk densities less than 70 lb/ft.^{3} (1120 kg/m^3) are called light-weight. Aggregates weighing more than 130 lb/ft.^{3} (2080 kg/m^3) are called heavyweight.

2.3.3.2 Synthetic Aggregates (Lightweight Aggregates)

The lightweight aggregates can be either natural or synthetic materials. The natural lightweight aggregates include pumice, scoria, tuff, etc. The aggregate falling in synthetic category are thermally processed materials. Expanded clay, shale, and slate are the most common synthetic aggregates used in structural concrete. Another type of aggregates are those which are made from industrial by-products, i.e. blast-furnace slag and fly ash. Lightweight aggregate types are also described in Chap. 1 under lightweight concrete heading.

The lightweight aggregates carry low bulk specific gravity. The most important aspect of lightweight aggregate is the porosity. They have high absorption values, which requires a modified approach to concrete proportioning. For instance, slump loss in lightweight concrete due to absorption can be an acute problem, which can be alleviated by pre-wetting the aggregate before batching.

2.3.3.3 Recycled Concrete Aggregates (RCA)

These aggregates are made from recycled concrete from demolished building, concrete sidewalks, bridges, and pavements. The waste concrete materials are thoroughly screened in order to remove any metal, scrap, or other impurities. It is then crushed down to a smaller aggregate size so that it can be reused for other construction and landscaping purposes.

After removal of contaminants, crushed concrete can be used in many new constructions such as new concrete for pavements, shoulders, sidewalks, curbs and gutters, and bituminous concrete. Portland cement association (PCA) [23] specifies that it is generally accepted when natural sand is used, up to 30% of natural crushed coarse aggregate can be replaced with coarse recycled aggregate without significantly affecting any of the mechanical properties of the concrete. As replacement amounts increase drying shrinkage and creep will increase and tensile strength and modulus of elasticity will decrease; however compressive strength and freeze/thaw resistance are not significantly affected.

In Canada, the majority of this material is used in place of natural aggregate for unbound road base applications. There has been little use of RCA in concrete. The CSA-Canadian Standards Association [6] specifies under annex-O that RCA may be used as a partial or total replacement of coarse aggregate for non-structural applications such as sidewalks, curb and gutter and some pavements or concrete base and unshrinkable fill (also known as controlled low-strength material or CLSM), and other low-risk applications. CSA further specifies that if recycled concrete aggregate is to be used in concrete, particular attention should be given to assessing durability characteristics; deleterious materials; potential alkali-aggregate reactivity; chloride contamination; and workability characteristics of concrete manufactured with that material.

CSA describes three main categories of RCA that are used as either aggregate in new concrete or granular base material.

(a) Construction and demolition waste (CDW)

CDW consists of materials arising from activities such as the construction or demolition of buildings and civil infrastructure, road planning, and maintenance. CDW can be mainly composed of concrete, but might also be contaminated with other demolition materials.

(b) Reclaimed concrete material (RCM)

RCM consists of used materials, such as hardened, hydraulic cement concrete that has been obtained from variable sources like sidewalks and concrete roads and construction and demolition waste (CDW).

(c) Returned crushed concrete (RCC)

RCC is unused concrete material obtained from plastic concrete that has been returned directly to the ready-mixed concrete plant, allowed to harden, and processed by crushing. It can be used for the same applications as CDW and RCM.

When demolished concrete is crushed, a certain amount of mortar and cement paste from the original concrete remains attached to stone particles in recycled aggregate. This paste reduces the specific gravity and increases the porosity compared to similar virgin aggregates.

Because of the attached mortar, recycled aggregate has significantly higher water absorption than natural aggregate. Therefore, to obtain the desired workability of RCA, it is necessary to add a certain amount of water to saturate recycled aggregate before or during mixing, if no water-reducing admixture is applied. One option is to first saturate recycled aggregate to the condition: water-saturated surface dry, and the other is to use dried recycled aggregate and to add the additional water quantity during mixing. The additional water quantity is calculated on the basis of recycled aggregate water absorption in prescribed time [24].

Another description of aggregates is given in ASTM C294 which classifies aggregates as follows:

Silica minerals (e.g., quartz, opal, chalcedony, tridymite, cristobalite), feldspars, micaceous minerals, carbonite minerals, sulphides, ferromagnesian minerals, zeolites, tron oxides, and clay minerals: These mineralogical classifications are of help in recognising properties of aggregates.

2.3.4   Properties of Aggregates

(a) *Cleanliness, soundness, and strength*: All these are important properties for any aggregate. Aggregates are considered clean if they are free of excess clay, silt, mica, and organic matter, chemical salts, and coated grains. An aggregate is physically sound if it retains dimensional stability under temperature or moisture change and resists weathering without decomposition. Aggregates must be resistant to degradation due to cycles of freezing and thawing, and wetting and drying, which are a function of the aggregate lithology, strength, absorption, and porosity. To be considered adequate in strength, an aggregate should be able to develop the full strength of the cementing matrix. When wear resistance is important, the aggregate should be hard and tough.

(b) *Shape and Texture*: Shape is classified mainly in four headings: rounded, irregular, flaky or elongated, and angular. However, surface texture is classified under six headings: glassy, smooth, granular, rough or pitted, crystalline, and honeycombed or porous. Surface texture is a measure of smoothness or roughness of the aggregate.

Both shape and surface texture affect the workability and possibly the density and strength of concrete. The workability increases with the smoother and round aggregates. Concrete made with sharp angular (crushed) aggregate is considerably less workable, but angular particles interlock better and can result in higher compressive strength due to good aggregate and mortar bond. An aggregate with a rough, porous texture is better than smooth surface aggregate, as the former can increase the aggregate-cement bond resulting in better compressive and flexural strength of concrete. Elongated or flaky particles have a detrimental effect on the workability of concrete, resulting in the necessity of more highly sanded mixes and the consequent use of more cement and water. As per OPSS.PROV-1002 [22], the percentage of flat or elongated particles should not exceed 20%.

(c) *The size and grading* are also important properties of aggregate as they also affect the workability and strength of concrete mix. In general, the larger the maximum size of the aggregate, the smaller the cement requirement for a particular water/cement ratio. This is due to the fact that the workability of concrete increases with the increase in the maximum size of the aggregate. In a mass concrete work, the use of a larger size aggregate is beneficial due to the lesser consumption of cement. This will also reduce the heat of hydration and corresponding thermal stresses and shrinkage cracks. However, in structural concrete, the maximum size is usually restricted to 25 or 40 mm (1 in. or 1½ in.) because of the size of the concrete section and spacing of reinforcement.

As for as grading of aggregate is concerned, it should be such that the smaller particles in a mix fill the voids between the larger particles. The proper grading of an aggregate produces dense concrete and needs less quantity of fine aggregate and cement paste. It is therefore essential that the coarse and fine aggregates should be well graded to produce quality concrete, for which the grading requirements of BS-882 and ASTM-C33 are reproduced in Tables 2.8, 2.9, and 2.10, respectively, along with grading requirements specified by Ontario Provincial Standard Specifications (OPSS), developed jointly by the Ministry of Transportation (MTO) and Municipal Engineers Association (MEA) that are

Table 2.8 Coarse aggregate grading requirements as per BS-882 [25]

| Sieve size | | Percentage by weight passing BS sieve | | | | | | | |
| | | Nominal size of graded aggregates | | | Nominal size of single-sized aggregates | | | | |
mm	in.	40–5 mm (1½ to $^3/_{16}$ in.)	20 to 5 mm (¾ to $^3/_{16}$ in.)	14 to 5 mm (½ to $^3/_{16}$ in.)	40 mm (1½ in.)	20 mm (¾ in.)	14 mm (½ in.)	10 mm ($^3/_8$ in.)	5 mm ($^3/_{16}$ in.)
50.0	2	100	–	–	100	–	–	–	–
37.5	1½	90–100	100	–	85–100	100	–	–	–
20.0	¾	35–70	90–100	100	0–25	85–100	100	–	–
14.0	½	25–55	40–80	90–100	–	0–70	85–100	100	–
10.0	$^3/_8$	10–40	30–60	50–85	0–5	0–25	0–50	85–100	100
5.0	$^3/_{16}$	0–5	0–10	0–10	–	0–5	0–10	0–25	45–100
2.36	No. 7	–	–	–	–	–	–	0–5	0–30

Table 2.9 Coarse aggregate grading requirements as per ASTM-C33 [26]

| Sieve size | | Percentage by weight passing sieve | | | |
| | | Nominal size of graded aggregate | | | |
mm	in.	37.5–4.75 mm (1½ in to No. 4)	25.0–4.75 mm (1 in. to No. 4)	19.0–4.75 mm (¾ in. to No. 4)	12.5–4.75 mm (½ in. to No. 4)
75	3	–	–	–	–
63.0	2½	–	–	–	–
50.0	2	100	–	–	–
37.5	1½	95–100	100	–	–
25.0	1	–	95–100	100	–
19.0	3/4	35–70	–	90–100	100
12.5	1/2	–	25–60	–	90–100
9.5	3/8	10–30	–	20–55	40–70
4.75	No. 4	0–5	0–10	0–10	0–15
2.40	No. 8	–	0–5	0–5	0–5

produced in Tables 2.11, 2.12, and 2.13. However, lightweight aggregate shall meet the requirements of ASTM C 330/C 330M.

(d) *Free Moisture and Absorption of Aggregates*: The moisture content and absorption of aggregates are important in calculating the proportions of concrete mixes since any excess water in the aggregates will be incorporated in the cement paste and give it a higher water/cement ratio than expected. All moisture condi-

Table 2.10 Fine aggregate grading requirements as per BS-882 [25] and ASTM-C33 [26]

Sieve size		Percentage by weight passing sieve				
		BS-882				
			Additional limits			
BS #	ASTM #	Overall limits	Coarse	Medium	Fine	ASTM C33
10 mm	$\frac{3}{8}$ in.	100	–	–	–	100
5 mm	4 in	89–100	–	–	–	95–100
2.36 mm	8 in	60–100	60–100	65–100	80–100	80–100
1.18 mm	16 in	30–100	30–90	45–100	70–100	50–85
600 µm	30 in	15–100	15–54	25–80	55–100	25–60
300 µm	50 in	5–70	5–40	5–48	5–70	10–30
150 µm	100 in	0–15	–	–	–	2–10

Table 2.11 Fine aggregate grading requirements as per OPSS.PROV 1002 [22] MTO Test # LS-602

MTO sieve designation	Percentage passing
9.5 mm	100
4.75 mm	95–100
2.36 mm	80–100
1.18 mm	50–85
600 µm	25–60
300 µm	10–30
150 µm	0–10
75 µm	0–3 (natural sand) 0–6 (manufactured sand)

Fine aggregates shall have no more than 45% passing any sieve and retained on the next consecutive sieve. The fineness modulus shall be a minimum of 2.3 and a maximum of 3.1
MTO Ministry of Transportation Ontario

Table 2.12 Coarse aggregate grading requirements as per OPSS.PROV 1002 [22] Structural concrete, sidewalks, curb and gutter—MTO test # LS 602

Nominal maximum size	19.0 mm	16.0 mm	13.2 mm	9.5 mm	6.7 mm
MTO sieve designation (mm)	Percentage passing				
26.5	100	–	–	–	–
19.0	85–100	100	100	–	–
16.0	65–90	96–100	–	–	–
13.2	–	67–86	90–100	100	100
9.5	20–55	29–52	40–70	85–100	–
6.7	–	–	–	–	75–100
4.75	0–10	0–10	0–15	10–30	40–80
2.36	–	–	–	0–10	0–20

Table 2.13 Coarse aggregate grading requirements as per OPSS.PROV 1002 [22] concrete pavement or concrete base—MTO test # LS 602

Nominal maximum size	37.5 mm	19.0 mm	Combined
MTO sieve designation (mm)	Percentage passing		
53.0	100	–	100
37.5	90–100	–	95–100
26.5	20–55	100	–
19.0	0–15	85–100	35–70
9.5	0–5	20–55	10–30
4.75	–	0–10	0–5

tions are expressed in terms of oven dry unit weight. The different moisture conditions of aggregates are described as follows:

- *Oven-dry condition*: All free moisture, whether external surface moisture or internal moisture, driven off by heat.
- *Air dry*: No surface moisture, but some internal moisture remains.
- *Saturated-surface dry condition (SSD)*: Aggregates are said to be SSD when their moisture states are such that during mixing they will neither absorb any of the mixing water added, nor will they contribute any of their contained water to the mix. Note that aggregates in SSD condition may possess "bound water" (water held by physical chemical bonds at the surface) on their surfaces since this water cannot be easily removed from the aggregate.
- *Damp or wet condition*: Aggregate containing moisture in excess of the SSD condition. The free water, which will become part of the mixing water, is in excess of the SSD condition of the aggregate.

(e) *Specific Gravity and Bulk Density*: Specific gravity is the ratio of the weight of a given volume of aggregate to the weight of an equal volume of water. Water, at a temperature of 73.4 °F (23 °C), has a specific gravity of 1. An aggregate with a specific gravity of 2.50 would thus be two and one-half times as heavy as water. Specific gravity is important for several reasons.

Each aggregate particle is made up of solid matter and voids that may or may not contain water. Since the aggregate mass will vary with its moisture content, specific gravity is determined at a fixed moisture content. Four moisture conditions are defined for aggregates (as explained above in section d) depending upon the amount of water held in the pores or on the surface of the particles. Most aggregates have a specific gravity between 2.4 to 2.9. In general a low specific gravity indicates a porous, weak and absorptive aggregate while a high specific gravity represents good quality.

In Portland cement concrete the specific gravity of the aggregate is used in calculating the percentage of voids and the solid volume of aggregates in computations of yield. The absorption is important in determining the net water/cement ratio in the concrete mix. Knowing the specific gravity of aggregates is also critical to the construction of water filtration systems, slope stabilisation projects, railway bedding, and many other applications.

Bulk Density: The bulk density (previously unit weight and sometimes called dry-rodded unit weight) of an aggregate is the mass of aggregate divided by the volume of particles and the voids between particles. The bulk density is used in estimating quantities of materials and in some mixture proportioning calculations. It affects concrete's mix design, workability, and unit weight.

If the moisture content of the aggregate varies, its bulk density will also vary. For coarse aggregate, increasing moisture content increases the bulk density, but for fine aggregate increasing the moisture content beyond the saturated surface-dry condition can cause the bulk density to decrease. This is because thin films of water on the sand particles cause them to stick together so that they are not as easily compacted. The resulting increase in volume decreases the bulk density. This phenomenon is called bulking and is of little importance if the aggregates for a concrete mixture are batched by mass. However, if volumetric batching is used, bulking must be taken into account when moisture content varies. Other properties that affect the bulk density of an aggregate include grading, specific gravity, surface texture, shape, and angularity of particles [27].

(f) *Fineness Modulus (FM)*: FM is the sum of total percentages retained on each specified sieve divided by 100. The standard sieves are 6″, 3″, 1½″, ¾″, $\frac{3}{8}$″, No. 4, No. 8, No. 16, No. 30, No. 50, and No. 100. The FM is an index to the fineness or coarseness of an aggregate. ASTM C 33 and OPSS.PROV-1002 require the FM of fine aggregate to be between 2.3 and 3.1. Example 2.1 below describes the mathematical calculation of fineness modulus of an aggregate.

Example 2.1
Based on values in Table 2.7, the fineness modulus will be:

$$\frac{2+16+27+62+88+96}{100} = 2.91$$

The higher the FM, the coarser the aggregate. Fine aggregate affects many concrete properties, including workability and finishability. Usually, a lower FM results in more paste, making concrete easier to finish. For the high cement contents used in the production of high-strength concrete, coarse sand with an FM around 3.0 produces concrete with the best workability and highest compressive strength. The fineness modulus of a fine aggregate is calculated from the sieve analysis. It is used for the purpose of estimating the quantity of coarse aggregate to be used in the mix design.

The American Concrete Institute (ACI) has developed Table 4.9 (Chap. 4). This table for various size coarse aggregates gives the volume of dry-rodded coarse aggregate per unit volume of concrete for different fineness moduli of sand. If the size of coarse aggregate and fineness modulus is known, the volume of dry-rodded coarse aggregate can be obtained from this table. The proportion of coarse and fine aggregate in a concrete mix depends upon the fineness modulus of the fine aggregate and the size of the coarse aggregate.

The volume relationship in the ACI table actually relates to the total surface area of the aggregate, or the water demand of the aggregate. For example, if the FM is constant, the volume of coarse aggregate increases with the size of the aggregate, or with the decrease in surface area of the coarse aggregate. Likewise, as the fineness modulus of the fine aggregate decreases for any one size of coarse aggregate the volume of the coarse aggregate increases. As the particle size of the fine aggregate decreases, the surface area increases.

2.4   Water

Water is an important constituent in concrete. Water for use in concrete and curing shall be obtained from an approved source and shall be of such a quality as not to affect setting time, strength, durability, reinforcement, and appearance of the hardened concrete.

A part of the water used in concrete is utilised for hydration of cement causing it to set and harden. The remaining water facilitates mixing, placing, and compacting of fresh concrete. This remaining water must be kept to the minimum, as too much water reduces the strength of concrete.

2.4.1   Mixing Water

Mixing water added for making concrete impacts concrete slump and is used to determine water-to-cementitious material ratio (w/c) of the concrete mixture. Mixing water for concrete should be clean and free from objectionable quantities of organic matter, silt, clay, acids, alkalis, and other salts and sewage. Generally, water fit for drinking is suitable for mixing concrete. As a rule, any water with a pH of 6.0–8.0, which does not test saline, is suitable for use. CSA A23.1-14 [28] specifies that water of unknown quality, including treated wash water and slurry water, shall not be used in concrete unless it produces 28-day concrete strengths equal to at least 90% of a control mixture. The control mixture shall be produced using the same materials, proportions, and a known acceptable water.

The use of seawater causes surface dampness, efflorescence, and staining. Such water should not be used where appearance of the concrete is of importance or where a plaster finish is to be applied. Seawater also increases the risk of corrosion of steel; accordingly its use in reinforced concrete is not recommended.

2.4.2 Curing Water

In general, water satisfactory for mixing is also suitable for curing purposes. However, it is essential that curing water be free from substances that attack hardened concrete. For example, water containing free carbon dioxide (CO_2) attacks concrete, which results in surface erosion. Iron or organic matter may cause staining. Curing with seawater may lead to corrosion of reinforcement.

The use of impure water for washing aggregates can adversely affect strength and durability if it deposits harmful substances on the surface of the particles. The quality of mixing and curing water is further discussed in Chap. 6.

2.5 Admixtures and Additives

Admixtures are materials added to concrete in order to modify its properties. Pigments and other materials, which impart their own properties to the concrete, are termed additives. The American Concrete Institute defines chemical admixtures as "A material other than water, aggregates, hydraulic cement, and fiber reinforcement, used as an ingredient of a cementitious mixture to modify its freshly mixed, setting, or hardened properties and that is added to the batch before or during its mixing".

Since admixtures may also have detrimental effects, their suitability for a particular concrete should be carefully evaluated before use, based on a knowledge of their main active ingredients on available performance data and on trial mixes. It should be remembered that admixtures are not intended to replace good concreting practice and should not be used indiscriminately.

2.5.1 Admixtures

Admixtures are classified according to their function such as air entraining, water reducing, retarding, accelerating, and plasticisers (super-plasticisers). All other varieties of admixtures fall into the specialty category whose functions include corrosion inhibition, shrinkage reduction, alkali-silica reactivity reduction, workability enhancement, bonding, damp proofing, and colouring.

2.5.1.1 Accelerators

These admixtures are used to accelerate the rate of setting and hardening of concrete. As they increase the rate of early strength development, they are useful for quick repairs and where early removal of formwork is required for reuse. Since they accelerate the hydration of cement and increase the rate of evolution of heat, they are very useful for concreting in cold weather. The strength development of concrete can also be accelerated by other methods such as using type III or type HE (high-early-strength) cement and lowering the water/cement ratio.

Calcium chloride ($CaCl_2$) is the most commonly used chemical in accelerating admixtures, especially for non-reinforced concrete. Besides accelerating strength gain, calcium chloride causes an increase in drying shrinkage, potential reinforcement corrosion, discoloration (a darkening of concrete), and potential for scaling. However, it may be noted that calcium chloride is not an antifreeze agent. When used in allowable amounts, it will not reduce the freezing point of concrete by more than a few degrees. Hence concrete protection from freezing by this method is not advisable.

The other common admixtures used as accelerators are aluminium chloride, sodium carbonate (washing soda), potassium carbonate, ferric salts, calcium formate, calcium nitrate, calcium nitrite, sodium thiocyanate, and combinations of these.

2.5.1.2   Retarders

These admixtures are used to delay the setting and hardening of concrete. They are useful particularly in hot weather and mass concreting where high temperatures can reduce the normal setting and hardening times. It is also used in ready-mixed concrete to gain more time for delivering concrete to far sites. The materials used as retarding admixtures can be sugar, carbohydrate derivatives, soluble zinc salts, soluble borates [29], and tartaric acid and salts. In general, using set retarders, some of the benefits achieved can be summarised as:

- Improved workability
- Reduced segregation
- Flexibility in scheduling of placing and finishing operations
- Higher delivery time
- Helping eliminate cold joints
- Reduced thermal cracking in mass concrete
- Increased compressive and flexural strengths

2.5.1.3   Air-Entraining Agents

These admixtures are used to increase the resistance of concrete to frost and deicing salts by introducing tiny air bubbles with diameters less than 1 mm (0.04 in.) into the hardened concrete paste. These minute air bubbles are distributed uniformly throughout the cement paste. Entrained air can be produced in concrete by use of an air-entraining cement, by introduction of an air-entraining admixture, or by a combination of both methods. An air-entraining cement is a Portland cement with an air-entraining addition inter-ground with the clinker during manufacture. However, an air-entraining admixture is added directly to the concrete materials either before or during mixing.

Dosage rates of air-entraining admixtures generally range from 15 to 130 mL per 100 kg of cementitious material. Higher dosages are sometimes required depending on the materials and mixture proportions. For example, concrete containing fly ash or other pozzolans often requires higher doses of air-entraining admixture to achieve the same air content compared to a similar concrete using only Portland cement.

Materials in use that have the property of entraining air are natural wood resins, animal fats and oils, sulphonated soaps, some synthetic detergents such as salts of organic acids and sulphonated hydrocarbons, salts of proteinaceous material, fatty and resinous acids and their salts, etc. These admixtures have proved very useful in the construction of concrete road and runway pavements in cold weather. They also improve the workability and cohesiveness of fresh concrete and reduce bleeding and segregation.

2.5.1.4 Workability Agents/Water Reducers/Plasticisers

These admixtures are used to increase the workability of concrete without an increase in water content. By using these admixtures, water/cement ratio can be lowered that produces concrete of higher strength. The effectiveness of water reducers on concrete is a function of their chemical composition, concrete temperature, cement composition and fineness, cement content, and presence of other admixtures.

The materials used for improving the workability of concrete are lignosulphonates, hydroxylated carboxylic acids, sulphonated melamine, and inorganic materials, such as zinc salts, chlorides, and phosphates.

Water-reducing agents are further grouped as mid-range and high-range admixtures (super-plasticisers).

Mid-range water-reducing admixtures provide significant water reduction (between 6 and 12%) for concretes with slumps of 125–200 mm (5–8 in.) without the retardation associated with high dosages of normal water reducers. Normal water reducers are intended for concretes with slumps of 100–125 mm (4–5 in.). Mid-range water reducers can be used to reduce stickiness and improve process of finishing, pumping, and placing of concretes containing silica fume and other supplementary cementing materials. The most common materials used are lignosulphonates and polycarboxylates.

High-range water-reducing admixtures or super-plasticisers can greatly reduce water demand and cement contents and make low-water/cement-ratio, high-strength concrete with normal or enhanced workability. A water reduction of 12–30% can be obtained through the use of these admixtures [30]. The reduced water content and water/cement ratio can produce concretes with (1) ultimate compressive strengths in excess of 70 MPa (10,000 psi), (2) increased early strength gain, (3) reduced chloride-ion penetration, and (4) other beneficial properties associated with low water/cement ratio concrete. The most common materials used are sulphonated naphthalene condensates, sulphonated melamine condensates, modified lignosul-phonates, etc [31].

2.5.1.5 Waterproofing/Water-Repelling Agents

Waterproofing admixtures for concrete are also called as water-resisting admixtures and permeability-reducing admixtures. Concrete cannot be made completely imper-meable by the use of admixture. A concrete having proper mix design, low water/cement ratio, and sound aggregate, fully compacted and adequately cured, will result in impervious concrete without any admixture. However, the resistance of concrete to the absorption of fluids can be improved by adding water-repelling agents.

ACI 212 [31] divides permeability-reducing admixtures into two categories: permeability-reducing admixture for hydrostatic conditions (PRAH), and permeability-reducing admixture for non-hydrostatic conditions (PRAN). PRAHs are used in concrete that is exposed to water under pressure and are sometimes called waterproofing admixtures. Materials include hydrophilic crystalline chemi-cals that react with cement and water to grow pore-blocking deposits or polymer globules that pack into pores under pressure.

PRANs are used for applications that are not subject to hydrostatic pressure, and are sometimes called damp-proofing admixtures. Most PRANs contain water-repellent chemicals that shed water and reduce water absorption into the concrete. Most common water repellents used are animal fats, vegetable oils, resin, petroleum oils and waxes, soaps, etc. Other PRANs are finely divided solids such as talc, ben-tonite, colloidal silica, and silicates. These fillers reduce water migration through pores, although not to the same degree as a PRAH, and are sometimes called densifiers.

2.5.1.6 Gas-Forming Agents

The most commonly used gas-forming agent is aluminium powder. It is used in minute quantities (0.005–0.2% of the weight of cement) [32], which reacts with free lime in the cement to produce hydrogen gas. This action causes a slight expansion in plastic concrete and reduces voids caused by normal settlement, which occur dur-ing the placement of concrete, thus preventing bleeding.

Gas-forming agents are used to produce lightweight concrete. The gas improves effectiveness and homogeneity of grouted concrete and hence is useful for filling joints and holes in concrete and grouting for foundation bolts etc. These materials are also used in larger quantities to produce autoclaved cellular concretes. The amount of expansion that occurs is dependent upon the amount of gas-forming material used, the temperature of the fresh mixture, the alkali content of the cement, and other variables. Where the amount of expansion is critical, careful control of mixtures and temperatures must be exercised. Gas-forming agents will not overcome shrinkage after hardening caused by drying or carbonation.

2.5.1.7 Bonding Admixtures

These admixtures are added to Portland cement mixtures to improve the bonding properties of fresh concrete to hardened concrete. Flexural strength and resistance to chloride-ion ingress are also improved. They are added in proportions equivalent to 5–20% by mass of the cementing materials, the actual quantity depending on job conditions and type of admixture. The ultimate result obtained with a bonding admixture will be only as good as the surface to which the concrete is applied. However the required surface must be dry, clean, sound, free of dirt, dust, paint, and grease, and at the proper temperature.

Bonding agents are mostly polymer emulsions (latexes). Polymer includes polyvinyl chloride, polyvinyl acetate, acrylics, etc. When a bonding agent is sprayed on a concrete surface, the pores in the concrete absorb the water and allow the resin particles to set and bond. Bonding agents should not be confused with bonding admixtures. Admixtures are an ingredient in the concrete; bonding agents are applied to existing concrete surfaces immediately before the new concrete is placed. Bonding agents help "glue" the existing and the new materials together.

2.5.1.8 Corrosion-Inhibiting Admixtures

Corrosion inhibitors are used in concrete for parking structures, marine structures, and bridges where chloride salts are present. The chlorides can cause corrosion of steel reinforcement in concrete. Corrosion inhibitors improve the natural ability of concrete to protect embedded reinforcement by forming passivating oxide layer on the steel.

The most commonly used corrosion-inhibiting admixture is calcium nitrite. The other materials used are sodium benzoate, calcium lignosulphonate, sodium nitrite, phosphates, ester amines, etc. A certain amount of nitrite can stop corrosion up to some level of chloride ion. Therefore, increased chloride levels require increased levels of nitrite to stop corrosion.

2.5.1.9 Expansion-Producing Admixtures

These admixtures are used to reduce the effects of drying shrinkage. These admixtures expand during the period of hydration of the concrete or react with other constituents of the concrete to produce expansion.

Propylene glycol and polyoxyalkylene alkyl ether have been used as shrinkage reducers. Drying shrinkage reductions of between 25 and 50% have been demonstrated in laboratory tests [30]. These admixtures have negligible effects on slump and air loss, but can delay setting.

2.5.1.10   Pumping Aids

Concrete pumping is frequently interrupted by stoppages due to a blocking of the pipes, etc. To improve concrete pumpability, pumping aids are added to concrete mixtures. These admixtures increase viscosity or cohesion in concrete as well as gliding ability of fresh concrete and reduce dewatering of the paste while under pressure from the pump. Some pumping aids may increase water demand, reduce compressive strength, cause air entrainment, or retard setting time. These side effects can be corrected by adjusting the mix proportions or adding another admixture to offset the side effect.

Common materials used are organic and synthetic polymers, organic flocculents, organic emulsions of paraffin, coal tar, asphalt, acrylics, bentonite and pyrogenic silicas, and hydrated lime (ASTM C 141). Some admixtures that serve other primary purposes but also improve pumpability are air-entraining agents, and some water-reducing and -retarding admixtures.

2.5.2   Supplementary Cementitious Materials (SCMs)/Additives

SCMs are also referred to as mineral admixtures or additives. They may be used individually or in combination in concrete. They can be added to concrete mixes as a blended cement or as a separately batched ingredient at the ready-mixed concrete plant. The principal cementitious material in most concrete mixes is Portland cement. However, other SCMs are added to concrete as part of the total cementitious system, either as an addition or as a partial replacement of Portland cement.

Supplementary cementing materials are often added to concrete to make concrete mixtures more economical, reduce permeability, increase strength, or influence other concrete properties. The most common supplementary cementing materials used in today's concrete mixes are fly ash, silica fume, natural pozzolans, and slag (granulated blast-furnace slag).

2.5.2.1   Pozzolans

The pozzolanic materials in use are fly ash (coal fly ash) or pulverised fuel ash (pfa), natural pozzolans, and silica fume. Pozzolans are usually siliceous or siliceous and aluminous materials. Fly ash is a residue from the combustion of pulverised coal collected from the fuel gases of thermal power plants.

Fly Ash: There are two classes of fly ash: "F" is made from burning anthracite and/or bituminous coal, and "C" is produced from lignite or subbituminous coal.

However, in Canada, CSA defines three types of fly ash based on calcium oxide (CaO) content. Type F fly ash has a CaO content less than 15%, type CI (C intermediate) fly ash has a CaO content range that spans more than 15% and less than 20%, and type CH fly ash has CaO content that exceeds 20% [33]. Fly ash used in concrete should conform to CSA A3001 or ASTM C618.

Fly ash benefits [34]:

- Improved workability
- Decreased heat of hydration.
- Reduced concrete cost
- Improved resistance to sulphate attack
- Improved resistance to alkali-silica reaction (ASR)
- Higher long-term strength
- Decreased shrinkage and permeability

Fly ash concerns:

- Fly ash reduces the amount of air entrainment, and concrete mixtures high in fly ash often require more air-entraining admixture.
- Slow initial strength gain.
- Longer setting time.
- Fly ash variability.

Silica Fume: Silica fume is a by-product resulting from the reduction of high-purity quartz with coal and wood chips in an electric arc furnace during the production of silicon metal or ferrosilicon alloys. Silica fume is an extremely fine powder, with particles on average 100 times smaller than cement particles. Silica fume is pozzolanic and provides no hydraulic properties. However, it is highly pozzolanic and very effective when used as a blended ingredient with ordinary Portland cement [35]. Because it has a very fine particle size, silica fume results in an increased water demand, leading to the use of high-range water reducers to maintain or decrease the water-to-cementitious (w/cm) ratio of the mixture.

Silica fume is available as a densified powder. Silica fume used in concrete should also conform to CSA A3001 or ASTM C1240. CSA 3001 defines two types of silica fume: type SF and type SFI based upon silicon dioxide (SiO_2) content. Type SF silica fume has a SiO_2 content of at least 85%. Type SFI has a SiO_2 content of at least 75% [33].

These additives are used to increase the workability, impermeability, and resistance to chemical attack. Silica fume is a very effective pozzolan and, when combined with the significant decrease in permeability provided, silica fume is very effective at mitigating ASR (Alkali silica reaction) and sulphate attack. Since these additives retard the rate of setting and hardening of concrete, they are useful in mass concrete work.

Natural Pozzolans: Various naturally occurring materials possess or can be processed to possess pozzolanic properties. These SCMs are also covered under CSA A3001 and ASTM C618 (AASHTO M 295). Natural pozzolans are generally derived from volcanic origins. In Western Canada, natural pozzolans have been commercially produced on a limited basis. Metakaolin has been produced by controlled calcining of kaolinite clay in southwestern Saskatchewan. This product has been used at rates of 5–15% by weight of cementitious materials [36]. Other natural pozzolans include diatomaceous earth, opaline cherts and shales, pumicites, volcanic glass, zeolites, and rice husk ash.

2.5.2.2   Slag Cement (Granulated Blast-Furnace Slag)

It is a waste product obtained during the manufacture of pig iron. These additives are added to the concrete as a separate cementing material or as an ingredient of blended cements. Slag is changed to glassy sand-like substance and then it is ground known as ground granulated blast-furnace slag—GGBFS. Its composition is very similar to OPC but reacts with water slower than OPC. Slag cement is hydraulic and produces calcium silicate hydrate (CSH) as a hydration product.

Slag cement is not pozzolanic but it does consume calcium hydrate (CH) by binding alkalis in its hydration products. Therefore, although it is a hydraulic cement, it provides the benefits of a pozzolan. Most notably, although curing of any concrete is essential for achieving a quality product, it is even more critical with slag-cement-based concrete. Due to the lower reaction rate, especially at lower temperatures slag-cement-based concrete needs to be cured for more period than the normal OPC concrete. However, the slower reaction rate and associated heat evolution make slag cement an ideal ingredient for mass concrete placement where control of internal temperatures is critical to achieving durability. Slag cement is commonly used as a partial substitute for Portland cement in concrete and can be used at Portland cement replacement rates of up to 80% particularly in a mass concrete [35]. It is used to produce blast-furnace cement, the advantages of which are explained earlier in this chapter.

Slag cement benefits and concerns:

- Increased setting time compared to OPC concrete.
- Increased curing period compared to OPC concrete.
- Slow setting and lower heat evolution make slag cement an ideal ingredient for mass concrete placement.
- Up to 80% replacement of OPC with slag cement can be used for mass concrete.
- Slag cement is effective at mitigating alkali-silica reaction (ASR).

OPSS.PROV 1350 [37] specifies that the use of supplementary cementing material shall be restricted to the following proportions by mass of the total cementing material:

(a) Slag up to 25%.
(b) Fly ash up to 10%, except for silica fume overlays and high-performance concrete (HPC), where up to 25% is permitted.

(c) A mixture of slag and fly ash up to 25%, except that the amount of fly ash shall not exceed 10% by mass of the total cementing materials, in concrete other than silica fume overlays and HPC (High performance concrete).

2.5.2.3 Pigments

Pigments are added mostly to produce coloured concrete products, such as concrete bricks, concrete roofing tiles, pedestrian and vehicle paving, mortars, stucco, ready-mixed concrete, architectural panels, and many more precast and in situ products. Colouring pigments are normally used in concrete for architectural purposes and the best effect is produced when they are inter-ground with the cement clinker rather than added during mixing.

Pigments normally do not affect the concrete properties, but those based on carbon may cause some loss of strength at early ages. Although there are a number of methods for colouration of concrete, iron oxide pigments are the most commonly used by the construction industry. Suitable pigments used in concrete to produce various colours are:

1. Black iron oxide and carbon black produce grey to black colours.
2. Brown iron oxide and raw and burnt umber produce brown colour.
3. Red iron oxide produces red colour.
4. Chromium oxide or hydroxide produces green colour.
5. Hydroxide of iron is used to get yellow colour.
6. Ultramarine produces blue colour,
7. Titanium dioxide produces white colour. It will never make concrete made with grey cement truly white, but it will lighten it to a lighter shade of grey.

Iron oxide pigments are available in two classes: synthetic and natural. Any concrete pigment can be wholly composed of either natural or synthetic iron oxides, or it can be a blend of the two classes. There are some basic differences between synthetic and natural iron oxides, but in general they all offer good longevity and colour consistency from lot to lot. Pigments are recognised by ASTM 979.

Initially, naturally occurring mineral oxides were used in concrete and the colour control was not easy. However, nowadays preference is given to synthetic mineral oxides, which have proved more colour consistent with bright colours. In further development of concrete-colouring system, pigment producers and blenders have more recently produced a new generation of colouring pigments in which milled pigment and other ingredients are converted into microspheres. The microspheres readily disperse in water-based media in a similar way to instant coffee granules. A plasticiser is also included in the formulation, which significantly reduces the total water demand of the mix. Batch-to-batch uniformity and streak-free colour are guaranteed. The product is supplied in water-soluble bags and is added straight into the mixer truck [38].

Forms of Pigment

Mostly pigments are used in three forms described below [39]:

Powder Form

Powder pigments are the most common form of pigment used by all concrete producers. They are the least expensive form of pigments and are typically available in either 50 pound bags or batch size bags as required. Powder pigments have unlimited shelf life when stored properly. It is best to keep them dry, so they don't get lumpy and the bag doesn't disintegrate. For best results, they should be added early in the batching sequence in order to achieve maximum dispersion and colour development of the coloured concrete.

Liquid Form

Liquid colour basically consists of the powdered pigments suspended in a water-based dispersion. Liquefying the pigment simplifies the conveying process and therefore allows pigment additions to be automated through the use of dispensing equipment. Dispensing systems can also create verification and documentation of all pigment additions to the concrete mix.

Liquid pigment also mixes more readily with the concrete and therefore significantly faster colour development and better dispersion of colour take place in the mixer. Liquid colour costs more than powdered colour, but ultimately it helps produce more consistent coloured concrete. The use of liquid colour in precast and ready mix has been increasing at an increasing rate over the last 5–10 years.

Granular Form

Granular pigments are basically powder pigments joined together to make small balls. They have been used in the construction industry for approximately since 20 years. Granules can also be added automatically, and since they are dry they do not require the addition of any extra water to the mix. As stated earlier, this form of colour is only available for synthetic iron oxides, and is used mostly in concrete paver plants and some concrete masonry manufacturing facilities.

Bibliography

1. ASTM C150/C150M—17, Standard Specification for Portland Cement.
2. CSA A3001 Cementitious Materials for Use in Concrete.

3. ASTM C1157/C1157M—17, Standard Performance Specification for Hydraulic Cement.
4. CEMEX Mortars UK, Educational Guide to Cementitious Materials.
5. ASTM C595 (Revised 2017)- Standard Specification for Blended Hydraulic Cements.
6. Canadian Standards Association (CSA) A23.1–14 - Concrete materials and methods of concrete construction/Test methods and standard practices for concrete.
7. The Cement Association of Canada- Brief on Contempra (Portland-Limestone Cement): A New Lower Carbon Cement.
8. ACI 207.lR-87 (Revised 2005), Mass Concrete.
9. ACI 223–90/98 (Revised 2010), Standard Practice for the Use of Shrinkage-compensating Concrete.
10. ASTM C91–05 (Revised 2012) - Standard Specification for Masonry Cement.
11. CAN/CSA-A179–04 (R2014) - Mortar and Grout for Unit Masonry.
12. University of Saskatchewan Class Notes- Chapter 2- Portland, Blended and Other Hydraulic Cements.
13. St Marys Cement Toronto- Specification for Masonry Cement Type-S.
14. ASTM C1328 (Revised 2012)—Standard Specification for Plastic (Stucco) Cement.
15. API (American Petroleum Institute) Spec 10A (R2015)- Specification for Cements and Materials for Well Cementing.
16. Dr. Mark A. Shand; Premier Magnesia North Carolina—A Brief on Magnesia Cements North Carolina.
17. BS 4550-78 (Amended 1984), Part-3, Methods of Testing Cement-strength Tests.
18. ASTM C151-89 (Revised 2016), Standard Test Method for Autoclave Expansion of Portland Cement.
19. ASTM Cl86-86 (Revised 2005), Standard Test Method for Heat of Hydration of Hydraulic Cement.
20. Portland Cement Association (PCA)-Concrete Technology Today- Portland Cement, Concrete, and Heat of Hydration, Volume 18/Number 2, July 1997.
21. ASTM C109/C109M—16a, Standard Test Method for Compressive Strength of Hydraulic Cement Mortars (Using 2-in. or [50-mm] Cube Specimens).
22. Ontario Provincial Standard Specification-OPSS.PROV 1002-Material Specification For Aggregates-Concrete, April 2013.
23. Portland Cement Association USA, Recycled Aggregates.
24. Mirjana Malešev, Vlastimir Radonjanin, and Snežana Marinković, Article: Recycled Concrete as Aggregate for Structural Concrete Production- Journal: Sustainability 2010, 2, 1204–1225; doi:10.3390/su2051204, www.mdpi.com/journal/sustainability.
25. BS 882–92, Specification for Aggregates from Natural Sources of Concrete.
26. ASTM (American Society for Testing and Materials) C33–86 (Revised 2016), Specifications for Concrete Aggregates.
27. ACI Education Bulletin E1-99-Aggregates for Concrete-Developed by Committee E-701-Materials for Concrete Construction.
28. Canadian Standards Association- CSA- A23.2-14, Test methods and standard practices for concrete.
29. A.M. Neville, Properties of Concrete, Longman, London, 1997.
30. University of Saskatchewan Class Notes- Chapter 6- Admixtures of Concrete.
31. ACI 212.3R-91 (Revised 2016), Chemical Admixture for Concrete.
32. Road Research Laboratory, Ministry of Transport, Concrete Roads Design and Construction, published by Her Majesty's Stationary Office, printed by Lowe and Brydone Ltd, London, 1966.
33. Concrete Alberta- Concrete Tech Tip # 27- Supplementary Cementing Materials (SCM's).
34. Iowa State University- Institute for Transportation- Best Practices Workshop, Supplementary Cementitious Materials.
35. Lawrence L. Sutter, Professor Michigan Technological University USA- Tech Brief SUPPLEMENTARY CEMENTITIOUS MATERIALS, Best Practices for Concrete

Pavements- This Tech Brief was developed under Federal Highway Administration (FHWA) Washington, DC USA, http://www.fhwa.dot.gov/pavement.
36. Concrete Alberta- Concrete Tech Tip # 27, Supplementary Cementitious Materials (SCMs).
37. Ontario Provincial Standard Specification- OPSS.PROV 1350, Material Specifications for Concrete— Materials and Production- November 2013.
38. Iain Christie and Margaret Hatfield of Roy Hatfield Ltd, Colour Technology for Concrete, Concrete Society Journal, CONCRETE, Vol. 31, No. 4, April 1997.
39. Cathy Higgins and Jimmy Crawford- Colouring Concrete Using Integral Pigments- East Texas Precast.

Chapter 3
Physical Properties of Concrete

3.1 Introduction

The properties of concrete in two states, the plastic and the hardened state, are vital for an engineer. The process of batching, mixing, transporting, placing, compacting, and finishing of concrete in a plastic state seriously affects the properties of the hardened concrete. Hence, it is important that the concrete in its plastic state should have a proper workability, so that it can be transported, placed, fully compacted, and finished without segregation and bleeding.

For structural strength, the concrete in its hardened state must have the properties assumed for it in the design of the structure. The important properties of hardened concrete are strength, durability, impermeability, and volume stability.

3.2 Plastic Properties of Concrete

Concrete key properties in its plastic state are as follows:

1. Workability
2. Segregation
3. Bleeding
4. Plastic shrinkage
5. Air entrainment

3.2.1 Workability

It is the property of concrete which gives a measure of the ease with which it can be placed in position, compacted, and finished. The degree of workability necessary in a concrete mix depends entirely upon the purpose for which it is used and the

© Springer Nature Switzerland AG 2019
A. Surahyo, *Concrete Construction*,
https://doi.org/10.1007/978-3-030-10510-5_3

methods and equipment used in handling and placing it in the work. The workability of fresh concrete should be suitable for the conditions of handling and placing so that after compacting properly the concrete surrounds all reinforcement without segregation.

In their textbook, Mehta and Monteiro [1] note that workability is a composite property, with at least two main components:

1. *Consistency* which is described as the "ease of flow" of the concrete
2. *Cohesiveness* which is described as the "tendency not to bleed or segregate"

3.2.1.1 Factors Affecting Workability

The most important factors which influence the workability of concrete are water/cement ratio, size, grading, shape, and proportions of aggregate; the amount and qualities of cement and other cementitious materials; the presence of entrained air and chemical admixtures; and temperature. The workability increases with the following:

- Increase in water content
- Increase in cement content
- Increase in fineness of cement
- Increase in ratio of fine aggregate
- Increase in size of coarse aggregate
- Use of smoother or rounded aggregate
- Use of water-reducing (plasticiser) and air-entraining agents

However, increase in the above items is limited to certain levels based on required strength of concrete. The workability drops, resulting in slump loss, with the following:

- Use of dry, porous, or friable aggregates, as they absorb water from the mix and increase the surface area to be wetted
- Elevated concrete temperature, ambient air temperature, and wind velocity
- Use of high-early-strength cement
- Cement deficient in gypsum [2]
- Low cement content and coarsely ground cement
- High ratio of volumes of coarse aggregate to fine aggregate

3.2.1.2 Measurement of Workability

Workability is not itself a measurable property, but it can be assessed by means of tests to measure properties that vary with the workability. The most common tests are:

1. Slump test
2. Compacting factor test

3. Vebe test
4. Flow table test

Out of these four tests, the slump test, is used most widely on construction sites because of the simplicity of the apparatus required and the test procedure.

Slump Test

The slump test is the most well-known and widely used test method to characterise the workability of fresh concrete. The test measures consistency and is used on job sites to determine rapidly whether a concrete batch should be accepted or rejected. As per requirements of BS-1881 [3], ASTM-C143 [4], and CSA-A23.2 [5], slump test is made in a frustum of a cone, with 300 mm (12 in.) height, 200 mm (8 in.) diameter at the base, and 100 mm (4 in.) at the top. As per CSA-A23.2 [5], a test specimen mould shall be made of metal not thinner than 1.5 mm. This test method is considered applicable to plastic concrete having coarse aggregate up to 40 mm in size. Samples for determining acceptance of concrete for air and slump shall be taken according to CSA-A23.2-1C after approximately 10% of the load has been discharged. Primarily, the test is performed as follows:

- Clean and dampen the cone. Place it on a smooth, horizontal, and non-absorbent surface. Now fill the cone with concrete sample using a scoop, in three equal layers.
- Rod each layer exactly 25 times with a standard 16 mm (5/8 in.) slump rod.
- Spread the blows evenly over the whole area and make sure that the rod just penetrates the layer below.
- After rodding of the final layer, level the top with the rod, and clean off spillage from sides and base plate.
- Now carefully and slowly lift the cone straight up.
- Upturn the slump cone, place the rod at its top, and measure the distance between the underside of the rod and the highest point of the concrete. This slump is considered as a "true slump" (Fig. 3.1). Instead of highest point of the concrete, ASTM-C143 and CSA-A23.2 suggest to measure the distance from the centre point (average height) of the top surface of the concrete.
- If the slump is not true, then take a new sample and repeat the test. If again a true slump is not achieved, but instead it collapses (Fig. 3.2) or a shear slump (Fig. 3.3) is produced, then it is an indication of lack of cohesion of the mix, which might be due to increase in moisture content of aggregate or change in grading of aggregate.

The slump test is very useful on site because it is easy to check every concrete pour. Slumps up to 1 in. (25 mm) are treated as very low, 1–2 in. (25–50 mm) low, 2–4 in. (50–100 mm) medium, and 4–7 in. (100–175) as high. The slump test is not applicable to flowing or self-consolidating concrete. OPSS-1350 (1996) [6] recommends slumps as specified in Table 3.1 prior to the addition of super-plasticiser.

Figures showing the making of slump, and its three main types.

Fig. 3.1 True slump

Fig. 3.2 Collapse

Fig. 3.3 Shear

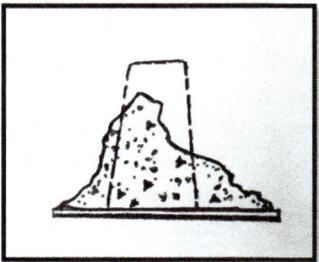

Compacting Factor Test

The test is useful for concrete of low workability. The test requires measurements of the weights of the partially and fully compacted concrete and the ratio of these two is known as the compacting factor. The test is most suitable for laboratory conditions.

The apparatus used (Fig. 3.4) consists of two hoppers with traps. Before starting make sure that the internal surfaces of the hoppers and the cylinder are smooth, clean, and damp. Now slowly pour the concrete sample in the upper hopper with a scoop until it is filled up to the top level. Open the trap door and allow the concrete to fall into the lower hopper. As soon as the concrete comes to rest, remove the trap door of the second hopper and allow the concrete to fall into the cylinder. The

Table 3.1 Placing concrete slump requirements as per OPSS.PROV-1350 [6]

Slump components	Slump (mm)
Reinforced concrete within vertical formwork such as abutments, columns, piers, walls and beams, and footings	80 ± 20
Reinforced or plain concrete in flat sections such as bridge decks, and approach slabs	70 ± 20
Slip-formed barrier walls	15 ± 10
Tremie concrete	150 ± 20

Fig. 3.4 Compacting factor apparatus

Hoppers

Cylinder

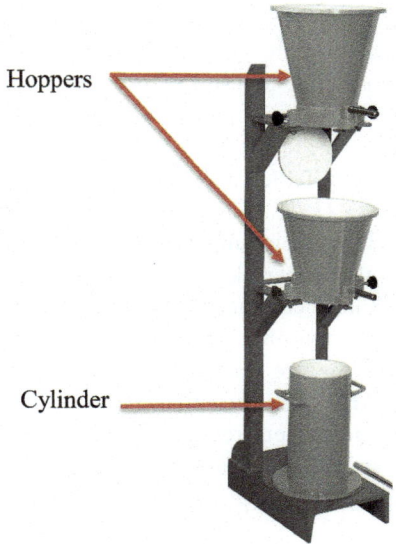

weight of concrete (w_1), which exactly fills the cylinder, is measured. The cylinder is then emptied and refilled with the concrete in four layers, each rodded or vibrated to get fully compacted concrete. If the weight of compacted concrete is taken as w_2 then

$$\text{Compacting factor} = w_1 / w_2$$

The compacting factor is expressed in decimal, as it is always less than one. Concretes having compacting factor 0.70 are treated as very stiff, 0.75 as stiff, 0.85 as stiff plastic, and 0.90 as plastic. However, concretes having compacting factor below 0.70 or above 0.98 are regarded unsuitable as per BS-1881 [3].

Vebe Test

Vebe test is used for very dry mixes, having low and very low workability. The apparatus used is shown in Fig. 3.5. Before starting, clean and dampen the inner surface of slump cone and place it inside the cylinder, which should be clamped properly. Now lower the funnel on to the cone and fill the cone in three layers. Rod each layer, like in the slump test, with a standard slump rod 25 times. After the top layer has been rodded, remove the cone slowly and measure the slump.

Now the disc-shaped rider is allowed to rest on top of the concrete, and the vibratory unit is switched on. The time, measured from the commencement of vibration until the surface of concrete first contacts the sides of the cylinder, is noted. This time, in seconds, is used as a measure of the consistency of the concrete. Concretes having a Vebe time of 32 to 18 s are treated as extremely dry, 18 to 10 s as very stiff, 10 to 5 s as stiff, 5 to 3 stiff plastic, and 3 to 0 s as plastic [7].

Flow Table Test

This test is suitable for flowing concrete having high to very high workability made with super-plasticising admixtures. The apparatus used is shown in Fig. 3.6. The slump cone used in this test has the internal dimensions: 200 mm (8 in.) height, 200 mm (8 in.) diameter of base, and 130 mm (5 in.) diameter of top.

Before starting, clean and dampen the inner surface of slump cone and table. Also make sure that the hinged top of the table can be lifted to the correct limit and is then free to fall to the lower stop. Now fill the cone with concrete sample in two equal layers, rodding each layer lightly ten times with the wooden tamping rod. After the top layer has been rodded, remove the cone slowly. Now slowly raise the table top by the handle till it reaches the upper stop, and then

Fig. 3.5 Vebe apparatus

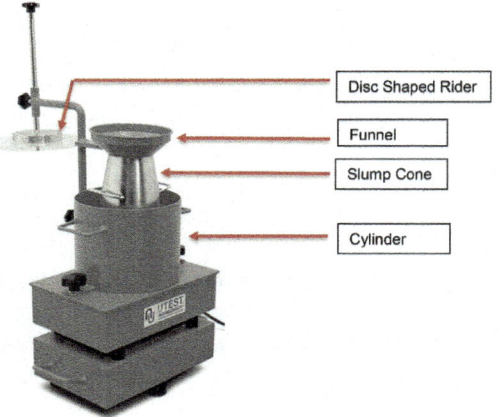

Fig. 3.6 Flow table

Slump Cone

Flow Table

allow the table top to fall freely to the lower stop. Repeat this process and give a total of 15 jolts by lifting and dropping the table top. Now measure the total diameter of the spread concrete. The mean of original base diameter of cone and this diameter in millimetres is taken as a measure of the flow or consistency of the concrete.

A value of 400 indicates a medium workability and 500 a high workability [3].

3.2.2 Segregation

Separation of coarse and fine particles in a mix is termed as segregation. It generally occurs due to use of extremely wet or dry mixes, over-vibration, transporting concrete mixes for long distances, and dropping concrete from excessive heights. The segregation adversely affects the properties of hardened concrete, results in production of porous and honeycombed concrete, and reduces its strength and durability.

Its extent can be controlled by using proper vibration, by taking care in handling and placing of concrete, and by the use of air entrainment and appropriate grading of aggregates. It is further discussed in detail in Chap. 7.

3.2.3 Bleeding

After concrete has been compacted, the force of gravity tends to pull the heavy solid particles downward, displacing the lighter water upward. This upward migration of water resulting in a layer of water on the upper surface of concrete is known as bleeding. It is actually a form of segregation. It generally occurs due to use of too much water or coarse cement, the particles of which settle more rapidly than those of the finer cement. It also appears due to over-vibration or too much tamping of the top of concrete. However, the major reason is the poor grading and inadequate consistency of the mix.

The bleeding results in a weak top surface, which becomes less durable than the bottom portion of concrete, and the strength of concrete becomes non-uniform and the top surface lacks the resistance to abrasion.

The bleeding can be reduced by decreasing the water content, using finer cement, increasing ratio of fine aggregate, and using air-entraining agents. Reduction in bleeding is also obtained by the addition of pozzolanas or other fine material or aluminium powder [8]. Bleeding can also be reduced by controlling over-vibration, by delaying the finishing of concrete for a little time until the bleeding water evaporates, and by using wooden floats.

3.2.4   Plastic Shrinkage

Plastic shrinkage is described in the following pages under the heading "Structural properties of concrete".

3.2.5   Air Entrainment

The introduction of air into concrete mixes by chemical admixture (or sometimes, air-entraining cement) has some distinct effects on the characteristics of both the plastic and the hardened concrete. Air-entrained concrete contains millions of minute air bubbles that are distributed uniformly throughout the concrete mix. Air-entraining concrete or aerated concrete and air-entraining admixtures are also discussed in previous two chapters.

In fresh concrete, the tiny air bubbles act as a lubricant in the mix, which improves its workability and increases its slump. The tiny air bubbles also act as fines, thereby cutting down the amount of sand needed. Because air entrainment increases slump, it is possible to decrease the amount of water to get higher strengths without affecting workability. Air also produces more cohesive concrete, and reduces bleeding and segregation in concrete.

In hardened concrete, air bubbles greatly enhance concrete's durability in moist freeze/thaw environments while making it more impermeable to water, including sulphate-laden groundwater. Other uses of air-entrained concrete include resistance to deicing salts and alkali-silica reactivity.

3.2.5.1   Measurement of Air Content in Concrete

Several methods are available for measuring the air content of fresh concrete. However, the following five techniques are commonly performed:

1. Pressure method
2. Volumetric method
3. Chace air indicator method
4. Gravimetric method
5. Air void analyser

Since pressure method is most commonly used on construction sites for determining air content of concrete in plastic state, only that will be described here in detail, whereas other tests will be explained briefly.

Pressure Method

The apparatus used for pressure test is shown in Fig. 3.7. This apparatus measures the air content of fresh concrete based on the pressure-to-volume relationship of Boyle's law. The Boyle's law states that the volume occupied by air is proportional to the applied pressure.

Mostly the test is performed as follows:

- Clean and dampen the bowl. Place it on a smooth, horizontal, and non-absorbent surface. Now fill the bowl with concrete sample using a scoop, in three equal layers.
- Rod each layer exactly 25 times with a standard 16 mm (5/8 in.) slump rod.
- Spread the bowls evenly over the whole area and make sure that the rod just penetrates the layer below. Tap sides of the bowl ten or more times to remove air bubbles and consolidate the mix.
- After rodding and tapping of the final layer, level the top with the rod, and clean off spillage from sides and base plate.
- Now place cover with air chamber and valve on bowl and clamp down both petcocks. Add water through one petcock until all trapped air under the cover is forced out through the other petcock. During this operation, the bowl should be tilted and tapped. When the water coming from second petcock is found free of bubbles, it means that the trapped air has been expelled.

Fig. 3.7 Measuring air content of fresh concrete by pressure method

- Now with the valve closed, air is pumped into the air chamber to the predetermined operating pressure. When the valve between the air chamber and bowl is opened, the air expands into the test chamber, and the pressure drops in proportion to the air contained within the concrete sample. The required air content is displayed on the attached pressure gauge.

The applicable standards for this test are ASTM C 231, AASHTO T-152, and CSA-A23.2-4C.

Volumetric Air Meter Method

This test method specifies the procedure for determining the air content of plastic concrete using the water displacement method. The bowl section of the apparatus is filled with concrete and consolidated and smoothed. The top section of the apparatus is then attached and filled with water, and a measured quantity of 70% isopropyl alcohol (de-foaming agent) is added to dispel the foam generated during agitation. The unit is then inverted, rolled, and agitated to remove the air.

The amount that the water column goes down, with the aid of a de-foaming agent, is a measure of the air content. The air content of the concrete is read directly from the sight tube. CSA-A23.2-7C [5] specifies that this test method works for all concretes, but is most commonly used for lightweight aggregate concrete.

The applicable standards for this test are ASTM C 173, AASHTO T-196, and CSA-A23.2-7C.

Chace Air Indicator Method

To some extent, this method is identical in concept to the volumetric air meter. To perform the test, a sample of cement mortar paste is placed in a container and alcohol is added to free the air. The change in the level of alcohol in the container stem indicates the air content. AASHTO T-199 provides that this test cannot be used as a substitute for pressure, volumetric, and gravimetric methods. It should not be used for determining the compliance of air content with the specifications.

Applicable standard: AASHTO T-199.

Gravimetric Method

In the gravimetric method, air content is determined based on differences in actual and theoretical (air-free) unit weights (density) of the concrete. The unit weight is determined by weighing a known volume of fresh concrete. The air content is calculated using two independent equations given in AASHTO T121 or ASTM C138.

This test method is sensitive to consolidation and strike-off of the concrete in the container. Determining the actual unit weight incorrectly, for instance, by failing to properly strike off the concrete after the mould is filled, can cause a relatively large error in air content. In that context, this method is less reliable in comparison to first two methods.

Applicable standards: AASHTO T121 and ASTM C138.

Air Void Analyser (AVA)

The AVA determines air void parameters in fresh samples of air-entrained concrete. The test apparatus determines the volume and size distributions of entrained air voids and thus allows an estimation of the spacing factor, the specific surface, and the total amount of entrained air.

The AVA is a relatively new piece of equipment that can be used to accurately evaluate the air void system of fresh, plastic concrete. One of the limitations of the AVA has been that it is a sensitive machine. It has been considered usable only in buildings, not in the field, because vibrations, such as those caused by wind or people movements, can have a significant effect on the AVA's results.

3.3   Structural Properties of Concrete

Concrete key properties in its hardened state are as follows:

1. Strength
2. Durability
3. Permeability and porosity
4. Shrinkage

3.3.1   Strength

Strength is the most important property of hardened concrete and the quality of concrete is often judged by its strength. It is usually determined by the ultimate strength of a specimen in compression; however, there are other strengths to consider besides compression, depending on the loading applied to the concrete. Flexure or bending, tension, shear, and torsion are applied under certain conditions and must be resisted by the concrete or by steel reinforcement in the concrete. Simple tests available for testing concrete in compression and in flexure are used regularly as control tests during construction. An indirect test for tension is available as the splitting tensile test, which can easily be applied to cylindrical specimens made on the job. Laboratory procedures can be used for studying shear and torsion applied to concrete; however, such tests are neither practical nor necessary for control, as the designer can evaluate such loadings in terms of compression, flexure, or tension.

The specifications or code designates mostly the compressive strength required of the concrete in several parts of the structure. In those cases in which strength specimens fail to reach the required value, further testing of the concrete in place is usually specified. This may involve drilling cores from the structure or testing with certain non-destructive instruments that measure the hardness of the concrete.

3.3.1.1 Compressive Strength

Concrete usually gains strength over a long period of time; hence the compressive strength at the age of 28 days is commonly used as a measure of this property. Concrete structures, except road pavements, are normally designed on the basis that concrete is capable of resisting only compression, whereas steel reinforcement takes care of the tension.

Cubes, cylinders, and prisms are the three types of compression test specimens used to determine the compressive strength on site. If curing conditions, methods of sampling, and casting are allowed to vary, the resulting test evaluations are worthless because it would be difficult to assess whether a low strength is due to poor-quality concrete or poor testing procedure. Standard cubes are usually of 150 mm (6 in.) × 150 mm (6 in.) size, and the cylinders are either 150 mm (6 in.) diameter by 300 mm (12 in.) high or 100 mm (4 in.) diameter × 200 mm (8 in.) high. The prisms used in France are 100 mm (4 in.) × 100 mm (4 in.) × 500 mm (20 in.) in size. The specimens are cast, cured, and crushed in laboratory as per standard method to determine the compressive strength.

CSA-A23.2-3C recommends that 100 mm diameter moulds should be filled in three equal layers and each layer should be rodded uniformly 20 times with a 10 mm diameter × 450–600 mm long hemispherically tipped steel rod (see Fig. 3.8). However, for 150 mm diameter moulds, each layer should be rodded 25 times with 16 mm diameter × 450–600 mm long steel rod. Since the specified mix used for the concrete curb shown in Fig. 3.8 is very tight (40 ± 20) the technician is using vibrator

Fig. 3.8 Preparing
concrete cylinders

instead of steel rod. Prepared test cylinders should be placed on a rigid horizontal surface in a controlled environment such as a curing box for a minimum 20 h. The temperature is to be maintained 20 ± 5 °C where the cylinders are stored. After setting for a minimum 20 h, cylinders should be moved properly to a laboratory for standard curing and testing.

The compressive strength of concrete is taken as the maximum compressive load it can carry per unit area. Usually cubes or cylinders prepared from the site of work are tested after 7 days and 28 days. The 7-day tests are a guide to the rate of hardening; the strength at this age for Portland cement concrete is mostly two-thirds of the strength required at 28 days. Twenty-eight-day strength test results are usually used for quality control and acceptance of concrete and to determine that the concrete mixture as delivered meets the requirements of the specified strength in the contract specifications. A test result is the average of at least two concrete samples made from the same concrete and tested on the same age of 28 days.

The rate of increase of strength with age is almost independent of the cement content. With ordinary Portland cement concrete (including supplementary cementitious material "slag", air content 5–8%, and retarder as a chemical admixture), mostly 100% of the specified strength is reached at 28 days and increases more over time subject to appropriate curing.

Compressive strength varies from less than 10 N/mm^2 (1500 lb/in.^{2}) for lean concretes to more than 55 N/mm^2 (8000 lb/in.^{2}) for special concretes. Higher strengths up to and exceeding 70 MPa (10,000 psi) are sometimes also specified for certain applications. The minimum compressive strength of reinforced concrete as specified by BS-5337 is to be considered as grade 25 (C25), whereas CSA23.3-04 provides that specified concrete compressive strengths used in design shall not be less than grade 20. However, in practice, a grade 30 mix is mostly used because of durability considerations as discussed in Chap. 1.

3.3.1.2 Tensile Strength

The tensile strength of concrete is usually considered about one-tenth of its compressive strength. The direct tensile strength of concrete is considered when calculating resistance to shearing force and in the design of cylindrical liquid-containing structures. However, it is difficult to measure concrete strength in direct tension and also the variation in results is high; therefore the direct tensile test is not standardised and rarely used.

The indirect methods adopted for determining the tensile strength are known as the splitting test. Splitting test is simple to perform and gives more uniform results than other tension tests. The strength determined is closer to the actual tensile strength of the concrete than is the modulus of rupture value. Under this test a standard test cylinder is loaded in compression on its side and by means of an equation a value of tensile strength can be computed.

3.3.1.3 Flexural Strength

Many structural components are subject to flexure, or bending. Road pavements, slabs, and beams are examples of elements that are loaded in flexure. The determination of flexural strength is essential to estimate the load at which the concrete members may crack. The flexural strength at failure, also expressed as the modulus of rupture in psi (MPa), is thus determined and used when necessary. Its knowledge is of importance in the design of concrete road or runway pavements, as road slabs rarely fail in compression. Flexural modulus of rupture is about 10–20% of compressive strength depending on the type, size, and volume of coarse aggregate used. An elementary example is a simple beam that is placed across two supports, and then is loaded by the testing machine until the concrete cracks and fails. Load can be applied to the centre point of the beam (ASTM C 293), or equally at the third points (ASTM C 78). When this beam is loaded, the bottom portion of the beam is in tension and the upper portion of beam is in compression. Failure of the concrete beam will be a tensile failure in the lower portion, as concrete is much weaker in tension than in compression. Hence steel bars are usually added in the lower part of the beam to support a much greater load because the steel bars have a high tensile strength.

The tests are described in BS-1881 and ASTM- C78 [7] and ASTM-C496 [9]. However Fig. 3.9 produced here shows the relation between compressive strength, tensile strength, and flexural strength.

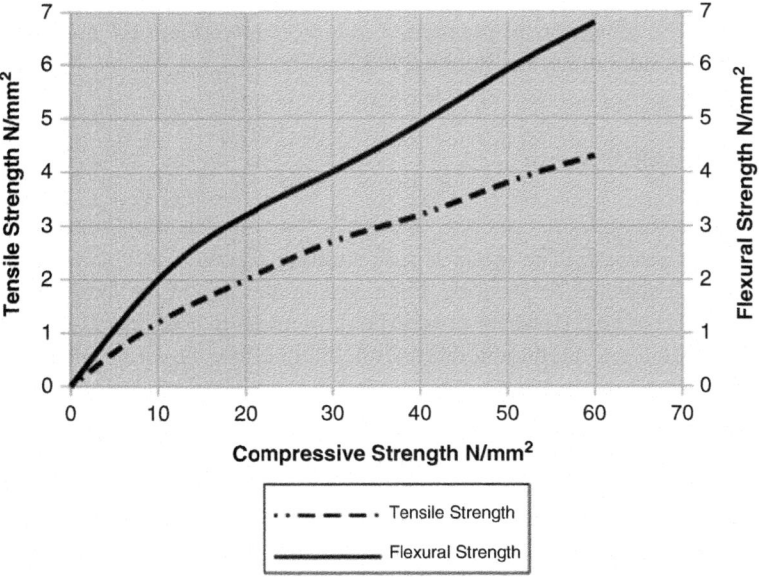

Fig. 3.9 Relation between compressive strength, tensile strength, and flexural strength [10]

3.3.1.4 Testing of Concrete in Hardened State

The main purpose of determining the strength of concrete in plastic state is to ensure that the required strength of concrete in actual structure is satisfactory. However, if the compression strength achieved is below the required value then either concrete in the actual structure is unsatisfactory or the concrete samples collected during concrete pouring are not truly representative of the concrete in the structure. To resolve such disputes, samples are sawed or cored from the suspected part of the hardened concrete and tested to determine the required strength of concrete in the structure. Non-destructive tests may also be performed for the purpose of poured concrete quality assurance. Some of the common tests are briefly described as follows:

Destructive Concrete Tests

Concrete core test: A rotary diamond drill is used to slice cylindrical cores from the completed structure. CSA-A23.2-14C recommends that a minimum core diameter of 100 mm be used whenever practicable. These cores, which may contain some steel, are soaked in water, capped, and examined in compression to get the quantity of the concrete strength in the original structure. The proportion of core height to diameter and the location, where the core is acquired, influence the strength. Additionally, core cylinder strength is generally lower than that of standard cylinders due to construction-site curing. There are also chances that damage may occur due to the vibration of the core drill. The strength is minimum at the top surface and raises with depth through the element. A proportion of core height to diameter of 2 provides a standard cylinder test. This test is described in CSA-A23.2-14 and BS EN 12504-1:2009.

The cores cut to determine the strength of concrete of the actual structure may also be used to find out segregation and honeycombing of concrete. In some cases, the beam specimens are also sawn from the road and airfield slabs for finding flexural strength.

Pull-Out Test: The concrete pull-out test procedure starts with the fixing of pull-out test equipment. Pull-out test equipment produces pull-out forces required to pull a steel insert out of concrete which was embedded during casting. The force measured is correlated to an equivalent compressive strength of the concrete on standard cylinders, through a correlation curve.

Such strength relationships are affected by the configuration of the embedded insert, bearing ring dimensions, depth of embedment, and type of aggregate (lightweight or normal weight). Before use, the relationships must be established for each test system and each new concrete mixture. Such relationships are more reliable if both pull-out test specimens and compressive strength test specimens are of similar size, consolidated to similar density, and cured. This test is described in CSA-A23.2-14, ASTM C-900-15, and BS EN 12504-3:2005.

Non-destructive Tests

The primary non-destructive tests for strength on hardened concrete areas are given below:

Rebound hammer (hardness) test: The most common non-destructive test is the rebound test. For the rebound hardness test, the Schmidt hammer also called a Swiss hammer is utilised. Under this test, a spring-loaded hammer is released to impact against a piston in contact with the concrete surface. The amount of rebound (rebound number) is documented on a scale and this highlights the strength of the concrete. It is a quick, non-destructive test that can be used for the determination of the approximate compressive strength of concrete in place.

The results of Schmidt test are usually affected by surface finish, moisture content, temperature, rigidity of the member being tested, depth of carbonation, and direction of impact (upward, downward, horizontal). ASTM C805 provides that different instruments of the same nominal design may give rebound numbers differing from 1 to 3 units. Hence, tests should be made with the same instrument in order to compare results.

Presently in the market there are many redesigned Schmidt hammers which provide better and more accurate results in comparison to traditional Schmidt hammers such as classic Schmidt hammer and silver Schmidt hammer.

Penetration Resistance Test: Penetration resistance test rapidly and accurately determines the compressive strength of the concrete. The strength is found out by firing (shooting) a steel probe or pin on the surface of the concrete with the known amount of force. The penetration is inversely proportional to the compressive strength of concrete in the standard test condition. This test is not affected by surface texture and moisture content like rebound hammer test; however, the mix proportions, strength of aggregates, and material properties are still important.

Windsor probe equipment is a famous penetration resistance measurement equipment for the concrete. The technique consists of accessories like a gunpowder-actuated driver and hardened alloy rod probe. It also consists of accessories like loaded cartridges and a depth gauge.

It uses a gunpowder-actuated driver to fire/drive a hardened alloy probe on the concrete. The depth of penetration depends on the strength of the concrete. Usually three sets of probe are driven and average value is taken to estimate the strength of concrete. Accuracy is about the same as that of the Swiss hammer, but small indentations are left in the surface of the concrete, which might be unsightly in some exposed concrete.

As regards pin penetration test, the device requires less energy than the Windsor Probe system. Being a lower energy device, sensitivity is reduced at higher strengths. Hence it is not recommended for testing concrete having strength above 28 N/sqm. This test is described in ASTM C803. Testing of hardened concrete is also described in BS EN 12390-3:2009.

Ultrasonic Pulse Velocity (upv): The upv method is mostly used for assessing the quality of concrete in a structure. ASTM C597-16 provides that this test method is applicable to assess the uniformity and relative quality of concrete, to indicate the presence of voids and cracks, and to evaluate the effectiveness of crack repairs. It is also applicable to indicate changes in the properties of concrete, and in the survey of structures, to estimate the severity of deterioration or cracking.

In general upv test can be used for:

- The homogeneity of a material
- The presence of voids, cracks, or other internal imperfections or defects
- Changes in the concrete which may occur with time (i.e. due to the cement hydration) or damage from fire, frost, or chemical attack
- The strength or modulus of a material
- The quality of the concrete in relation to specified standard requirements

Ultrasonic pulse velocity equipment consists of two ultrasonic transducers; one of the transducers acts as a transmitter and the other acts as a receiver. Ultrasonic pulse velocity is a stress-wave technique that propagates an ultrasonic wave through the material and measures the time needed for the wave to pass through. Optimal results are obtained when both sides of the structure are accessible. The wave travel time is used to infer the quality of the concrete. Shorter travel times indicate higher quality concrete and longer travel times indicate lesser quality concrete which could contain voids and cracks. In special applications, ultrasonic pulse velocity can be used to estimate the depth of fire-damaged concrete or estimate the depth of cracks.

To estimate compressive strength of in-place concrete, pulse velocity measurements are conducted on the structure, and then calibrated for the particular concrete being tested. In the calibration procedure, compressive strength tests are performed on five to ten cores extracted from selected locations. A correlation between pulse velocity and compressive strength is then established for the particular concrete [11].

Some of the factors which affect this test are surface smoothness, travel path of the pulse, temperature effects on the pulse velocity, moisture content, presence of steel reinforcing bars, and age of concrete.

3.3.1.5   Factors Influencing Strength

The main factors affecting strength of concrete are the following:

Water/Cement Ratio

It is the prime factor affecting the strength of concrete. The strength decreases with an increase in w/c ratio in a manner similar to that illustrated in Fig. 4.2 (XX). The strength of concrete is usually considered inversely proportional to the water/cement ratio. However, it must be understood that strength depends on the free w/c ratio, which is calculated on the basis of mix water, excluding the water absorbed by the aggregates. In fact, the total water in a concrete mix consists of the water absorbed by the aggregate to bring it to a saturated surface-dry condition and the free water available for the hydration of the cement and for the workability of the fresh concrete. The workability of concrete hence depends to a large extent on its free-water content. Similarly, the strength of concrete is better related to the free w/c ratio since on this basis the strength of the concrete does not depend on the absorption characteristics of the aggregates.

Cement

The rate of development of strength varies with the type of cement used, and is higher at early ages for rapid-hardening Portland cement and lower for low-heat Portland cement. Variations in the quality of cement are responsible for considerable variations in concrete strength.

The increase in cement content and fineness of cement particles increase the concrete strength, whereas the strength decreases as the cement content is decreased. Figure 3.10 shows the effect of cement content on strength.

Aggregate

The shape of aggregate indirectly affects the strength of concrete. For a given workability, a rounded aggregate requires a lower water/cement ratio than does an angular one, and thus strength will be higher. Similarly, a coarser aggregate grading will permit a lower water/cement ratio, and thus gives higher strengths for a given workability.

The aggregate size and cleanliness also affect strength and are discussed in Chaps. 2 and 6.

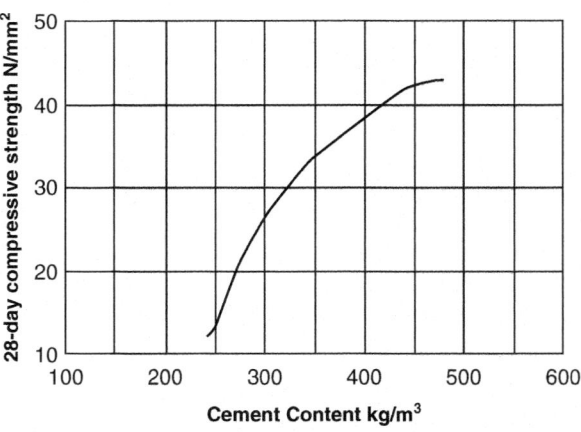

Fig. 3.10 Effect of cement content on strength [10]

Compaction

The degree of compaction obtained in the concrete greatly affects the strength. If full compaction is not achieved, porous and honeycombed concrete will be produced resulting in loss of strength. Only 5% of air voids may reduce the strength of concrete by about 30%. Hence, the concrete must be sufficiently workable in its plastic state, so that it could be fully compacted without bleeding and segregation. It is further explained in Chap. 7.

Curing

Curing also plays an important role in the development of strength. The strength and durability of concrete will be fully developed only if it is cured. Since the hydration of cement proceeds only in the presence of an adequate amount of water, moisture must be maintained in the concrete during the curing period. The development of concrete strength under various curing conditions is shown in Fig. 7.13, Chap. 7.

In general, the longer the period during which concrete is kept in water, or to prevent loss of water by other means the greater will be its final strength. Curing is further explained in Chap. 7.

Admixture

Admixtures can only affect concrete strength by modifying certain properties of concrete like rate of hydration or setting time, workability, and air entrainment.

Accelerating admixtures speed up the rate of development of strength at early ages. Retarding admixtures delay the setting time and are particularly used in mass concreting to overcome the heat evolution or when concreting is done in hot weather. Plasticisers are used to increase workability by decreasing the water/cement ratio resulting in a higher strength. Air entrainment in concrete causes a reduction in strength at all ages, and to achieve a required strength the cement content in the mix should be increased. However, it has also been noticed that air entrainment increases the workability of concrete mix; hence the addition of entrained air can be accompanied by a reduction in the water/cement ratio, and this consequently compensates for the loss of strength. Air entrainment also increases the resistance of concrete to frost damage.

Steps of Concrete Preparation

If concrete materials are not properly proportioned and mixed into a homogeneous mass, it will result in poor-quality and low-strength concrete. After proper mixing, the concrete should be handled, transported, placed, and compacted in such a manner that bleeding and segregation should not occur, which would result in honeycombing, i.e. porous and poor-quality low-strength concrete. If full compaction is not achieved, the resulting voids produce a marked reduction in concrete strength. The time interval between mixing and placing the concrete should be reduced to the minimum possible. Adequate curing is essential for the development of the strength of concrete.

3.3.2   Durability

Durability is one of the most important properties of hardened concrete. The durable concrete is one which can withstand the conditions for which it has been designed without deterioration. PCA (Portland Cement Association) defines durability as the ability of concrete to resist weathering action, chemical attack, and abrasion while maintaining its desired engineering properties.

To ensure a high degree of durability, it is essential that clean, sound materials and the lowest possible water content are used in the concrete, together with thorough mixing. Good consolidation during placement of the concrete is important, including proper curing and protection of the concrete during the early hardening period. Another property that helps ensure durability is the water-to-cementitious ratio (w/c). The term w/c generally refers to the weight of water divided by the weight of cementitious material. Cementitious materials include Portland cement, slag, silica fume, fly ash, and any other material having cementitious properties.

The durability of a concrete could be affected by external factors like environment, freezing and thawing, wetting and drying, abrasion, and chemical attack and by internal factors like interaction between the constituent materials such as alkaliaggregate reaction, volume changes, and corrosion. Thus, different concretes require different degrees of durability depending on the exposure environment and properties desired.

A good air void system is also essential to having a durable concrete when the concrete is exposed to freeze/thaw conditions. Air voids in concrete by adding air entrainment relieve the pressure caused by frozen water in concrete. Concrete having a total air void content of about 6.5% seems optimal. A mix having 6.5% total air voids will have approximately 1.5% entrapped air voids and 5.0% entrained air voids [12]. Entrapped air is the larger bubbles formed in the mixing process and does not provide much protection against freeze/thaw action. The entrained air bubbles are smaller and more closely spaced. These small bubbles give protection against freeze/thaw. Concrete with entrained air will have a lower strength than the same mix without entrained air, but the concrete can attain strengths required for most purposes by an increase in the cementitious factor of the mix or by reducing the water content.

Deicing agents, if added, accelerate freeze/thaw cycles. The soil and groundwater may have high sulphate content, which is potentially harmful to the concrete. Use of type MS or type HS cements is therefore recommended to protect concrete against sulphate attack. All these topics are discussed in Part II of the book.

3.3.3 Permeability and Porosity

One of the main characteristics influencing the durability and strength of concrete is its permeability to the potentially deleterious substances like ingress of water, oxygen, and carbon dioxide.

Permeability is that property of concrete which permits liquids to pass through it, whereas *porosity* is the property in which liquids can penetrate into it by capillary action, and depends on the total volume of the spaces occupied by air or water between the solid matters in the hardened concrete. A higher permeability or porosity leads to deterioration of concrete.

Excess water leaves voids and cavities after evaporation, and if they are interconnected water can penetrate or pass through the concrete. The air voids result due to incomplete compaction and the water voids are due to water left behind in the concrete after compaction. Air voids must be reduced as much as possible by compacting the concrete adequately. Water voids must be reduced by keeping the water/cement ratio as low as possible.

Porous aggregates increase the permeability of concrete, the use of which should be avoided. Poor curing also causes an increase in permeability as it leads to shrinkage cracks around the large aggregate particles. In order to have dense concrete, well-graded aggregate has to be used.

Admixtures are sometimes used to make concrete less permeable, but in general good mix design with appropriate water/cement ratio followed by careful placing, compacting, and curing are the best methods of ensuring high resistance to penetration of water. ACI-318-08 (Revised 2014) suggests that (Table 7.1) in order to have low-permeable concrete, the structural concrete should have a w/c ratio of not more than 0.50 for exposure to freshwater and not more than 0.40 for exposure to seawater.

3.3.4 Shrinkage

The contraction or decrease in volume that occurs in concrete when it dries and hardens is termed as shrinkage.

The decrease in volume is mainly the result of moisture loss caused by drying and hydration as well as the chemical changes that result in the carbonation of cement hydration products [13]. The volume changes due to shrinkage results in cracks, which must be avoided as it will affect adversely the durability and integrity of the structure.

Shrinkage is mainly of three kinds:

3.3.4.1 Plastic Shrinkage

Plastic shrinkage is the first change to occur. Plastic shrinkage is caused by volume loss due to the hydration reaction and by evaporation. The shrinkage that takes place before concrete has set is called plastic shrinkage. When concrete is in plastic state it will shrink if it is subjected to loss of water by evaporation, which may occur due to air and concrete temperature, humidity, and wind velocity. The loss of water can also occur by suction of underlying dry concrete or soil. As per recommendations of ACI 305:R [14], evaporation rates greater than 0.5 kg/h/m^2 (0.1 lb/h/ft.^{2}) of the exposed concrete surface have to be avoided in order to prevent plastic shrinkage.

To control the evaporation of water immediately after casting, the concrete should be covered with wet hessian cloth (burlap) or polyethylene sheets or by spraying with water or curing compound. If plastic shrinkage cracks appear in the fresh concrete, the cracks can be closed by striking each side of the crack with a float or by reworking the concrete surface after about 15 min of the finishing time.

3.3.4.2 Autogenous Shrinkage

In a set concrete, the shrinkage which occurs due to the result of a chemical reaction (hydration) and ageing within the concrete is known as autogenous shrinkage. As the name implies, it is self-produced by the hydration of cement. In fact, the hydration of cement results in the generation of heat and a reduction in the volume of the hydration cement paste system (cement paste and mix water). This reduction is about 8.0% of the original volume of cement paste system [15] and is known as autogenous shrinkage and mostly occurs due to loss of water used up in hydration. Since this type of shrinkage occurs within a concrete mass, that is, without contact with the ambient medium, it is often called self-desiccation shrinkage.

Using less cement can decrease this type of shrinkage. The other factors which affect the rate and magnitude of this shrinkage are chemical composition of cement, initial water content, temperature, and time.

3.3.4.3 Drying Shrinkage

Shrinkage that takes place after the concrete is set and hardened is called drying shrinkage. The common cause of cracking in concrete is restrained drying shrinkage. It is caused due to loss of water from hardened concrete. After curing, when concrete starts drying, it first loses water from its voids and capillary pours, and then the water is drawn out of its cement gel, resulting in shrinkage of the hardened concrete.

In general, drying shrinkage is directly proportional to the water/cement ratio and inversely proportional to the aggregate/cement ratio. Drying shrinkage can be reduced by increasing the amount of aggregate and reducing the water content. The development of both autogenous and drying shrinkage can be delayed by wet curing, and prolonged curing can help avoid the development of cracking. In addition to appropriate curing, for reducing drying shrinkage, it is also necessary to prevent evaporation of water from the surface of the concrete. The water evaporation can be controlled as discussed in plastic shrinkage. Surface crazing on walls and slabs is an example of drying shrinkage. The cracks, caused due to shrinkage cracking, can be controlled by using contraction joints and proper steel detailing. Shrinkage cracking may also be reduced by using shrinkage-compensating concrete. In general, the factors affecting drying shrinkage are produced in Table 3.2.

For concrete that can dry completely (except water-retained structures) and where the shrinkage is unrestrained, the linear coefficient is approximately 0.00025 at 28 days and 0.00035 at 3 months, and at the end of 12 months 0.0005. If concrete member is restrained so that a reduction in length due to shrinkage cannot take place, tensile stresses are caused. A coefficient of 0.0002 may be considered to a stress of 3.5 N/mm² or 500 lb./in.². In cases like a building floor, it is important to minimise the action of these stresses by proper curing and providing joints. Shrinkage is considered in the calculation of deflections and the design of fixed arches.

3.3.5   Modulus of Elasticity

Concrete behaves nearly elastic under load; this deformation increases with the applied load and is commonly known as elastic deformation. However, under constant load, the concrete continues to deform; that is, strain increases with time and this deformation is termed as creep, which will be discussed later.

The elastic property of concrete is measured with modulus of elasticity which is the ratio of load per unit area, i.e. stress to the elastic deformation per unit length, i.e. strain. In other words:

$$\text{Modulus of elasticity} = \text{stress} / \text{strain}$$

The modulus of elasticity, also known as Young's modulus, is an important factor in the design of most concrete structures. The modulus is used for calculation of deformation, deflection, or stresses under normal working loads. It increases with increase in cement content and age and with a reduction in the water/cement ratio, as does the strength. The value of the modulus of elasticity ranges between 500 and 1600 times the compressive strength, but the average modulus of elasticity (Ec) for

Table 3.2 Factors affecting drying shrinkage

Factors	Reduced shrinkage	Increased shrinkage
Cement type	Types I, II	Type III
Aggregate size	1½ in. (38 mm)	¾ in. (20 mm)
Aggregate type	Quartz	Sandstone
Cement content	550 lb/yd³ (325 kg/m³)	700 lb/y³ (415 kg/m³)
Slump	3 in. (75 mm)	6 in. (150 mm)
Curing	7 days	3 days
Placement temperature	60 °F (16 °C)	85 °F (29 °C)
Aggregate state	Washed	Unwashed

(Reprinted with permission from Concrete Repair & Maintenance, Illustrated by Peter Emmons, © 1993 R.S. Means Co, Inc. Kingston, MA 02332, USA [16])

a 1:2:4 concrete is 24 KN/mm^2 or 3.5×10^6 lb/in.^{2}. However, an approximate relationship can be used to estimate the modulus of elasticity for normal-weight concrete as provided by ACI-318 (2008) [17]:

$$Ec = 33w_c^{1.5} \sqrt{fc'} \text{ lb / in.}^2$$

where w_c = weight of concrete (lb/ft.^{3}) and $\sqrt{fc'}$ = the compressive strength determined by the test cylinders at 28 days.

Hence for normal-weight and normal-density concrete we have

$$Ec = 57,000 \sqrt{fc'} \text{ lb / in.}^2$$

or $4500 \sqrt{fc'}$ MPa for SI (metric) units

CSA-A23.3-04 [13] also proposes that the modulus of elasticity, Ec, of normal-density concrete with compressive strength between 20 and 40 MPa may be taken as $4500 \sqrt{fc'}$ MPa.

The modulus of elasticity of structural lightweight concrete is generally 20–50% lower than for normal-weight concrete of equal strength.

3.3.5.1   Modular Ratio

In reinforced concrete, the modulus of elasticity is used in the form of the modular ratio, i.e. the ratio of the modulus of elasticity for the reinforcing steel (Es) to the modulus of elasticity for the concrete (Ec). Hence:

$$\text{Modular ratio} = Es / Ec$$

Now, for 1:2:4 concrete considering Ec = 24 KN/mm^2 or 3.5×10^6 lb/in.^{2} and Es = 200 KN/mm^2 or 30×10^6 lb/in.^{2} modular ratio will be Es/Ec = 200/24 = 8.5.

The modular ratio is not constant for all grades of concrete. It varies with the grade of concrete. However, as recommended in BS-8110, the generally accepted arbitrary value for the modular ratio is taken as 15 by considering a fix value of Ec as 14 KN/mm^2 or 2×10^6 lb/in.^{2} for concretes of all proportions.

3.3.5.2   Poisson's Ratio

When concrete is subjected to an axial compression, it contracts in the axial direction and expands laterally. Poisson's ratio is defined as the ratio of the lateral strain to the associated axial strain. The value for Poisson's ratio ranges from 0.1 to 0.3. It increases as the water/cement ratio is increased and decreases as the cement content is decreased. For serviceability calculations, BS-8110 recommends to consider value of 0.2. As regards static modulus of elasticity test and dynamic modulus test, the static value is generally lower than the dynamic value, the former being frequently taken as 0.15 and the latter as 0.24. The dynamic value has particular application in roadwork, where transient loads are applied. However, in general, Poisson's ratio is used in the advanced structural analysis of flat-plate floors, roof shells, and mat foundations.

3.3.6 Creep

After the concrete has changed from the plastic to the hardened state, it is subject to changes in its volume and dimensions due to changes in temperature as well as creep. Creep of concrete is the continuous deformation with time, which takes place under conditions of sustained loading. It is therefore defined as the time-dependent increase in strain under sustained loading. Creep strain is much larger than the elastic strain on loading. When the sustained load is removed, there is an immediate elastic recovery, which is followed by a gradual decrease in strain over a period of time.

The extent of the creep is proportional to the stress applied, and the tendency of concrete to creep gradually decreases as the strength increases. It increases with increasing water/cement ratio and decreases with increase in relative humidity. The type of aggregate also influences the tendency to creep; a sandstone concrete, for instance, creeps much more than a granite concrete. The different types of cement also influence creep because of the different rates of gain in concrete strength; for example, concrete made with rapid-hardening Portland cement shows less creep than concrete made with ordinary Portland cement and loaded at the same age.

The effect of creep of concrete is not often considered directly in reinforced concrete design; however, it is taken into account when calculating deflections. Characteristic values for creep, expressed in deformation per unit length, for 1:2:4 concrete loaded at 28 days with a sustained stress of 4 N/mm^2 or 600 lb/in.^{2} are 0.0003 at 28 days after loading, and 0.0006 at 1 year. Creep is usually expressed in terms of creep coefficient C_t. ACI committee 209 provides the following expression to determine creep coefficient which is applicable to normal-, semi-low, and low-density concretes:

$$C_t = \frac{t^{0.6}}{10 + t^{0.6}} \times C_u Q_{cr}$$

where t = time in days after loading, C_u = ultimate creep coefficient, and Q_{cr} = correction factor.

C_u varies between 1.30 and 4.15. In the absence of specific creep data for local aggregates and conditions, the average value suggested for C_u is 2.35.

3.3.7 Thermal Properties

The important thermal properties, which may be significant in the performance of concrete, are thermal conductivity, specific heat, thermal diffusivity, and coefficient of thermal expansion. The first two are required in mass concrete to which insulation is applied to control the loss of heat by evaporation, conduction, and radiation. However, thermal diffusivity is an index of the facility with which the temperature can change and the coefficient of thermal expansion is of great importance in ordinary structural works.

3.3.7.1 Thermal Conductivity

It is a measure of ability of the concrete to conduct heat and is measured in British thermal units (BTU) per hour per square foot area of the body when the temperature difference is 1 °F per foot thickness of the body. The average value of thermal conductivity of normal-weight 1:2:4 gravel concrete at normal temperature is 10 British thermal units inches per square foot per hour per F (1.44 W/m^2/°C). Thermal conductivity depends upon the composition of concrete and varies with the density and porosity of the material. The structural concrete of normal aggregate conducts heat more readily than lightweight concrete.

3.3.7.2 Specific Heat

It gives the heat capacity of concrete. It increases with the moisture content of concrete. The specific heat values for ordinary concrete are between 0.2 and 0.28 BTU/lb/F. Specific heat varies with temperature. Hydrated cement paste has a low specific heat and a very definite curve of variation with temperature. Since water has a specific heat of 1.0, the water content is relatively effective in raising the specific heat of concrete. The specific heat of ice at 0 °F is 0.4 and for other non-metallic construction materials is mostly between 0.16 and 0.24.

3.3.7.3 Thermal Diffusivity

It is a constant and is a measure of the rate at which temperature changes will take place within the mass of hardened concrete. Its variability over a range of 0.020–0.080 ft.^{2}/h (0.00186–0.0074 m^2/h) is controlled largely by the composition of the mass and is very similar in characteristics to the thermal conductivity. This property is directly proportional to the thermal conductivity and inversely proportional to the specific heat multiplied by the density. It may be determined from the formula

$$a = k / cp$$

where a = thermal diffusivity (diffusion constant), k = thermal conductivity, c = specific heat, and p = density in pounds per cubic foot.

3.3.7.4 Thermal Expansion

Thermal expansion of concrete depends mostly on the type of aggregate and the amount of cement used. Concrete made with siliceous aggregates such as flint gravels, quartzite, and many stones expands more than that made with calcareous aggregates such as calcite and limestone. Rich mixes expand more than lean ones, as the coefficient of expansion of the cement paste is greater than that of the aggregate.

The expansion caused in concrete due to chemical reactions between its ingredients may cause buckling. Expansion due to heat of hydration of cement can be reduced by keeping cement content as low as possible, using type IV cement (low-heat Portland cement) and chilling the aggregates, water, and concrete in the forms. Expansion due to increase in air temperature may be decreased by producing concrete with a lower coefficient of expansion, usually by using coarse aggregates with a lower coefficient of expansion.

The coefficient of thermal expansion is required in the design of chimneys, tanks containing hot liquids, and exposed or long lengths of construction. Provisions must be made to resist the stresses due to changes in temperature or to limit the strains by providing joints. Hardened concrete has a coefficient of thermal expansion ranging from 7 to 12×10^{-6} per °C (4 to 7×10^{-6} per °F). A coefficient of 10×10^{-6} mm/mm per °C (5.5×10^{-6} in./in. per °F), which is about the same as that of mild steel, is commonly used. Roughly, the amount of movement caused by temperature changes is obtained by multiplying the average thermal expansion of concrete by the length of the structure and by the degree change in temperature. A change of 100 °F (38 °C) in a 100 ft. (30.5 cm) length of concrete will change the overall length by 7/8 in. (22 mm).

Bibliography

1. Mehta, P.K. and Monteiro, P.J.M., CONCRETE - Microstructure, Properties & Materials, McGraw Hill, 3rd edition, 2006.
2. ACI 309. 1 R-93 (Revised 2008), Behavior of Fresh Concrete During Vibration.
3. BS 1881, Part 103-1993 Testing Concrete- Method for Determination of Compacting Factor (Replaced by BS EN 12350-4-2009).
4. ASTM C143-90a (Revised 2015a), Test Method for Slump of Portland Cement Concrete.
5. Canadian Standards Association- CSA- A23.2-14, Test methods and standard practices for concrete.
6. Ontario Provincial Standard Specification- OPSS 1350, Material Specification for Concrte- Materials and Production, 1996.
7. ASTM C78-84 (Revised 2018), Test for Flexural Strength of Concrete.
8. A.M. Neville – Properties of Concrete; Pearson Education Limited England- 5th Edition 2011.

9. ASTM C496-90 (Revised 2017), Test for Splitting Tensile. Strength of Cylindrical Concrete Specimens.
10. M.L. Gambhir, Concrete Technology, TATA McGraw Hill, India, 1989.
11. CTL Group- Ultrasonic Pulse Velocity- Checking Concrete's Pulse- Skokie, Illinois 60077 USA- www.ctlgroup.com
12. Minnesota Department of Transportation USA, "Properties and Mix Designations"- Concrete Manual, September 2003.
13. Cement Association of Canada- CSA- A23.3-04- Concrete Design Handbook, Canada 2006.
14. ACI 305R-91 (Revised 2010), *Hot* Weather *Concreting.*
15. P.C Aitcin, A.M. Neville & P. Aacker, Integrated View of Shrinkage Deformation. ACI journal, Concrete International, Vol. 19, No. 11, November 1997.
16. Peter H. Emmons, Concrete Repair & maintenance Illustrated, R. S. Means Company, Inc. USA, 1993.
17. ACI 318-08 (Revised 2014): Building Code Requirements for Structural Concrete and Commentary.

Chapter 4
Proportioning of Concrete Mixes

4.1 Introduction

Concrete mixtures are proportioned to have required workability and to assure that the hardened concrete will have the required properties. A concrete mix having excessive coarse aggregate lacks sufficient mortar to fill the void system, resulting in a non-cohesive and harsh mix. The strength and impermeability of harsh mixes will be less than those for a properly proportioned mix. On the other hand, excessive amounts of fine aggregate will increase the cohesion and the mix will turn sticky and difficult to move. This will need more water content resulting in increased drying shrinkage and cracking.

In addition to the above, the cement content also affects the workability of a concrete mix. High cement content mixes are generally sticky and sluggish. Furthermore, the lower w/c ratio reduces the workability of rich mixes. Hence, it is important that the concrete proportions should be selected to provide suitable workability, density, strength, and durability for the particular application at the lowest cost.

The common method of expressing the proportions of the materials in a concrete mix is in the form of parts, or ratios of cement, fine aggregate, and coarse aggregate with cement being taken as unity, e.g. 1:2:4 mix. The proportions in which the cement, fine aggregate, and coarse aggregate are mixed may be either by volume or by weight. In volumetric ratios 1:2:4 means one part by volume of cement, two parts by volume of fine aggregate, and four parts by volume of coarse aggregate. The amounts of water and admixtures, if any, are expressed separately.

Generally, for all important concrete works, proportioning is usually done by weight. In measuring by weight, if Portland cement has a nominal weight of 1440 kg/m³ or 90 lb/ft.³, then 1:2:4 means 1440 kg of Portland cement to 2 m³ of fine aggregate to 4 m³ of coarse aggregate or 90 lb of cements to 2 ft.³ of fine

© Springer Nature Switzerland AG 2019
A. Surahyo, *Concrete Construction*,
https://doi.org/10.1007/978-3-030-10510-5_4

aggregate to 4 ft.^{3} of coarse aggregate. If the basis of batch of concrete is a 50 kg or 1 cwt bag of cement, this mix is equivalent to 50 kg of cement to 0.07 m^3 of fine aggregate to 0.14 m^3 of coarse aggregate or 112 lb of cement to 2.5 ft.^{3} of fine aggregate to 5 ft.^{3} of coarse aggregate.

The other forms of expressing the proportions or ratios to specify concrete are as cement/aggregate ratio, water/cement ratio, and fine aggregate/coarse aggregate ratio, usually by weight.

Such methods have certain merits in terms of simplicity of expression; however, they do not account for the varying characteristics of the constituent materials and usually result in under- or over-rich mixes. Hence nowadays, in order to satisfy strength and durability requirements, designed mixes are mostly used in which limiting values of properties like minimum strength or grade of concrete, maximum water/cement ratio, minimum cement content, and maximum size of aggregate are specified.

4.2 Normal Concrete Mix Design

Basically, the problem of designing a concrete mix consists of selecting the correct proportions of cement, fine and coarse aggregate, and water to produce concrete having the specified properties. However, proportions calculated by any method must always be considered subject to revision on the basis of experience with trial batches. A number of different approaches to design a mix have been proposed. In general, most of the available mix design methods are based on empirical relationship, charts, and graphs developed from extensive experimental investigations. Some of the common mix design methods for medium- and high-strength concrete are:

1. Trial-and-error method of mix design
2. Fineness modulus method of mix design
3. Road note No. 4 method of mix design
4. The British mix design method
5. The ACI mix design method
6. The mix design according to Indian Standard (IS)

In general, for designing concrete mixes, a basic step is to select water/cement ratio required to produce a given mean compressive strength. Because of the variations in concrete strength, concrete mixes are designed for a mean compressive strength (required average strength), sufficiently higher than the specified design compressive strength by an amount termed as margin. OPSS 1350 [1] proposes that the concrete mix shall be designed in conformance with the general requirements of ACI 211.1 [2]. Here, we will only consider British and ACI mix design methods.

4.2.1 The British Mix Design Method

The British mix design method based on Building Research Establishment (BRE) Ltd., UK [3], will be discussed in detail here. However, initially we will highlight in brief the requirements of BS-5328 [4], which provides guidance on four types of concrete mixes: designed, prescribed, standard, and designated. For type of designed mixes, it suggests that the grade of concrete should be specified with limiting mix parameters, like maximum w/c ratio, minimum cement content, and minimum grade to satisfy strength and durability requirements of the work. Meeting these requirements, Table 4.1 provides guidance on mix design limits for durability of concrete made with normal-weight aggregates of 20 mm (¾ in.) nominal maximum size, in various exposure conditions. Based on values given in Table 4.1, mix proportions are determined by the concrete producer/ready-mix concrete supplier with the help of previous record, and by making various trial mixes.

Now let us discuss BRE mix design method. It has replaced the mix design method of Road Note No. 4 and is suitable for design of normal-weight concrete mixes. The method is based on data obtained by the BRE, the Transport Research Laboratory, and the British Cement Association. The stepwise procedure of mix design is as follows:

Step 1: Target Mean Compressive Strength

In case mean compressive strength is not given, then it is calculated as below:

$$fm = fc + M$$

Table 4.1 Guidance on mix design limits proposed by BS-5328 [4] for durability of concrete made with normal-weight aggregates of 20 mm (¾ in.) nominal maximum size

Condition of exposure	Type of concrete	Maximum w/c ratio	Minimum cement content (kg/mm³)	Minimum grade
Mild	Reinforced	0.65	275	C30
	Prestressed	0.60	300	C35
Moderate	Reinforced and prestressed	0.60	300	C35
Severe	Reinforced and prestressed	0.55	325	C40
Very severe	Reinforced and prestressed	0.55	325	C40
Most severe	Reinforced and prestressed	0.45	400	C50
Abrasive	Reinforced and prestressed	0.50	350	C45

Notes:
1. Where concrete is subjected to freezing while wet, air entrainment should be used. Using air-entrained concrete the grade may be reduced by 5
2. For aggregates other than 20 mm, minimum cement content can be adjusted, like +40 kg/m³ for 10 mm aggregate and −30 kg/m³ for 40 mm aggregate
3. To achieve low w/c ratio, quantity of water can be reduced by using chemical admixtures

where fm = the target mean strength, fc = the specified compressive/characteristic strength, and M = the margin = $K \times s$, where K = a constant and s = the standard deviation.

The values of K and s generally depend on the variation in concrete strength. K increases as the proportion of defectives is increased. The values recommended for K are as follows:

K for 10% defectives = 1.28
K for 5% defectives = 1.64
K for 2.5% defectives = 1.96
K for 1% defectives = 2.33

For the value of "s", it is usually required that the standard deviation for a mix design should be calculated from at least 20 results as it will provide better judgement for estimating the true standard deviation. Usually this value is taken as 4 N/mm² for concrete with a characteristic strength of 20 N/mm² or above as shown in Fig. 4.1. However, in case of fewer results, a standard deviation of 8 N/mm² should be used for concrete having characteristic strength of 20 N/mm² or more as shown in Fig. 4.1.

Step 2: Free Water/Cement Ratio

Knowing the type of coarse aggregate, type of cement, and age, obtain the value of compressive strength for a free w/c ratio of 0.50 from Table 4.2. This strength value is now plotted in Fig. 4.2 and a curve is drawn from this point and parallel to the printed curves until it intercepts a horizontal line passing through the ordinate representing the target mean strength.

The corresponding value for free w/c ratio can then be read from the abscissa. This value of w/c ratio is now compared with the maximum w/c ratio specified for the durability in Table 4.1 and the lower of the two values is adopted.

Step 3: Free Water Content

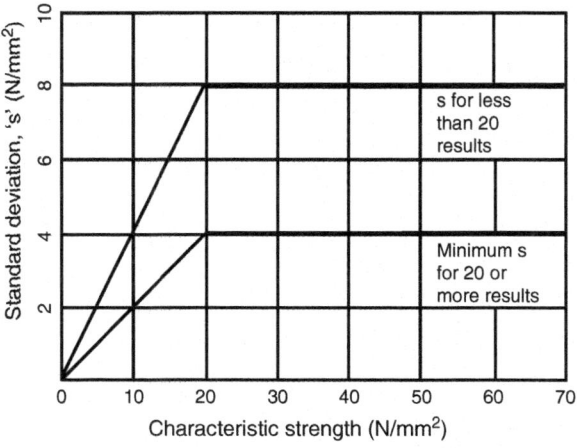

Fig. 4.1(XX) Relationship between standard deviation and characteristic strength © BRE, UK [3]

Table 4.2 (YY) Approximate compressive strengths of concrete mixes made with a free w/c ratio of 0.50 [3]

Type of cement	Type of coarse aggregate	Compressive strengths (N/mm² [psi])			
		Age (days)			
		3	7	28	91
Ordinary Portland cement or sulphate-resisting Portland cement	Uncrushed	22 (3200)	30 (4400)	42 (6100)	49 (7100)
	Crushed	27 (3900)	36 (5200)	49 (7100)	56 (8100)
Rapid-hardening Portland cement	Uncrushed	29 (4200)	37 (5400)	48 (7000)	54 (7800)
	Crushed	34 (4900)	43 (6200)	55 (8000)	61 (8900)

© BRE, UK

Fig. 4.2 (XX) Variation of compressive strength with free w/c ratio © BRE, UK [3]

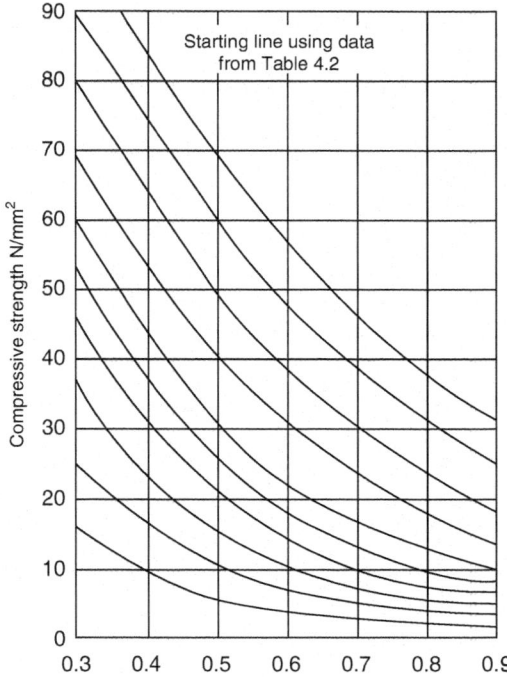

Based on the type and size of aggregate and appropriate workability (specified in terms of slump or Vebe time), the water content is calculated from Table 4.3.

Step 4: Cement Content

Knowing the water content and w/c ratio, the cement content is calculated as:

$$\text{Cement content} = \text{free water content} / \text{free w/c ratio}$$

Table 4.3 (YY) Approximate free water contents (kg/m³) required to give various levels of workability [3]

Maximum size of aggregate	Type of aggregate	Water content kg/m³			
		Slump: in mm and Vebe			
		0–10	10–30	30–60	60–180
		>12	6–12	3–6	0–3
10	Uncrushed	150	180	205	225
	Crushed	180	205	230	250
20	Uncrushed	135	160	180	195
	Crushed	170	190	210	225
40	Uncrushed	115	140	160	175
	Crushed	155	175	190	205

This calculated cement content, however, should not be less than the minimum cement content required as per Table 4.1.

Step 5: Total Aggregate Content

This step involves the calculation of total aggregate required. The wet density of concrete is obtained from Fig. 4.3 depending upon the relative density of the combined aggregate and free water content. If information is not available for relative density of aggregate then an approximate value of 2.8 is assumed for uncrushed aggregate and 2.9 for crushed aggregate.

Now total aggregate content (saturated and surface-dry condition) is calculated as

$$\text{Total aggregate content} = \text{the wet density of concrete} \left(kg / m^3 \right)$$
$$- \text{cement content} \left(kg / m^3 \right) - \text{free water content} \left(kg / m^3 \right).$$

Step 6: Fine and Coarse Aggregate Contents

Step 6 involves separately calculating the required fine and coarse aggregate content. Based on known free w/c ratio, maximum size of aggregate, and slump and grading of the fine aggregate (defined by its percentage passing a 600 μm sieve), the proportion of fine aggregate is calculated from Fig. 4.4a–c. The contents of fine and coarse aggregate, based on the found total aggregate value as per step 5, are then calculated as follows:

$$\text{Fine aggregate content} = \text{total aggregate content} \times \text{proportion of fines}$$

$$\text{Coarse aggregate content} = \text{total aggregate content} - \text{fine aggregate}$$

The coarse aggregate content can be subdivided if single-sized 10, 20, and 40 mm materials are to be combined. In this case BRE suggests the ratio to be considered: 1:2 for combination of 10 and 20 mm material, whereas 1:1 and 5:3 for combination of 10, 20, and 40 mm material.

Step 7: Trial Mixes

Fig. 4.3 (XX) Estimated weight density of fully compacted concrete © BRE, UK [3]

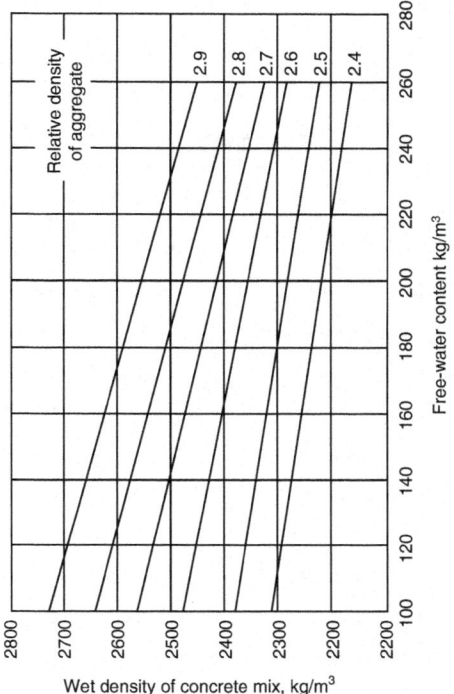

Finally, the trial mixes of the calculated proportions are prepared and checked for their 28-day strength, workability, cohesiveness, finishing properties, etc. If any one of these properties is unsatisfactory, suitable adjustments are made to obtain the final proportions satisfying the design requirements. For example, lack of cohesiveness can be corrected by increasing the fine aggregate content with making necessary deduction in coarse aggregate content. Similarly, if at the designed water content the workability of the trial mix appears below that required, additional water can be added to obtain the required workability.

4.2.2 The ACI Mix Design Method

The ACI mix design method is suitable for normal- and heavy-weight concretes having 28-day cylinder compressive strength of 45 N/mm². However, the procedure for selection of mix proportions given in this section is applicable to normal-weight concrete based on ACI 211.1.91 [2]. Use of admixture, silica fume, etc. in concrete is beyond the scope of this method. Steps for design as per ACI 211.1.91 [2] are enlisted below.

Step 1: Slump

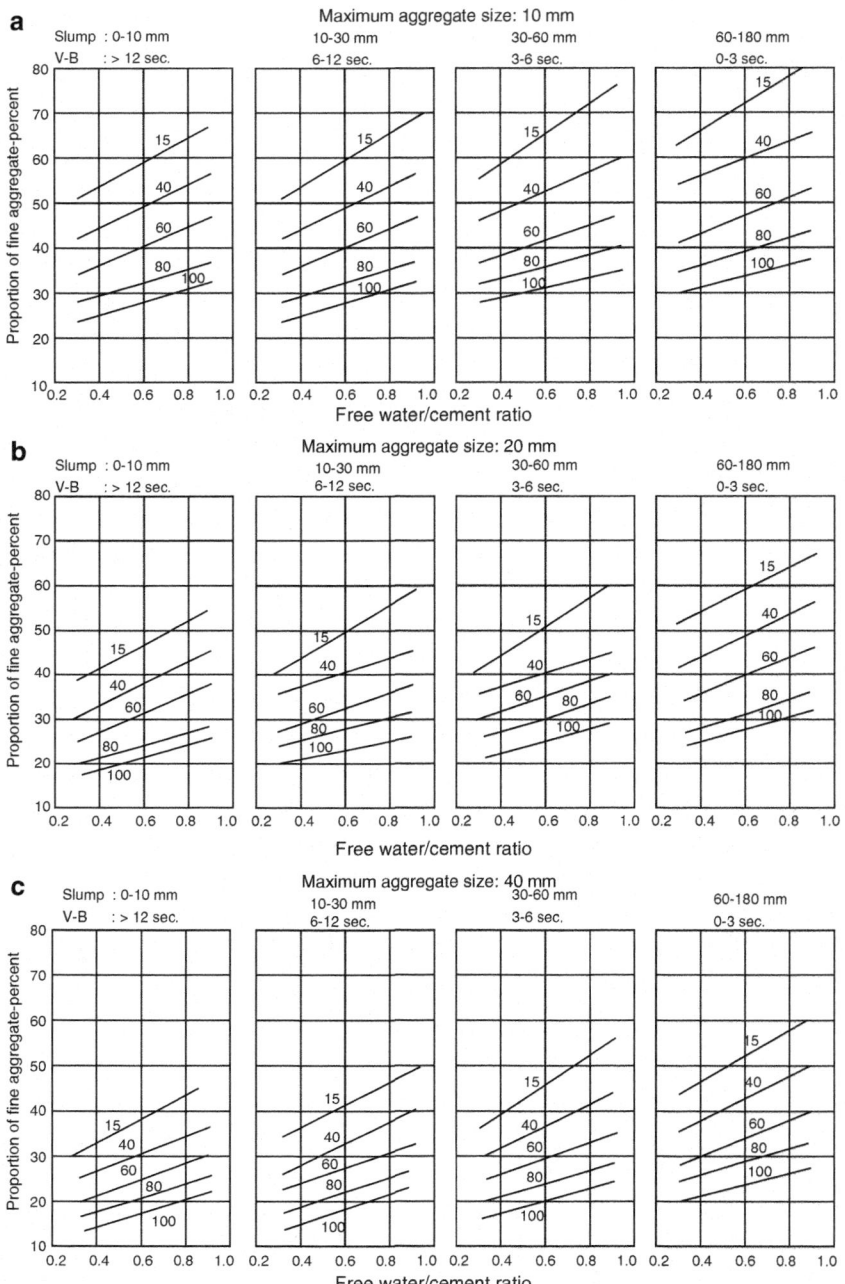

Fig. 4.4 (XX) (**a**) Recommended proportions of fine aggregate according to percentage passing 600 μm sieve. Numbers of each graph present the percentage of fines passing 60 μm sieve © BRE, UK [3]. (**b**) Recommended proportions of fine aggregate according to percentage passing 600 μm sieve. © BRE, UK [3]. (**c**) Recommended proportions of fine aggregate according to percentage passing 600 μm sieve. © BRE, UK [3]

Table 4.4 Recommended slumps for various types of construction [2] as per ACI 211-91

Types of construction	Slump, mm (in.)	
	Maximum	Minimum
Reinforced foundation: walls and footings	75 (3)	25 (1)
Plain footings, caissons, and substructure walls	75 (3)	25 (1)
Beams and reinforced walls	100 (4)	25 (1)
Building columns	100 (4)	25 (1)
Pavements and slabs	75 (3)	25 (1)
Mass concrete	50 (2)	25 (1)

The value of slump appropriate for the concrete mix is selected from Table 4.4. Slump may be increased when chemical admixtures are used.

Step 2: Required Average Compressive/Target Mean Compressive Strength

In case average compressive strength is not given, then based on specified compressive strength, it is calculated as below:

$$fc'r = fc' + 1.34Ks$$

where $fc'r$ = required average compressive strength, fc' = specified compressive strength, K = factor from Table 4.5 for increase in standard deviation if the total number of tests is less than 30, and s = standard deviation.

Now for calculation of standard deviation, ACI specifies that the field test records used to calculate "s" should have performed at least 15 consecutive compressive strength tests, from a single group with the same mix proportions. In this case, i.e. for a single group of consecutive test results, "s" will be calculated as below:

$$s = \left[\frac{\sum_{1}^{n}(X_i - X)^2}{n-1} \right]^{1/2}$$

where s = standard deviation, n = number of test results considered, X = average of all n test results considered, and Xi = individual test results.

However, for two groups of consecutive test results "s" will be calculated as.

$$s = \left[\frac{(n_1-1)(s_1)^2 + (n_2-1)(s_2)^2}{(n_1+n_2-2)} \right]^{1/2}$$

where s = standard deviation for the two groups combined; s_1 and s_2 = standard deviations for groups 1 and 2 respectively, calculated as per equation developed for single group above; and n_1 and n_2 = number of test results in groups 1 and 2, respectively.

Table 4.5 K-factor for increasing the standard deviation for number of tests considered as per ACI-301 [5]

Total number of tests considered	K-factor for increasing standard deviation
15	1.16
20	1.08
25	1.03
30 or more	1.00

Table 4.6 Required average compressive strength when data is not available to establish a standard deviation as per ACI-301 [5]

Specific strength fc'	Required average compressive strength $fc'r$
Less than 3000 psi	fc' + 1000 psi
3000–5000 psi	fc' + 1200 psi
Over 5000–10,000 psi	fc' + 1400 psi
Over 10,000–15,000 psi	fc' + 1800 psi

In case field test data is not available to establish a standard deviation, the required average compressive strength is calculated from Table 4.6.

Step 3: Effective w/c ratio

For the required average compressive strength, effective w/c ratio is determined from Table 4.7.

Step 4: Water Content

For the known slump as per step 1 and required normal maximum size of aggregate, the water content is determined from Table 4.8.

Step 5: Cement Content

Knowing the w/c ratio and water content from steps 3 and 4, respectively, cement content is calculated.

Step 6: Coarse Aggregate Content

Now from Table 4.9, for a maximum size of aggregate and fineness modulus of sand, the dry bulk volume of coarse aggregate per unit volume of concrete is determined. The content of coarse aggregate is then calculated as

Coarse aggregate content = dry bulk volume × given density of coarse aggregate

Step 7: Fine Aggregate Content

The fine aggregate content is determined by subtracting the sum of volumes of coarse aggregate, cement, and water content from unit weight (density) of concrete, which is determined from Table 4.10.

Step 8: Trial Mixes

Table 4.7 Relationship between w/c ratio and average compressive strength of concrete based on ACI-211-91 [2]

Average compressive strength at 28 days		Effective w/c ratio (by weight)	
MPa	Psi	Non-air-entrained concrete	Air-entrained concrete
–	6000	0.41	–
40	–	0.42	–
35	5000	0.48	0.40
30	–	0.54	0.45
–	4000	0.57	0.48
25	–	0.61	0.52
–	3000	0.68	0.59
20	–	0.69	0.60
15	–	0.79	0.70
–	2000	0.82	0.74

Table 4.8 Approximate mixing water and air content requirements for different slumps and nominal maximum sizes of aggregates as per ACI-221-91 [2]

Slump, mm (in.)	Water content kg/m³ (lb/yd³) for indicated nominal maximum sizes of aggregate						
	10 mm (3/8 in.)	12 mm (½ in.)	20 mm (¾ in.)	25 mm (1 in.)	40 mm (1½ in.)	50 mm (2 in.)	75 mm (3 in.)
Non-entrained concrete							
25–50 (1–2)	205 (350)	200 (335)	190 (315)	180 (300)	165 (275)	155 (260)	130 (190)
75–100 (3–4)	225 (385)	215 (365)	205 (340)	195 (325)	180 (300)	170 (285)	145 (210)
150–175 (6–7)	240 (410)	230 (385)	215 (360)	200 (340)	190 (315)	180 (300)	160 (270)
Approximate entrapped air content %	3	2.5	2	1.5	1	0.5	0.3
Air-entrained concrete							
25–50 (1–2)	180 (305)	175 (295)	165 (280)	160 (270)	150 (250)	140 (240)	125 (205)
75–100 (3–4)	200 (340)	190 (325)	185 (305)	175 (295)	165 (275)	155 (265)	135 (225)
150–175 (6–7)	215 (365)	205 (345)	200 (325)	185 (310)	175 (290)	165 (280)	155 (260)
For different exposure conditions recommended average total air content %							
Mild exposure	4.5	4.0	3.5	3.0	2.5	2.0	1.5
Moderate exposure	6.0	5.5	5.0	4.5	4.5	4.0	3.5
Extreme exposure	7.5	7.0	6.0	6.0	5.5	5.0	4.5

Table 4.9 Dry bulk volume of coarse aggregate per unit of volume of concrete as per ACI-211-91 [2]

Maximum size of aggregate, mm (in.)	Volume of oven-dried rodded coarse aggregate per unit volume of concrete for different fineness moduli of fine aggregate			
	2.40	2.60	2.80	3.00
10 ($3/8$)	0.50	0.48	0.46	0.44
12 (½)	0.59	0.57	0.55	0.53
20 (¾)	0.66	0.64	0.62	0.60
25 (1)	0.71	0.69	0.67	0.65
40 (1½)	0.75	0.73	0.71	0.69
50 (2)	0.78	0.76	0.74	0.72
73 (3)	0.82	0.80	0.78	0.76
150 (6)	0.87	0.85	0.83	0.81

Table 4.10 First estimate of unit weight (density) of fresh concrete as per ACI-211-91 [2]

Maximum size of aggregate, mm (in.)	First estimate of concrete weight, kg/m³ (lb/yd³)	
	Non-air-entrained concrete	Air-entrained concrete
10 ($3/8$)	2285 (3840)	2190 (3690)
12 (½)	2315 (3890)	2235 (3760)
20 (¾)	2355 (3960)	2280 (3840)
25 (1)	2375 (4010)	2315 (3900)
40 (1½)	2420 (4070)	2355 (3960)
50 (2)	2445 (4120)	2375 (4000)
75 (3)	2465 (4160)	2400 (4040)
150 (6)	2505 (4230)	2435 (4120)

Now the calculated mix proportions should be checked by means of trial batches, and tested. Only sufficient water should be used to produce the required slump regardless of the amount assumed in selecting the trial proportions. The concrete should be checked for unit weight, proper workability, freedom from segregation, and finishing properties. The final proportions may be obtained by appropriate adjustments.

Example 1: American Method

Design a concrete mix as per American method with the specified requirements as follows:

1. Required average compressive strength, at 28 days   25 MPa
2. Slump required   75–100 mm
3. Nominal maximum size of coarse aggregate   20 mm
4. Dry rodded mass of coarse aggregate   1600 kg/m³
5. Cement used   OPC (type I)
6. Fineness modulus of fine aggregate   2.6

7. Bulk specific gravity of coarse and fine aggregate   2.6

(with negligible absorption and moisture content)

SOLUTION:

Step 1: Slump

The slump is required to be 75–100 mm.

Step 2: Required average/target mean compressive strength

The required average compressive strength is given as 25 MPa.

Step 3: W/C ratio

Based on the given compressive strength 25 MPa, the w/c ratio from Table 4.7 (without air entraining) = 0.61.

Step 4: Water content

Having slump = 75–100 mm and maximum size of aggregate = 20 mm.

Using Table 4.8, mixing water quantity (non-air-entrained concrete) = 205 kg/m^3.

Step 5: Cement content

Calculated w/c ratio = 0.61 and calculated water quantity = 205 kg/m^3

Cement content = water content/w/c ratio = 205/0.61 = 336 kg/m

Step 6: Coarse aggregate content

Maximum size of coarse aggregate   = 20 mm

Fine modulus of sand   = 2.6

Therefore, dry bulk volume of coarse aggregate as per Table 4.9 = 0.64 per unit volume of concrete.

Hence coarse aggregate content = dry bulk volume × given density of coarse aggregate   = 0.64 × 1600 = 1024 kg/m^3

Step 7: Fine aggregate content (weight/mass method)

Now fine aggregate = unit weight (density) of concrete − (volume of coarse aggregate + cement content + water content). Hence from Table 4.10, with maximum size aggregate 20 mm:

Density = 2345 kg/m^3

Hence   fine   aggregate   content = 2345 − (1024 + 336 + 205) = 2345 − 1565 = 780 kg/m^3

RESULT: The estimated quantities in kg/m^3 are found:

Cement   = 336

Fine aggregate   = 780

Coarse aggregate   = 1024

Added water   = 205

Example 2: British Method

Design a concrete mix as per British method with the specified requirements:

1. Characteristic compressive strength at 28 days 30 N/mm^2
2. Defective rate   5%
3. Slump required   30–60 mm
4. Maximum aggregate size   20 mm

5. Aggregate type   Uncrushed
6. Cement used   OPC (ordinary Portland cement)
7. Type of exposure   Moderate

Step 1: Target mean compressive strength

fm = fc + Ks

Now fc = 30 N/mm^2

As no previous control data is specified, K for 5% defectives = 1.64

s = 8 N/mm^2 from Fig. 4.1

Step 2: Free w/c ratio

Cement used = OPC, type of coarse aggregate = uncrushed, and age = 28 days.

Using the above data, the compressive strength for free w/c ratio of 0.50 from Table 4.2 = 42 N/mm^2.

Now plotting this value in Fig. 4.2 against the value of target mean strength 43 N/mm^2

W/C ratio = 0.49

Now as per Table 4.1, for moderate exposure conditions:

Maximum free w/c ratio = 0.60

Hence the lower w/c ratio 0.49 will be taken.

Step 3: Free water content

Type of aggregate = uncrushed, size of aggregate = 20 mm, and specified slump = 30–60 mm.

Therefore from Table 4.3, free water content = 180 kg/m^3.

Step 4: Cement content

Cement content = free water content/free w/c ratio = 180/0.49 = 367 kg/m^3.

Minimum cement content specified for moderate exposure conditions as per Table 4.1 = 300 kg/m^3.

Hence cement content calculated as 376 kg/m^3 is correct.

Step 5: Total aggregate content

Free water content as per above = 180 kg/m^3 and relative density of aggregate = 2.6.

Concrete density from Fig. 4.3 = 2375 kg/m^3.

Now total aggregate content = wet density of concrete − cement content − free water content

= 2375 − 367 − 180 = 1828 kg/m^3

Step 6: Fine and coarse aggregate contents

For grading of fine aggregate, consider percent passing 600 μm sieve = 60%

Free w/c ratio = 0.49, maximum size of aggregate = 20 mm, and slump = 30–60 mm.

Hence proportions of fine aggregate as per Fig. 4.4b = 30–35, say 32%.

Now fine aggregate content = total aggregate content × proportions of fines

= 1828 × 32/100 = 585 kg/m^3.

Now coarse aggregate content = total aggregate − fine aggregate = 1828 − 585= 1243 kg/m^3.

Now in case if single-sized 10 and 20 mm coarse aggregates are used, the coarse aggregate content is proportioned as 1:2.

This will give us 414 kg/m^3 of 10 mm single-sized and 829 kg/m^3 of 20 mm single-sized aggregates.

RESULT: The estimated quantities in kg/m^3 are found:

Cement	= 367
Fine aggregate	= 585
Coarse aggregate	= 1243 (414-10 mm and 829-20 mm)
Added water	= 180

Bibliography

1. Ontario Provincial Standard Specification- OPSS 1350, Material Specification for Concrete-Materials and Production, 1996.
2. ACI 211.1-91 (Reapproved 2009), Standard Practice for Selecting Proportions for Normal, Heavy Weight and Mass Concrete.
3. D.C. Teychenne, J.C. Nicholls, R.E. Franklin, D.W. Hobbs, second edition amended by B.K. Marsh, Design of Normal Concrete Mixes, Building Research Establishment (BRE) Ltd UK, 1997.
4. BS 5328, Concrete: Part-1, Guide to Specifying Concrete-1991 (Amended 1995); Part-2, Methods for Specifying Concrete Mixes- 199 1 (Amended 1995); Part-3, Specification for the Procedures to be Used in Producing and Transporting Concrete-1990 (Amended 1992); Part-4, Specification for the Procedures to be used in Sampling, Testing and Assessing Compliance of Concrete-1990 (Amended 1995).
5. American Concrete Institute (ACI), Publication SP-15(95), Field Reference Manual, ACI 301-96 (Revised 2016).

Chapter 5
Sustainable Concrete

5.1 Introduction

The focus on sustainability continues to grow within the construction industry. Sustainability generally means having no net negative impact on the environment. Because concrete is the most widely used material worldwide, concrete industries have the environmental and societal responsibility to contribute to sustainable development. The concrete industry is a significant contributor to air pollution and also a consumer of vast quantities of natural materials, including water. It is general understanding that manufacturing 1 ton of Portland cement produces about 1 ton of CO_2 and other greenhouse gases (GHG) [1]. Worldwide, the cement industry produced about 1.4 billion tons CO_2 in 1995, which caused the emission of as much CO_2 gas as 300 million automobiles [2, 3].

According to the Canada Green Building Council [4], buildings generate:

- Up to 35% of all greenhouse (GHG) gases
- 35% of landfill waste as a direct result of construction and demolition activities
- Up to 70% of municipal water consumption in and around buildings

Worldwide, over ten billion tons of concrete are being produced each year. In the United States, the annual production of over 500 million tons implies about two tons for each man, woman, and child [1]. Such volumes require vast amounts of natural resources for aggregate and cement production.

The concept of sustainable concrete (green concrete) construction is that the energy and resource consumption due to the construction and operation of a concrete structure must be minimised. Relating to concrete structures, this can be achieved by the use of the material in the most efficient way considering its strength and durability within the service life of the structure. Sustainable development is about balancing human needs with the earth's capacity to meet them. Concrete offers a wide range of capabilities to help achieve this balance. Sustainable structures can cost less to build and maintain, use less energy, use fewer building materials, emit less greenhouse gas into the atmosphere, and are good for the planet and the social environment.

© Springer Nature Switzerland AG 2019
A. Surahyo, *Concrete Construction*,
https://doi.org/10.1007/978-3-030-10510-5_5

5.2 Achieving Sustainable Concrete

There are three components of sustainability: environment (issues such as greenhouse gases and global warming), economy, and society (well-being of the society). To meet its goal, sustainable development must provide that these three components remain healthy and balanced. The main focus of this chapter is to address environmental issues concerning concrete construction.

In the construction of concrete structures and use of construction materials, the critical elements of environmental impact are the utilisation of resources, the embodied energy, and the generation of waste materials including water usage. Like other building materials, concrete has embodied energy: it takes energy to manufacture and construct a concrete building or structure. However, concrete's lower embodied energy from the manufacturing stage to end of life can be handled more efficiently. Accordingly, in order to have sustainable concrete project, the following parameters need to be taken care of:

1. Minimise energy and CO_2 footprint
2. Minimise potable water use
3. Minimise waste
4. Increase use of recycled content

5.2.1 Minimise Energy and CO₂ Footprint

CO_2 emissions contribute about 70% of the potential global warming effect of emissions of greenhouse gases caused by human activities; however, the cement industry alone generates about 7% of it. Concrete is used only second to water on a volume consumption basis. Since concrete is used in large quantities in construction, it is important to reduce energy consumption and CO_2 emission in its manufacturing process.

Portland cement is usually manufactured by heating a mixture of limestone, shells, and chalk combined with other ingredients in a kiln to a high temperature up to about 1425–1650 °C and then inter-grinding the resulting clinker with gypsum to form a fine powder. The reaction between limestone and other ingredients to produce clinker results in production of CO_2. Furthermore, the fuel used in the kiln and the electricity in the grinding mills themselves produces some amount of gaseous waste, principally CO_2 and CO. These gases are nontoxic and are released to the atmosphere, where they contribute to global warming. The carbon footprint of concrete can be lowered by using supplementary cementitious materials (SCMs) such as fly ash, slag cement, and silica fume; by using higher strength concrete; and through the use of alternative fuels during the cement manufacturing process.

5.2.1.1 Using Supplementary Cementitious Materials

Proper use of SCMs can reduce the environmental footprint and contribute beneficially to the fresh and hardened properties of concrete. It is important to consider the impact on the properties of the concrete when determining the optimum amount of SCMs in a concrete mixture, which can affect water demand, curing times, durability, aesthetics, and other factors. It has been estimated that 18% replacement of Portland cement results in a 17% reduction of the CO_2 emissions and that, if just 30% of cement used globally were replaced with SCMs, the rise in CO_2 emissions from cement production could be reversed [5].

As per CSA-23.1-14 [6], use of high volumes of supplementary cementing materials (SCM) in producing concrete can meet the requirements of LEED (Leadership in Energy and Environmental Design) or similar green building accreditation systems. High-volume supplementary cementing material (HVSCM) concrete contains a level of SCM above that typically used for normal construction. CSA-23.1-14 describes the following two categories of HVSCM:

HVSCM-1: $FA/40 + S/50 \geq 1.00$
HVSCM-2: $FA/30 + S/40 \geq 1.00$

where FA = fly ash (types F, CI, or CH) content of the concrete (% mass of total cementing materials) and S = slag content of the concrete (% mass of total cementing materials).

As described in Chap. 2, supplementary cementing materials (SCMs) include pozzolanic materials, such as fly ash, silica fume, and natural pozzolans and ground granulated blast-furnace slag (GGBFS), referred to as slag. While the use of concrete with high levels of SCM offers many advantages, both technical and otherwise, such concrete displays different characteristics (in both plastic and hardened concrete) from plain hydraulic cement concrete and requires special consideration in the design and production stages.

Typical replacement levels for SCM vary depending on the nature of the material, the type of construction, and the placement conditions and traditionally fall in the range of 15– 25% for fly ash (Class-F) and 15–40 % for fly ash (Class-C) by mass of cementitious material and 30–45% for slag [39]. Concrete with higher levels of these materials, especially fly ash and to a lesser extent slag, to replace larger proportions of hydraulic cement in concrete mixtures has a record of good performance and durability. There is precedent for using concrete with up to 60% fly ash replacement and 75% slag. Such mixtures have been used successfully in industrial and heavy civil construction [40]. Currently, interest in using HVSCM concrete has expanded to commercial, institutional, and residential construction.

For HVSCM-1 and HVSCM-2 concrete mix design, CSA-23.1-14 [6] recommends that the laboratory trial mixes, followed by full-size batch tests, shall be made to demonstrate that the materials, mix formula, and production techniques chosen will produce concrete meeting the requirements for the job. The following properties, as applicable to the work, shall be evaluated in the trial:

- Workability
- Air content
- Finishability
- Setting time
- Temperature development
- Hardened air-void parameters
- Strength
- Durability

For reinforced concrete surfaces exposed to air and not directly exposed to precipitation, with depths of cover less than 50 mm, CSA provides that the water-to-cementing material ratio shall not be greater than 0.40 for HVSCM-1 concrete and not greater than 0.45 for HVSCM-2 concrete.

This requirement is intended to minimise the risk of corrosion of embedded steel due to carbonation of the concrete cover.

5.2.1.2 Using Blended Cements

Cements with large quantities of inter-ground limestone have been in use in many parts of the world, such as CEM IIA-L (up to 20% limestone) and CEM IIB-L (up to 35% limestone) in the European Union (refer Table 2.5). In Canada, the CSA A3001 has also recognised a new class of Portland-limestone cements (GUL, MHL, HEL and LHL, as mentioned in Chap. 2) with up to 15% inter-ground limestone. Use of these cements has a direct effect on reducing point-source CO_2 emissions at cement plants by approximately 10% [7].

Similar Portland-limestone cements have also been used in the United States since around 2005 under the ASTM C1157 performance standard for hydraulic cements. In 2012, the ASTM C595 standard for blended cements adopted a new category of cements with up to 15% limestone. To increase their acceptance in the construction market, the CSA Portland-limestone cements are required to meet the same set time and strength development properties as Portland cement. They also have been found to work well together with slag additions at the concrete plant, and sometimes perform better than with Portland cements [7].

5.2.1.3 Enhancing Energy Efficiency

Cement manufacturing is a very emission-intensive process. Approximately half (50%) of the emissions come from the chemical reaction that converts limestone into clinker, the active component in cement. Another 40% comes from burning fuel, and the final 10% is split between transportation and electricity. The energy required for the chemical reaction, which occurs at temperatures above 1400 °C, is typically provided by burning coal or petroleum coke, two of the most carbon-intensive fossil fuels.

Energy efficiency helps to protect the environment. When less energy is used, less energy needs to be generated by power plants, thus reducing energy consumption and production. This in turn reduces GHGs and improves the quality of the air.

The manufacture of Portland cement is the third most energy-intensive process, after aluminium and steel manufacture. In fact, for each metric ton of Portland cement, about 5½ million BTU of energy is needed [41]. Although cement production is energy inefficient, major initiatives have reduced energy consumption and the most significant has been the replacement of wet production facilities with dry processing plants.

Wet and Dry Production Facilities

Five different processes are used in the Portland cement industry to accomplish the pyroprocessing step: the wet process, the dry process, the semidry process, the dry process with a preheater, and the dry process with a preheater/precalciner. In the wet process and long dry process, all of the pyroprocessing activity occurs in the rotary kiln.

Wet cement production involves mixing raw materials (limestone and clay or loam) with water in order to produce slurry. Further in the process, water is evaporated from the homogenised mixture and this step in the production requires significant amounts of energy. The raw meal (dried slurry) is subjected to high temperatures in a rotary kiln, where the reaction of calcination takes place (its final products are lime and CO_2). The lime is further influenced by the temperatures of 1400–1500 °C resulting in clinker. The clinker production ends with the cooling phase in a cooler. The final stage of cement production is fine crushing of clinker with calcium sulphates (gypsum or anhydrite) and with possible additions of other minerals (blast-furnace slag, natural pozzolanas, fly ash, silica fume, or limestone) based on the requirements.

In case of dry cement production, the raw materials are mixed without water and therefore the evaporation process can be omitted. The latter technology could reduce the energy consumption from the "wet" to "dry" process by over 50%.

Advanced dry process system consists of the kiln and a suspension preheater. The raw meal (dried slurry) is fed in at the top of the preheater tower and passes through the series of cyclones in the tower. Hot gas from the kiln and, from the clinker cooler, is blown through the cyclones. Heat is transferred efficiently from the hot gases to the raw meal. Typically, 30–40% of the meal is decarbonated before entering the kiln. A development of this process is the "precalciner" kiln. Most new cement plants are of this type. The principle is similar to that of the dry process preheater system but with the major addition of another burner, or precalciner. With the additional heat, about 85–95% of the meal is decarbonated before it enters the kiln [8]. These systems allow more thermal processing to be accomplished efficiently in the preheater.

Preheater and precalciner kiln systems often have an alkali bypass system between the feed end of the rotary kiln and the preheater tower to remove the undesirable volatile constituents. Otherwise, the volatile constituents condense in the preheater tower and subsequently recirculate to the kiln. Buildup of these condensed materials can restrict process and gas flows. The alkali content of Portland cement is often limited by product specifications because excessive alkali metals (i.e. sodium and potassium) can cause deleterious reactions in concrete. In a bypass system, a portion of the kiln exit gas stream is withdrawn and quickly cooled by air or water to condense the volatile constituents to fine particles. The solid particles, which are removed from the gas stream by fabric filters and ESP's, are then returned to the process [9].

The semidry process is a variation of the dry process and are often used as intermediate steps in the conversion to dry processes. In the semidry process, the water is added to the dry raw mix in a pelletiser to form moist nodules or pellets. The pellets then are conveyed on a moving grate preheater before being fed to the rotary kiln. The pellets are dried and partially calcined by hot kiln exhaust gases passing through the moving grate.

All Ontario six plants use a dry process with preheaters, and most have precalciners. This upgraded plant design has also been a significant factor behind the 21% reduction in energy intensity over the past two decades in the Canadian cement sector [10].

Use of Fuel Sources

Cement kilns use a large variety of fuel sources to provide the energy required to produce the high temperatures necessary for the clinker formation. The kiln's combustion-generated CO_2 emissions are directly related to fuel use. The most common fuel sources for the cement industry are:

- Coal
- Fuel oil
- Petroleum coke

In addition, some cement plants use natural gas. The most frequently used fuels and their energy content are shown in Table 5.1 [11].

The Carbon War Room suggests that the cement industry should aim for a coal replacement rate of 50% by 2020 as part of its CO_2 emission reduction strategy. This compares with a current replacement rate of 30% in Quebec Canada, which has

Table 5.1 Typical data on energy content and CO_2 emission for frequent fuels [11]

Fuel	Energy content (MJ/kg)	CO_2 emission factor (kg/MJ)
Coal	32	0.103
Fuel oil	40	0.077
Natural gas	36	0.056
Petroleum coke	34	0.073–0.095

successfully used lower carbon fuels in cement kilns for more than 20 years, and the national average replacement rate of 10%. Canada's cement industry has a lower substitution rate country-wise than many other countries, and it falls below the global average. In 2011, Canadian cement producers derived 90% of thermal energy for production from conventional fuels, compared to a global average of 87%. By comparison, cement producers in the European Union derived 66% of thermal energy from conventional fossil fuels, with rates as low as 34% in Austria and 38% in Germany. In the United States, the average in 2011 was 84% [10].

In order to replace coal and other fossil fuels, alternative fuels are being adopted by many countries. Alternative fuels or waste fuels include scrap tires, waste oils, plastics, paper residues, waste solvents, and other wastes that have high-energy content. Cement kilns are well suited for waste combustion because of their high process temperature and because the clinker product and limestone feedstock act as gas-cleaning agents. Used tyres, wood, plastics, chemicals, and other types of waste are co-combusted in cement kilns in large quantities.

The amount of CO_2 generated by waste fuel is considered to be zero as shown in Table 5.2 [11]. This is based on the argument that the CO_2 generated by waste fuels would be released into the atmosphere by natural degradation, and during the natural process the energy content would not be applied in any manufacturing process.

The EPA recognises tyre-derived fuel as a Best Management Practice and encourages industries to recover the energy from this waste stream while offering the added benefit of reducing the landfill for scrap tires. Tyre-derived fuel (TDF) has approximately 20% more BTUs than a comparable weight of coal, and since tyres are manufactured in part with natural latex which literally grows in trees TDF has a lower greenhouse gas impact than coal. The steel-reinforcing belts in a tyre are a recycled material source for some of the iron needed in cement production. The use of tyres as fuel can actually reduce certain emissions in cement production [12]. The Rubber Manufacturers Association reported in 2013 that an estimated 44,300,000 scrap tyres were diverted from landfills for energy recovery in cement kilns [13].

Plants in Belgium, France, Germany, the Netherlands, and Switzerland have reached average substitution rates of from 35% to more than 70%. Some individual plants have even achieved 100% substitution using appropriate waste materials. However, very high substitution rates can only be accomplished if a tailored pre-treatment and surveillance system is in place. Municipal solid waste, for example, needs to be pre-treated to obtain homogeneous calorific values and feed characteristics. The cement industry in the United States burns 53 million used tyres

Table 5.2 Typical data on energy content and CO_2 emission for waste fuels [11]

Fuel	Energy content (MJ/kg)	CO_2 emission factor (kg/MJ)
Scrap tyres	21	NA
Plastics	33	NA
Waste oil	38	NA
Paper residues	6	NA
Waste solvents	18–23	NA

per year, which is 41% of all tyres that are burnt. About 50 million tyres, or 20% of the total, are still used as landfill [14].

5.2.2 Minimise Potable Water Use

Water resources are being exhausted day by day due to various uses. Shortage of water is perhaps the most critical environmental problem in several countries [15, 16]. Freshwater accounts for only 2.5% of the Earth's water, and most of it is frozen in glaciers and ice caps. The remaining unfrozen freshwater is mainly found as groundwater, with only a small fraction present above ground or in the air [17]. According to the United States Census Bureau, the world population is projected to reach nine billion by 2043, or an increase of 50% relative to 1999 [18]. Thus, it is expected that the water demand will have an increasing trend, leading to water recycling and conservation [18, 19] becoming a necessity. Therefore, potable water should be conserved to serve life-sustaining needs rather than infrastructure needs [20].

The need of a sustainably developed and environmental friendly concrete industry is aggravated by population growth and scarcity of water. World Business Council for Sustainable Development has stressed for the actions to be taken to offset the industry footprint on water, and mostly at local level, where individual facilities and activities have a direct impact, by using a risk-based approach. Primarily, companies have to understand and manage the quantities of water withdrawn, as well as the quality and quantity of water released, with particular attention in water-stressed areas [21].

Water is used in the concrete production process not only in the preparation of the concrete mixtures, but also in the washing of aggregates, wetting of aggregate stockpiles, and cleaning of trucks and equipment. Almost any water suitable for drinking is acceptable for use in concrete. To improve water conservation, recently approved practices in concrete production include replacing of some of the potable water with water reclaimed from previous concrete production, industrial processes, and other water sources typically not used for human consumption. Technologies for using increasing amounts of "grey water" (obtained from washing concrete production equipment and trucks) in concrete mixtures are rapidly becoming more common and accepted, although grey water with high solid contents (i.e. >15 lb/yd^3 [8.9 kg/m^3]) [22] can significantly impact concrete mixture water demand, setting time, compressive strength, and permeability [23].

Rainwater and surface runoff water can be used as a water conservation method by recycling these water resources in construction instead of using potable water. Furthermore, mixtures with less water should be developed with new technologies to create mortar and concrete containing a minimal amount of water. Concrete manufacturing conserves potable water use. Admixtures are used as part of the operations and manufacturing processes to reduce the water requirement of the mix and non-potable (grey) water. Admixtures can enhance sustainability by improving the workability of the concrete, reducing water demand, and improving durability. Modern paving concrete makes extensive use of chemical admixtures, most commonly air-entraining, water-reducing, set-retarding, and set-accelerating admixtures.

Reclaimed water from sewage treatment plant has also been used in concrete by some of the countries. Sewage is the wastewater released by residents, businesses, and industries in a community. It is 99.94% water, with only 0.06% of the wastewater dissolved and suspended solid material [24]. Sewage is treated to remove organic matter, solids, nutrients, disease-causing organisms, and other pollutants from wastewater. A typical sewage treatment system consists of preliminary treatment, primary treatment, secondary treatment, tertiary filtration, and disinfection. Finally, the disinfected water is discharged in a water body.

Several researches [20] around the world have studied the use of reclaimed water in concrete, with various levels of success. The reuse of recycled water from the recycling of unset/discarded concrete as mixing water for concrete is common practice in almost all ready-mixed concrete plants in Germany. The disposal of such wastewater is no longer being environmentally accepted. The recycled water consists primarily of the mixture of water, cement, and fines that remain after removal of the aggregate, but it also includes the wash water used for washing and cleaning the returning mixer trucks, concrete pumps, and other equipment, as well as the precipitation water collected in the production areas [25, 26]. Overall, it was found that concretes made with recycled water are durable and exhibit the similar properties as concretes made with drinking water or freshwater [25, 27].

The feasibility of using reclaimed wastewater [20] in concrete mixtures has also been studied in Indonesia. The reclaimed wastewater is lower in quality than potable water. Researchers have shown that concrete with improved initial compressive strength could be made with reclaimed wastewater used partially or totally in lieu of the mixing water [28]. The use of potable and treated waters was also tested in Saudi Arabia, and setting time and compressive strength were evaluated for the concrete. Pore solutions extracted from the mortar specimens were analysed for alkalinity and chloride content. Results showed that the treated water tested in this study qualifies to be used in making concrete [29].

An initial laboratory investigation conducted using reclaimed wastewater in concrete mixtures by University of Wisconsin-Milwaukee, USA, team found that no significant differences exist between mortar cubes made of potable water versus sewage treatment plant water [20]. Two main parameters were evaluated during this experiment: flow/workability and compressive strength. The average flow for mortar cubes made of potable water and reclaimed water was 98.1% and 89.5%, respectively. Although there was reduced flowability/workability of the mortar with reclaimed water, negative impact of the use of reclaimed wastewater on the mortar cubes was not noticeable.

Regarding the compressive strength, mortar cubes with sewage treatment plant water have shown improvement in strength during 3–28 days according to Fig. 5.1. These results suggest that the organic content present in the sewage treatment plant water may be acting as a dispersing agent, improving the dispersion of particles of cement and reducing clumping.

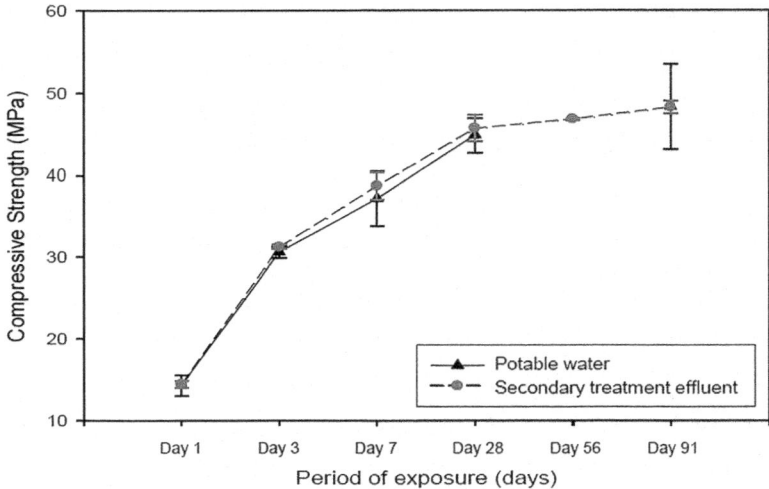

Fig. 5.1 Comparison of compressive strengths of mortar cubes made with potable and wastewater [20]

Lehigh Hanson Canada (cement manufacturer) has specified that it minimises its use of water by reusing and recycling water at every opportunity through on-site filtration and treatment of the process water; much of the concrete products are made with recycled process water. When discharges are unavoidable, wastewater is treated and monitored to ensure that it will not have significant impacts downstream.

Concrete can also be made more sustainable by better distributions of aggregate. Well-graded or reasonably graded combined aggregates in mixes can reduce the amount of cement required and the amount of water needed.

5.2.3   Minimise Waste

The world is literally covered in concrete and with an annual production of 3 tons per person globally it is the second most consumed material on earth, after water [30]. Concrete is an excellent construction material for long-lasting and energy-efficient buildings; however, there is no doubt it is also a significant source of greenhouse gas emissions. That's why researchers and innovators are exploring "greening concrete", or reducing the amount of waste concrete on construction sites and encouraging money-saving methods to recycle concrete collected at demolition sites. Accordingly it has been demanded from the civil engineering researchers, sustainable construction advocates, concrete manufacturers, and innovators for zero-waste movement.

Waste management is an essential aspect of sustainable concrete. Managing waste means eliminating waste where possible; minimising waste where feasible;

and reusing materials which might otherwise become waste. Solid waste management practices have identified the reduction, recycling, and reuse of wastes as essential for sustainable management of resources [31]. Waste concrete when disposed contributes to global greenhouse gases (GHGs) and when buried it results in land pollution.

Planning reduction of concrete waste prior to construction or demolition is important for achieving set priorities and goals. There are many procedures through which wastage of concrete materials or minimizing volume of materials in production can be achieved. Two such procedures are explained hereunder.

5.2.3.1 Improved Mechanical Properties

An increase in mechanical strength and similar properties leads to a reduction of materials needed. For example, doubling the concrete strength for strength-controlled members cuts the required amount of material in half.

With cement contributing to a significant portion of concrete's environmental impact, a common specification strategy is to reduce the cement content of a mix design to the lowest possible level. Contrary to the strategy above, high-percentage cement mix designs can be a solution for a lower carbon footprint in some applications. The use of high-strength concrete for certain design elements can result in a significant reduction of cross section, may eliminate the need for multiple elements, or may provide significantly longer service life.

To support the above idea, Portland Cement Association (PCA) [32] has provided a simplified example of a 40-storey building with a floor plate supported by with 16 columns (15' floor to floor height) per floor. Utilizing a 4000 psi mix design (with 440 lb of cement/yd^3) requires columns with a cross-sectional dimension of 36″ × 36″. Raising the cement content to 856 lb/yd^3 yields a compressive strength of 9000 psi. The column cross-sectional area is reduced to 24″ × 24″ as shown in Table 5.3.

Table 5.3 4000 psi mix design versus 9000 psi mix design

Description	4000 psi concrete	9000 psi concrete
Total cementitious materials in pounds per cubic yard	550	865
Supplementary cementitious in pounds per cubic yard	110 (fly ash)	40 (silica fume)
Portland cement in pound per cubic yard	440	825
Column cross section in inches	36 × 36	24 × 24
Concrete per column (15 ft.) in yard	5.00	2.22
Portland cement per column in pounds	2200 lb	1833 lb
Volume of cement reduction		16%
Volume aggregate reduction		55%

This results in a 16% net reduction for cement and a 55% reduction for aggregates, lowering the greenhouse gas footprint for these elements as well as providing an increase in net rentable floor area of 3120 ft^2.

5.2.3.2 Improved Concrete Durability

Durability is the key to sustainability. One of the solutions for preserving natural resources is through increasing the service life of concrete by increasing concrete durability. "Service life" means the period of time a building can be expected to withstand normal conditions, if properly maintained [33]. The basis for calculating the durability of concrete construction has been developed over many decades. Accurate durability assessment is the key to establishing a reliable way of ensuring an adequate service life and establishing it in a standard.

Concrete components provide a long service life due to their durable and low-maintenance surfaces. Concrete can resist weathering action, chemical attack, moisture, and abrasion while maintaining desired engineering properties. These characteristics of concrete make it sustainable in multiple ways: it avoids contributing solid waste to landfills, and it reduces the depletion of natural resources, and the generation of air, water, and solid waste from replacement materials. When properly designed, concrete structures can be reused or repurposed several times in the future [32].

The design service life for concrete structures is set at a minimum of 50 years and up to 200 years. There is 95% likelihood that the designed service life will be reached. In practice, this means that, depending on the design parameters, the average actual service life is substantially greater than the design service life, often more than double [33]. The designer should aim for the optimum overall package with regard to service life, insuring that the various design options and components act in harmony with each other. Thus in designing and building concrete structures, the service life can be influenced by the following choices:

- Strength grade and water/cement ratio
- Selection of constituent materials
- Mixing, transporting, placing in forms, compacting, and curing
- Concrete cover on reinforcement
- External factors such as weather conditions, effects of salts from soil, ground- or seawater, freezing and thawing, wetting and drying, leaching, and abrasion
- Air entrainment
- Permeability and porosity of concrete
- Method of construction
- Density of the concrete and maintenance

Internal concrete structures are, in principle, everlasting as there are no mechanisms that will damage indoor concrete in normal conditions. Their service life is assumed to be 200 years [33]. Durability of concrete is further described in Chap. 3.

Long-term use and serviceability are the key to a building's sustainability. However, for long-term use and serviceability, along with durable concrete structure, it should also be resilient—this is the latest approach of many organisations in different parts of the world. It is also considered that a building is not green if it's not resilient. As Louis Gritzo, Vice President of Research for FM Global states: "… a building that burns down is not green, regardless of how it was originally constructed … A green building must be resilient and sustainable in the long-term". It is this common-sense approach that is driving the resilience discussion [34].

Resilience is described by NRCMA (National Ready Mixed Concrete Association) as the ability of a system and its component parts to anticipate, absorb, accommodate, or recover from the effects of a hazardous event in a timely and efficient manner, including through ensuring the preservation, restoration, or improvement of its essential basic structures and functions [35].

Concrete can be incorporated in several key aspects to make projects more durable and disaster resistant. For example, concrete wall, floor, and roof systems offer an unsurpassed combination of structural strength and wind resistance. In addition, hardened exterior finishes for walls and roofs of a home or business provide the best combination of strength and security. Buildings and structures with resilient design and materials are not only better able to recover following disasters, such as hurricanes or fires, but also the new "green" buildings. Builders, architects, and designers have come to recognise that more durable public buildings, private homes, and businesses, often built with concrete to resist damage from natural disasters, also reduce the impact entire communities have on our planet [36].

5.2.4 Increased Use of Recycled Material

The last and final parameter for achieving sustainable concrete project is the increased use of recycled material. The use of recycled materials in concrete reduces greenhouse gasses, frees up landfill space, and reduces raw material consumption. After a concrete structure has served its original purpose, the concrete can be crushed and recycled into aggregate for use in new concrete or as a backfill or road base. However, care needs to be taken for possible reuse of concrete aggregate in concrete because of a possible adverse cumulative level of salts, alkali reaction, or other impurities.

In its report "Recycling Concrete" the CSI (Cement Sustainability Initiative) argues that the recycling of concrete can reduce natural resource exploitation and waste going to landfill. The report asks for an ultimate goal of "zero landfill" of concrete. The recovery of concrete falls between standard definitions of reuse and recycling. Recovering concrete has two main advantages: it reduces the use of new virgin aggregate and the associated environmental costs of exploitation and transportation, and it reduces landfill of valuable materials [37].

5.2.4.1 Recycled Concrete Aggregate (RCA)

Aggregates, constituting 60–75% of concrete by volume, are basically sand, gravel, or crushed stone which have low embodied energy. Recycled concrete from demolished buildings and pavements and reclaimed aggregate from concrete production can be used to replace a portion of new aggregate in concrete. In a 2008 report, Federal Highway Administration, USA, noted that 11 states currently use recycled concrete aggregate in new concrete [38]. These states report that concrete containing recycled aggregate can perform equal to concrete containing natural aggregates. Applications such as foundation slabs and insulated concrete-form walls are also well suited for recycled aggregate incorporation.

The Construction Materials Recycling Association, USA, estimates that approximately 140 million tons of concrete is recycled annually. This reduces the amount of material that is landfilled and the need for virgin materials in new construction. Concrete pieces from demolished structures can also be reused in stacked landscaping walls and gabion walls or as riprap for shoreline protection [32].

The use of recycled concrete as granular base has been increasing rapidly. For example, at Toronto's Pearson airport, 145,000 tons of concrete from old terminals and pavements was crushed on site and recycled for use in 500 mm thick granular base layers under new apron and taxiway pavements. Over 75,000 tons of RCA was used for this purpose, thus saving approximately 4000 truckloads of virgin granular base from being hauled more than 50 km from quarries to the airport and a similar number of truckloads of old concrete being hauled away to landfill. In an experimental trial with the Ministry of Transportation in Ontario (MTO), RCA was also used to replace 15 and 25% of the coarse aggregate in sidewalks in Nov. 2011. These sidewalks are in excellent condition, with no difference in the severity of isolated pop-outs after five winters [7].

Recycled concrete aggregates (RCA) are also discussed in Chap. 2.

5.2.4.2 Returned Fresh Concrete

Returned concrete is the unused fresh wet ready-mixed concrete that is returned to the plant in the concrete truck as surplus. This can be small amounts of concrete leftover at the bottom of the drum in the truck, or a whole truck not used by the customer on the construction site. The returned concrete to manufacturing plants can result in wasted material or would need disposal resulting in unnecessary environmental impact. Returned concrete can be successfully recycled and incorporated into new batches.

Typically the small quantities of returned waste concrete can be recovered by washing and reuse in concrete production or, if it has already hardened, it can be crushed and reused as aggregate. If the quantities returned are high, it is often standard practice to let it harden, crush it, and use as landfill. Sometimes, if the concrete is still wet, precast concrete products such as concrete block are made.

However, the good news is that the ASTM has released new Standard (ASTM C1798/C1798M-16e1) for Returned Fresh Concrete for Use in a New Batch of Ready-Mixed Concrete. This standard recognises unused concrete in a fresh state as a potential ingredient for a new concrete batch, as mentioned by ASTM member Rich Szecsy, President, Texas Aggregates and Concrete Association. In other words, recycled fresh concrete can be treated as a raw-material component just like water, aggregates, and cement. According to Szecsy, prior to this, manufacturers had some options for repurposing or recycling fresh concrete but such material was most often disposed of in a landfill. Owners and end users will benefit from the new standard, since it will help lower the environmental impacts of construction. However, ASTM specifies that "returned fresh concrete in a quantity of less than 450 kg (1000 lb) or 0.2 m^3 (0.25 yd^3) shall not be subject to this specification". The new standard can be purchased online on ASTM website.

5.2.4.3 Other Recycled Materials

There are many recycling industry by-products that can be used to produce sustainable concrete. Used foundry sand can replace regular sand in Portland cement concrete. Post-consumer glass can be used as a partial replacement of fine aggregate in regular and flowable concrete. Concrete containing residual solids from pulp and paper mills and paper-recycling plants exhibits improved resistance to chloride-ion penetration and freezing and thawing without loss of strength. The addition of recycled materials can improve the quality of concrete while reducing the amount of waste deposited in landfills. The subject of use of recycled material in concrete is vast and it can cover a wide range of by-products or waste materials. However, as mentioned earlier use of all such materials requires special consideration in the design and production stages.

Further research and acquiring of knowledge continue because there is a strong need to manufacture concrete in a more sustainable manner.

Bibliography

1. Christian Meyer, Columbia University- Concrete Materials and Sustainable Development in the United States.
2. Malhotra, V.M. (2000). "Role of supplementary cementing materials in reducing greenhouse gas emissions", in *Concrete Technology for a Sustainable Development in the 21st Century*, O.E. Gjorv and K. Sakai, eds., E&FN Spon, London, 2000.
3. Naik, T.R. (2007). "Sustainability of the cement and concrete industries." In *Proc. Int. Conf: Sustainable construction materials and Technologies*, Y.M. Chun, P. Claisse, T.R. Naik, E. Ganjian, eds., 11-13 June 2007 Coventry.: Taylor and Francis, London, ISBN 13: 978-0-415-44689-1,p19-25.
4. Canada Green Building Council- CaGBC National Office, 47 Clarence Street, Suite 202, Ottawa, Ontario K1N 9K1, info@cagbc.org.

5. Julia G. Tapali, Sotiris Demis and Vagelis G. Papadakis, University of Patras, Greece-Sustainable concrete mix design for a target strength and service life- Computers and Concrete, Vol. 12, No. 6 (2013) 755-774.
6. Canadian Standards Association, CSA-23.1-14
7. R. D. Hooton and M. Anson-Cartwright- Improving Concrete Sustainability Though Durability Design- 2016 International Concrete Sustainability Conference, National Ready Mixed Concrete Association Canada.
8. Understanding Cement- copyright © 2005-2018 by WHD Microanalysis Consultants Ltd.
9. Environmental Protection Agency USA (EPA)- AP 42 Section 11.6- Portland Cement Manufacturing- Upgraded September 2016.
10. Alternative Fuel Use in Cement Manufacturing- Implications, opportunities and barriers in Ontario, Workshop Written by The Pembina Institute and Environmental Defence- Funded by Cement Association of Canada and Holcim (Canada) Inc, May 2014.
11. Hernane G. Caruso- Reduction of CO2 Emissions from Cement Plants- A thesis presented to the University of Waterloo Canada, 2006.
12. Blumenthal, M., "The Use of Scrap Tires in Rotary Cement Kilns, Scrap Tire management Council, Washington DC, 1996.
13. Rubber Manufacturers Association, 2013 Scrap Tire Management Summary, Nov 2014, http://www.rma.org/download/scrap-tires/market-reports/US_STMarket2013.pdf
14. ClimateTechWiki- A clean Technology Platform, supported by Government of Netherlands-Energy Efficiency and Saving in the Cement Industry.
15. Okun, D. (1994). "The role of reclamation and reuse in addressing community water needs in Israel and the West Bank." In Proceedings of the First Israeli-Palestinian International Academic Conference on Water, Water and Peace in the Middle East, J. Isaac and H. Shuval, eds., Zurich, Switzerland, December 10-13, 1992. New York: Elsevier, 329-338.
16. Environmental Protection Agency (EPA) (2004). "Guidelines for water reuse technical issues in planning water reuse systems." Washington, DC: U.S Agency for International Development. EPA/625/R-04/108, chapter 3.
17. United Nations Educational Scientific and Cultural (UNESCO), World Meteorological Organization (WMO), and International Atomic Energy Agency (IAEA) (2006). "The state of the resource." In Water, A shared responsibility, the United Nations Water Development, Water Development Report 2. Chapter 4. (http://www.unesco.org/water/wwap/wwdr/wwdr2/pdf/wwdr2_ch_4.pdf) (Sep. 27, 2009).
18. U.S. Census Bureau (USCB) (2009). "World Population: 1950–2050." (http://www.census.gov/ipc/www/idb/worldpopgraph.php) (Sept. 27, 2009).
19. Sethuraman, P. (2006). "Water reuse and recycling—a solution to manage a precious resource?"(http://www.frost.com/prod/servlet/market-insight-top.pag?docid=90081832) (Sep. 21, 2009).
20. Marcia Silva and Tarun R. Naik- Sustainable Use of Resources—Recycling of Sewage Treatment Plant Water in Concrete. Second International Conference on Sustainable Construction Materials and Technologies, June 2010.
21. World Business Council for Sustainable Development www.wbcsd.org- Letter from the CSI members' CEOs: A continued commitment to sustainability.
22. Federal Highway Administration, FHWA-HIF-16-013- Tech Brief, Strategies for Improving Sustainability of Concrete Pavement- Tech Brief Developed by Mark B. Snyder (Consultant), Tom Van Dam (NCE), Jeff Roesler (University of Illinois, Urbana-Champaign), and John Harvey (University of California, Davis), April 2016.
23. Lobo, C. and G. M. Mullings. 2003. "Recycled Water in Ready Mixed Concrete Operations." Concrete in Focus. Spring 2003. National Ready Mixed Concrete Association, Silver Spring, MD.
24. Karen Mancl- Wastewater Treatment Principles and Regulations, Ohio State University Extension USA- February 2016.

25. Rickert, J. and Grube, H. (2003). "Influence of recycled water from fresh concrete recycling systems on the properties of fresh and hardened concrete." *VDZ, Concrete Technology Reports.* Düsseldorf, Germany.
26. Rickert, J. and Grube, H. (2000). "Analysis of recycled water component." *VDZ, Concrete Technology Reports.* Düsseldorf, Germany.
27. Chini, S. A. and Mbwambo, W. J. (1996). "Environmentally friendly solutions for the disposal of concrete wash water from ready mixed concrete operations." In Proceedings of CIB W89 Beijing International Conference.
28. Tay, J. and Yip, W. (1987). "Use of Reclaimed Wastewater for Concrete Mixing." *Journal of Environmental Engineering,* 113(5), 1156-1161.
29. Saricimen, H., Shameem, M., Barry, M., and Ibrahim, M. (2008). "Testing of treated effluent for use in mixing and curing of concrete."(e-printshttps://eprints.kfupm.edu.sa/1745/) (Sep. 21, 2009).
30. Michelle Volkmann, Concrete Ideas for Reducing Concrete Waste- Sustainable City Network, January 2015.
31. Tom Napier, Research Architect, U.S. Army Corps of Engineers, Engineer Research and Development Center/Construction Engineering Research Laboratory-*Construction Waste management*- Whole Building Design Guide, National Institute of Building Science Updated: 10-17-2016.
32. Portland Cement Association (PCA)- Using Concrete as a Sustainable Solution for Buildings- Continuing Education, Course # ARnov2015.2.
33. European Concrete- Sustainable Benefits of Concrete Structures- Editor: Jean-Pierre Jacobs 1050 Brussels, Belgium
34. Jeff Sieg and Larry Rowland, Why You Should Care About Resilience- For Construction Pros. com, September 9, 2015.
35. National Ready Mixed Concrete Association (NRMCA), Pathway to Resilience—A Guide to Developing a Community Action Plan, 2014.
36. Portland Cement Association (PCA)- Resilient Construction-Stronger, Safer Communities,2017.
37. World Business Council for Sustainable Development www.wbcsd.org- Cement Sustainability Initiative (CSI)- Concrete Recycling, 2014-15.
38. FHWA, *Recycled Concrete Aggregate*, Federal Highway Administration National Review, Federal Highway Administration, Washington, D.C., USA, 2008.
39. Kosmatka, S.H., Kerkhoff, B., Panarese, W.C - Design and Control of Concrete Mixtures, 14th Edition, Portland Cement Association, USA 2008.
40. Malhotra, V.M, Mehta, P.K – High Performance High Volume Fly Ash Concrete- Supplementary Cementitious Materials for Sustainable Development, Inc. Ottawa, Canada.
41. Naik, T. R., and Kraus, R. N. (1999)- "The Role of Flow able Slurry in Sustainable Developments in Civil Engineering." Proc., ASCE Conf. on Materials and Construction— Exploring the Connection, ASCE, Reston, Va., 826–834.

Part II
Practical Problems and Solutions

Reinforced concrete is one of the most durable materials that a designer can choose for almost any type of building or structure. Good durability can be achieved, but it demands careful attention in design, supervision, and workmanship during construction.

Concrete, in fact, has a wide range of mechanical, chemical, and physical properties, which in turn are adversely affected by factors like inadequate specifications, poor construction methods, load condition, constituent materials, and the local environment. Inadequate knowledge of factors influencing the behaviour of concrete has resulted in deterioration and failures in concrete structures, which have now become a growing worldwide concern.

There are chemical and physical interactions which has a significant effect on the durability and hence on the service life of the concrete. Many structures deteriorated much before the stipulated time causing a lot of economic damage and public inconvenience. Durability is thus attracting more and more world's attention and is becoming more and more important.

The durability problems may be due to inadequate quality control as well as improper selection of the materials, which is increasing due to the inadequate use of chemical and mineral admixtures in concrete. The present concrete is not the same like before when pure Portland cement, aggregates of good quality and non-polluted water was used. As stated by Bryant Mather [1986]; "I worked for 20 years in the concrete research laboratory to learn that no two pieces of concrete are alike. In the next 20 years I learnt that not one piece of concrete is alike." Proper selection of constituent materials of concrete plays a very important role in producing good quality concrete having the specified properties.

The most common forms of distress and deterioration of reinforced concrete are cracking, crazing, scaling, spalling, disintegration, discolouration, honeycombing, and efflorescence. Splitting of concrete as a result of reinforcing steel corrosion is common.

There are many causes of concrete deterioration/failures and can be categorized as under:

- Incorrect selection of constituent materials.
- Poor construction methods.
- External factors.
- Chemical attack.
- Corrosion of reinforcement.
- Hot and cold weather concreting.
- Errors in designing and detailing.

Chapter 6
Incorrect Selection of Constituent Materials

6.1 Introduction

The properties of concrete depend on the quantities and qualities of its constituent materials. Constituent materials used should satisfy structural performance and durability of the finished structure.

Constituent materials used to produce a concrete mix should be selected to meet the environmental or soil conditions, where the concrete is to be placed. For example, when concrete is to be subjected to different exposures like freezing and thawing, wetting and drying, heating and cooling, and sulphate attack, special type of ingredients—low-alkali cement, pozzolans, blast-furnace slag, sulphate-resisting cement, aggregate selected to prevent harmful expansion due to alkali-aggregate reaction, or aggregate composed of hard minerals including suitable admixtures—need to be properly selected to increase the resistance of concrete.

Selecting appropriate materials of suitable composition and processing them correctly are therefore essential to achieve concrete resistance to deleterious effects of water, aggressive solutions, reactive aggregates, extremes of weather conditions, etc.

6.2 Cement Selection

Since cement is the most active component of concrete, its selection and proper use are important in obtaining appropriate and economical concrete with required properties. Many types of cement are available in the market. Most are used for general purposes, whereas some are used in specific applications like for resistance to sulphate attack, reduced heat evolution, and early setting and hardening. If one fails to select cement properly to meet the special requirements, serious consequences may result. However, unnecessary use of such special purposes cement is not only

© Springer Nature Switzerland AG 2019
A. Surahyo, *Concrete Construction*,
https://doi.org/10.1007/978-3-030-10510-5_6

uneconomical but may also degrade other more important properties. Hence, it should be ensured that proper amounts and types of cement are obtained to meet the structural and durability requirements.

In order to select cement for mass concreting like in a dam, the generation of heat must be taken into account. In such cases, preference should be given to low-heat Portland (type IV) cement, as the use of ordinary cement would result in unacceptable large-temperature gradient within the concrete. Use of pozzolan as a replacement further delays and reduces heat generation. If the cement content is limited to 235 lb/yd^3 (140 kg/m^3), the temperature rise for most concretes will not exceed 19 °C (35 °F) [1].

On the other hand, rapid-hardening Portland (type III) cement that produces a much early strength is totally unsuitable for large masses of concrete, as its increased rate of hydration or setting is accompanied by a high rate of heat evolution. However, this cement is very useful in cold weather or where high-early strength is required for further construction or when formwork is to be removed early for reuse.

Concreting in hot weather leads to problems in mixing, placing, and curing, and adversely affects the properties and serviceability of the concrete. Concrete placed at elevated temperatures results in rapid hydration of cement, evaporation of water from mix, and evaporation of curing water, which leads to shrinkage cracking and lower strength. In such conditions, the use of cements with increased rate of hydration and use of high-compressive-strength concretes, which require higher cement contents, must be avoided. Use of normally slower setting type II Portland cement (modified cement) or low-heat Portland (type IV) cement as already briefed or slag cement (type IS) will produce better results.

Water-soluble sulphates occur in some soils. These sulphates of calcium, magnesium, and sodium can attack the cement matrix to give reaction products, which have an increased volume, and thus cause expansion. This, in turn, can lead not only to spalling and surface scaling, but also to more serious disintegration. In such areas good-quality, well-compacted concrete of low permeability should be used. However, considerable resistance to sulphate attack can be obtained by using sulphate-resisting Portland (type V) cement. Moreover, in very severe exposures, suitable pozzolan or slag along with type V cement has proved more satisfactory.

In cold countries, water freezing within the pores of concrete can cause disruption. Susceptibility to such attacks is greatest with poor-quality concrete used in wholly exposed positions, such as road curbs and slabs, dams, and reservoirs. Good-quality, low-permeability concrete, used in most building or road situations, is less vulnerable to frost action. However, frost resistance can be increased greatly by use of air-entraining Portland cement. In addition to this, use of rapid-hardening Portland (type III) cement and high alumina cement is very helpful in cold weather.

Since high-alumina cement concretes resist a number of agents that attack Portland cement concretes, high-alumina cement may be used in a particularly aggressive environment in the pH range from 3.5 to 4 [2].

Like this, many other types of cement have been developed for special uses, i.e. white and coloured Portland cements, masonry cement, antibacterial cement, expanding cements used for making shrinkage-compensating concrete, and hydrophobic cement which contains water-repellent film around its grain, and can be stored under unfavourable conditions of humidity for a long period of time without any significant deterioration. The other kinds of cement like waterproof and water-repellent Portland cements produce a more impermeable fully compacted concrete than ordinary Portland cement. These cements are suitable for use in water-retained structures. One more recently developed cement is known as macro-defect-free (MDF) cement, which is made from aluminous cement and has a very high compressive strength of 100–300 MPa (14,500–43,500 psi).

For some properties, the amount of cement is more important than its characteristics. In order to achieve durable concrete, Table 4.1 may be referenced which specifies minimum cement content required under various exposure conditions.

However, based on present researches, the engineers are motivated to use higher levels of supplementary cementitious materials (SCMs) on the basis of economic and environmental grounds, especially fly ash and to a lesser extent slag, to replace larger proportions of hydraulic cement in concrete mixtures. The environmental advantages are linked to the desire to use more recycled materials in "green buildings" and to reduce greenhouse gas emissions by lowering the use of hydraulic cement, the production of which consumes large quantities of energy and releases substantial quantities of carbon dioxide, a greenhouse gas. While the use of concrete with high levels of SCMs offers many advantages, such concrete displays different characteristics (in both plastic and hardened state) from plain hydraulic cement concrete and requires special consideration in the design and production stages.

6.3   Aggregate Selection

Since about 75% by volume of concrete mix is occupied by aggregate, its quality is of considerable importance, and has a significant effect on concrete quality. The aggregate selected should be free from impurities and deleterious materials. Organic impurities may occur either in the aggregate or in the mixing water. They are harmful because they retard the setting of the cement, and lead to reduced durability. These impurities generally consist of decayed vegetable matter and appear in the form of humus or organic loam, and are likely to be present in fine aggregate. Suspected aggregates should be washed with clean water; sometimes two or three washings may be necessary.

Coal, gypsum, hard-burned lime or dolomite, and glasses are the main deleteriously reactive substances, the contamination of which with concrete aggregates must be avoided. Coal usually contains sulphur compounds that cause sulphate attack. The presence of coal particles also results in surface staining of concrete. Gypsum and other sulphates like pyrite and marcasite should be avoided as they also give rise to sulphate attack. Hard-burned lime and hard-burned dolomite, if present in aggregates, will react with water in concrete and carbon dioxide from the air, and form carbonates and hydroxides, which result in swelling and pop-outs. Most of the glasses are highly siliceous. Silica found in aggregates can, in the presence of water, react with alkalis derived from cement and, in so doing, may cause expansion and subsequent damage to the concrete. The reaction is known as alkali/aggregate reaction. It is explained in detail in Chap. 9. The risk of damage is slight when used with most aggregates and cements. Where long-term experience of aggregates is not available, and a potential risk is suspected, specialist tests like quick chemical test or mortar-bar test are necessary to show whether alkali/aggregate reaction is likely. These tests are described in ASTM-C2-89 [3] and ASTM-C227-90 [4], respectively.

Similarly, unsound particles in aggregates with low density like shale, clay lumps, and wood must also be avoided as they lead to pitting and scaling. If present in large quantities, these particles may adversely affect the strength of concrete. The presence of mica in fine aggregate also affects the compressive strength of concrete adversely. Hence, if mica is present, a suitable allowance for the possible reduction in strength of concrete should be made.

The main cause of concrete distress is reinforcing steel corrosion. The presence of chlorides causes this corrosion. The use of fine poorly graded sands contaminated by chlorides must be avoided. To remove deleterious substance, the sand must be washed a number of times. There are also well-established limits to control the amounts of chloride ions expressed in percentage of weight of cement. The chloride issue is further explained in Chap. 9.

Aggregate porosity is an important property, as it affects the strength of concrete due to water absorption and permeability of the aggregate. An aggregate with high porosity will produce less durable concrete, particularly when subjected to freezing and thawing. Aggregates with high absorption or porosity may also produce concrete with high shrinkage. Hence, aggregates with low porosity and low shrinkage properties should be used. Aggregates containing quartz, dolomite, granite, and some basalt can generally be classified as low-shrinkage-producing aggregates. However, aggregates containing sandstone, shale, slate, and some types of basalt are known as high-shrinkage-producing aggregates. Moreover, the water added to concrete mix must be adjusted to take account of water to be absorbed by aggregate porosity.

The grading and size of aggregate are also important because of their effect on workability. The proper grading of aggregate will demand a lower water/cement ratio, which results in an increase of strength. Good grading saves cement content. It helps prevent segregation during placing and ensures a good finish. The aggregate size also affects the strength. For given mix proportions, the concrete strength decreases as the maximum size of aggregate is increased. For proper sizes of aggregate and grading limits, the tables provided in Part I of this book might be referred.

In selection of aggregate for fire resistance, preference should be given to lightweight aggregates. These aggregates have greater fire resistance, because they have a lesser tendency to spall, and the concrete also suffers a lower loss of strength with a rise in temperature. Lightweight aggregates are useful for making lightweight aggregate concrete due to their lower specific gravity. Carbonate aggregates like dolomites and calcites are generally more resistant to fire than siliceous aggregates.

However, aggregates containing quartz such as granite, sandstone, and quartzite are more susceptible to fire damage.

Where acid resistance is required, siliceous aggregates such as quartzite and granite are found generally acid resistant. However, as the cement paste of concrete also reacts with acid, a concrete with carbonate aggregates (under mild acid conditions) is found more acid resistant than concrete with siliceous aggregates.

While selecting aggregates, mostly in concrete roads and floors, importance is given to properties like toughness and hardness of aggregate due to the polishing effects of traffic and the internal abrasive effects of repeated loadings. Aggregate toughness is its resistance to failure by impact and this is normally determined from the aggregate impact test (BS-812: Part 112 [5]). Aggregate hardness is the resistance of an aggregate to wear and is normally determined by an abrasion test (BS-812: Part 113 [6]). As per BS-882 [7], the maximum aggregate impact value should be 25% when aggregate is used in heavy-duty concrete floor finishes, 30% when used in concrete pavement wearing surfaces, and 45% when it is to be used in other concretes. The maximum aggregate abrasion value should be 30% when aggregate is used for wearing surfaces and 50% when used for non-wearing surfaces. However, CSA A 23.2-14 [8] provides that the abrasion loss shall not be greater than 35% when the coarse aggregate is used in concrete paving or for other concrete surfaces subjected to significant wear and maximum 50% when coarse aggregate is used for non-wearing surfaces.

Mineral fillers which are considered a very fine aggregate are fine powders (mineral fillers are not a supplementary cementing materials) manufactured or produced from crushing coarse aggregate. If mineral filler is proposed to be used for producing concrete, CSA A23.1-14 [9] specifies that the potential alkali-aggregate reactivity (AAR) of mineral filler shall be assessed through standard tests. Limestone fillers should not be used in sulphate environments for any S class cements (Table 2.4). Deleterious reactions have been found with mineral fillers containing carbonates (e.g. limestone and dolomite) used in concrete exposed to sulphate environment.

6.4 Water Selection

6.4.1 Mixing Water

Water, as an ingredient of concrete, greatly influences many of its significant properties, both in the freshly mixed and hardened state. Potable water can be used in concrete without any testing or qualification; however, many waters that are not fit for drinking are also suitable for use in concrete based on acceptance criteria recommended by ASTM C1602 for water to be used in concrete and as discussed in Chap. 5.

The strength and durability of concrete are reduced due to the presence of impurities in the mixing water. Excessive impurities in mixing water affect setting time and concrete strength and also cause efflorescence, staining, corrosion of reinforcement, volume changes, and reduced durability. Water draining from some mines and some industrial water may contain or form acids which attack concrete. Effluents from sewerage works, gas works, and paint, textile, sugar, and fertilizer industries are harmful to concrete. Hence, use of such water should be avoided.

Tests show that water containing excessive amounts of dissolved salts reduces compressive strength by 10–30% of that obtained using freshwater [10]. In addition, water containing large quantities of chlorides tends to cause persistent dampness and surface efflorescence; hence it should not be used where surface finish is of importance. Use of saline water for mixing and curing concrete should also be avoided as it increases the risk of corrosion of the reinforcement, which leads to cracking, and spalling of concrete. As mentioned in Chap. 2, CSA A23.1-14 [9] specifies that water of unknown quality, including treated wash water and slurry water, shall not be used in concrete unless it produces 28-day concrete strengths equal to at least 90% of a control mixture. The control mixture shall be produced using the same materials, proportions, and a known acceptable water.

6.4.2 Curing Water

As explained above saline water should not be used for curing. Use of curing water containing carbon dioxide should also be avoided. Flowing pure water, formed by melting ice or by condensation or water draining from mountains, contains free CO_2, which dissolves $Ca(OH)_2$ available in concrete, and causes surface erosion. Pure water containing no minerals can leach calcium hydroxide and other hydration products out of concrete. Calcareous rather than siliceous aggregate gives best results under these conditions. In addition to this, the surface of concrete exposed to pure water should have good drainage and be smooth and dense. Iron and organic matter in the water are responsible for staining or discolouration, and especially when concrete is subjected to prolonged wetting even a very low concentration of these can cause staining. Water, which is satisfactory for mixing concrete, should only be used for curing.

The effects of some of the impurities in water on the properties of concrete are given below:

1. Inorganic Salts

    The strength and durability of concrete are adversely affected if excess quantities of inorganic impurities are present in mixing water. The major ions usually present in natural waters are calcium, magnesium, sodium, potassium, carbonate, bicarbonate, sulphate, chloride, and nitrate.

    Salts like sodium iodate, sodium phosphate, sodium arsenate, and sodium borate reduce the initial strength of concrete to a great extent. Carbonates of sodium and potassium may cause extremely rapid setting, and in large concentrations reduce the concrete strength.

Zinc chlorides retard the setting of concrete to such an extent that no strength tests are possible on 2nd and 3rd days. The effect of lead nitrate is completely destructive [10].

Seawater generally contains about 3% of sodium chloride and about ½ (half) percent of magnesium sulphate and magnesium chloride. These salts reduce the ultimate strength of concrete. However, the major concern is the risk of corrosion of reinforcing steel due to chlorides. In general, the risk of corrosion of steel is more when the reinforced concrete member is exposed to humid air than when continuously submerged under water, including seawater. However, it is generally unadvisable to use seawater for mixing. CSA A-23.1-14 [9] specifies that the concentration of sulphates in mixing water should not exceed 3000 mg/L (ppm). BS-3148 [11] recommends that water containing a combined total of not more than 2000 mg/L of these common ions is generally suitable as mix water. However, individually when chloride ion content does not exceed 500 mg/L, or SO_3 ion content does not exceed 1000 mg/L, the mix water is harmless.

2. Algae

Algae, if present in mixing water, combines with cement and reduces the bond between aggregates and cement paste. It also results in air entrainment, causing loss of concrete strength.

The Portland Cement Association further warns that the use of material that contains algae water or aggregates can disrupt the hydration process, reduce strength, or cause poor bonding to occur between the aggregates and paste.

3. Sugar

A small amount of sugar in concrete mixing water will greatly retard the setting of the concrete. About 0.05% of sugar by weight of cement will delay the setting time by about 4 h [12]. Sugar solutions are detrimental, particularly to fresh concrete. Only about 3% concentration may gradually corrode concrete.

Less than 500 ppm of sugar in mix water generally has no adverse effect on strength, but if the concentration exceeds this amount, tests for setting time and strength are required.

4. Oil Contamination

Many oils, if present in mixing water, will adversely affect the quality of concrete. Vegetable oils usually contain small amounts of fatty acids, which result in slow deterioration of concrete surfaces. Animal and fish oils have corrosive effects. Glycerin, which is soluble in water, attacks concrete slowly by reacting with and dissolving calcium hydroxide. Petroleum oils are not reactive to concrete, even though light oils may penetrate good concrete [13].

5. Acids and Alkalis

Water-containing acids or alkalis are usually unsuitable for concrete construction. Mountain or spring waters are usually acidic because of their reaction with carbon dioxide present in the atmosphere. Carbonic acid, which is the result of this reaction, gives rise to corrosion of reinforced steel. Similarly, when sulphur dioxide gas combines with water, it forms sulphurous and sulphuric acids. Both acids corrode steel reinforcement. Organic acids, such as tannic acid, can have significant effect on strength at higher concentrations. Use of acid waters with pH values less than 3.0 should be avoided.

Most alkalis like calcium, ammonium, and barium hydroxides are harmless. However, if mixing water containing sodium hydroxide is used, this may result in physical damage of concrete due to reaction between sodium hydroxide and carbon dioxide from the air. Such water is better tested before use.

Water that contains alkali carbonates and bicarbonates may affect the setting time of the cement and the strength of concrete. Their presence may also be harmful if there is a risk of alkali-aggregate reaction. BS-3148 [11] suggests that their combined total should not exceed 100 mg/L in a mix water. However, CSA A23.1-14 [9] provides that the maximum concentration of alkalis (NaO + 0.658 K_2O) in mixing water should be 600 mg/L.

The use of water with organic impurities, like coloured water or water with a pronounced odour, should also be avoided. In general, water which is fit for drinking purposes is usually treated fit for concrete construction. The pH value of water suitable for concrete construction shall generally be between 6 and 8. Moreover, BS-3148 also suggests that the selected mix water would be tested to determine its effect on setting and hardening characteristics of concrete, and that the initial setting time of its concrete test blocks should not differ by more than 30 min from the control test blocks, whereas the average compressive strength of the concrete test cubes should not be less than 90% of the average strength of the control test cubes.

6. Chloride and Sulphates:

   Water containing large amount of chlorides tends to cause persistent dampness and surface efflorescence. The presence of chlorides in concrete containing embedded steel can lead to its corrosion.

   Water containing sulphates results in expansive reactions and deterioration of concrete. Mixing water rich in sulphate ions results in an internal sulphate attack. Sulphates react with the aluminium causing internal expansion.

   It was further observed that sulphate ions retard setting times of concrete. This may affect workability of cement paste and hence its placement period. However, sulphates have mild effect on corrosion of steel in concrete. Effects of chlorides and sulphates on concrete are further discussed in the following chapters.

7. Silt or Suspended Particles

   Silt, clay, and dust may form a coating on aggregate particles, resulting in weakened bond between the aggregate and the cement paste. Excessive amounts of these fine materials may also increase water demand of the concrete, resulting in loss of concrete strength and an increase in its permeability.

   About 2000 ppm of suspended clay or fine rock particles can be tolerated in mixing water. However, mixing water with high content of suspended solids should be allowed to stand in a setting basing before use as it is undesirable to introduce large quantities of clay and slit into the concrete.

8. Wash Waters (Recycled Waters)

Concrete producers mostly generate process water by cleaning mixers and plant components, also referred to as wash water or slurry water. Wash water is also generated when returned concrete is washed out in concrete reclaimer systems. These systems collect wash water with the cement and aggregate fines in the form of a slurry that can be reused as a mixing water in concrete. However, ASTM-C1602 provides that chemical limits for wash water used as mixing water should have maximum concentration of sulphate (SO_4) as 3000 ppm, alkalis as ($Na_2O + 0.658$ K_2O) 600 ppm, and total solids as 50,000 ppm.

The use of slurry water to replace the same amount of tap water will influence the effective w/c of the concrete due to the high solid content in the slurry water. The use of water with solid content exceeding the 50,000 ppm limit by up to 13,400 ppm (pH 12, SG-specific gravity1.03) has been found to reduce slump but no significant effect on unit weight or semi-adiabatic temperature rise. Increase in the percentage of slurry water resulted in reduction of compressive and flexural strength, and the elastic modulus. Compressive strength was not less than 90% of concrete with tap water. Increase in the proportion of slurry water as a replacement of tap water also had a negative effect on the drying shrinkage and the resistance to acid attack [16].

The NRMCA [17] [study found that slurry water with significantly higher solid content than ASTM C94 limit resulted in increased water demand and accelerated setting time. The effects were more pronounced with an increase in the age of the slurry past 1 day and an increase in the solid content. Mixtures that had higher water contents due to the use of higher solid recycled slurry had an associated reduction in strength and increase in drying shrinkage and rapid chloride permeability. Hydration stabilising admixture (HSA) was effective in overcoming the negative effects of age and higher solid content in the wash water.

Extreme caution is needed when considering the use of recycled water slurries with higher solids than the specified limit. The effect of such waters varies with increased solid contents as well as the age of the slurry.

9. Industrial Wastewaters:

Industrial wastewaters may be used as mixing water in concrete as long as they only cause a very small reduction in compressive strength, generally not greater than 10–15%. Wastewaters from paint factories, coke plants, and chemical and galvanising plants may contain harmful impurities. Thus such wastewaters should not be used as mixing water without testing.

10. Reclaimed or Sanitary Wastewater

The sanitary sewage may be safely used as mixing water after treatment or dilution of the organic matter. Reclaimed water is treated wastewater from which solids and certain impurities are removed. It is typically used for non-potable applications such as for irrigation, dust control, fire suppression, concrete production, and construction. It is further described in previous Chap. 5.

6.5 Admixture Selection

Admixtures should not be used without a full study of all consequences. Many admixtures affect more than one property of concrete, sometimes adversely affecting desirable properties. Calcium chloride used as an accelerator, i.e. for an early strength development of concrete, reduces the resistance of cement to sulphate attack but increases the risk of alkali-aggregate reaction, and increases shrinkage, creep, and risk of corrosion. The low limits, as per Tables 9.2 and 9.3, practically ban the use of any chloride-based admixtures in concrete containing embedded metal.

While selecting admixtures, their effect on water demand and required dosage rate must be examined. Some of the water-reducing and -retarding admixtures are generally used in small doses (2–7 oz/100 lb of cementitious materials) [14], and normally reduce mixing water requirements 5–8%, whereas water-reducing high-range admixtures, i.e. super-plasticisers, are used in large quantities (10–90 oz/100 lb of cementitious materials) [14], and reduce water requirements 12–25% or more. Hence, while using different water-reducing admixtures, their effects on water content must be taken into account when calculating the quantity for mix water.

While using admixtures, the proper dosage rate is an important factor because when admixtures are used in large quantities, side effects may occur. For example excessive amounts of super-plasticiser can cause shrinkage cracks in concrete. Similarly too much air entrainment may reduce strength of concrete. Some accelerators begin to show retarding properties if they are added to concrete in excess. The use of calcium-aluminate cement as an admixture may cause flash set depending on dosage rate.

The accelerating admixtures are used to reduce the initial and final setting times of concrete. The amount of reduction varies with the amount of accelerator used. The variation in dosage rate affects the proper function of accelerators as well, such as calcium nitrate beginning to show retarding properties at excessive dose (6% by weight of cement) [15]. Ferric chloride acts as a retarder at addition of 2–3% by weight, but turns to an accelerator at 5%. Similarly, admixtures such as oxalic acid, lactic acid, urea, and formaldehyde are used to accelerate the setting of Portland cement when low w/c ratios are used. However, when these compounds are used in excess, severe retardation is reported [15]. For proper dose of accelerators, the manufacturer's instructions must be followed.

In addition to adequate selection of admixtures, their proper use is also of great importance. The successful use of admixtures depends upon the use of appropriate methods of preparation and batching. Concrete properties and performance may be affected adversely if care is not taken in these areas. The preparation of admixtures may require diluting some concentrated solutions so that accurate batching or dispensing is ensured. In such conditions, the manufacturer's instruction should be followed if there is any doubt in the procedures adopted.

For dispensing of admixtures into a concrete batch, controlling the time and rate of the admixture addition along with accurate measurement of admixture quantity is a very important factor and must be monitored closely.

Sometimes, during mixing, changing the time at which the admixture is added can vary the degree of effectiveness of the admixture. In addition to this, if the admixture is not thoroughly mixed through the load, or is not consistent in its concentration from load to load, there can be significant problems with finishing, particularly in large pours. Poor mixing and adding of admixture in dry batches or directly in a mixer should be avoided; instead, the admixture should be added to the mixing water and introduced in two or three stages throughout the load.

It has also been noticed that two or more admixtures often are not compatible in the same solution. For example, a vinsol resin-based, air-entraining admixture and a water-reducing admixture containing a lignosulphonate should not be allowed to intermix before actual mixing into the concrete because of their instant flocculation and loss of efficiency of both admixtures [15]. Hence, intermixing of different admixtures before actual mixing into a mixer must be avoided unless the manufacturer recommends it. However, it is usually preferable to add the different admixtures into the batch at different times during mixing.

Mostly, chemical admixtures are available in liquid form. Care should be taken to protect liquid admixtures from freezing. Long-time storage of liquid admixtures in vented tanks should be avoided. Evaporation of a portion of the liquid will adversely affect the performance of admixture.

Proper understanding of the technical data provided by the supplier for the particular application is a must. The overdose of admixture should be avoided. Trial tests should be carried out using the actual constituents of the mix to be used. Adequate supervision is also required at the batching so as to ensure correct levels of dosage of the admixture.

Bibliography

1. ACI 207.IR-87 (Revised 2005), Mass Concrete.
2. ACI 225R-91 (Revised 2016), Guide to the Selection and Use of Hydraulic Cements.
3. ASTM C289-87 (Revised 2007- Withdrawn 2016) Standard Test Method for Potential Reactivity of Aggregates (Chemical Method).
4. ASTM C227-90 (Revised 2010), Standard Test Method for Potential Alkali Reactivity of Cement Aggregate Combinations (Mortar-Bar Method).
5. BS 812, Part 112-90, Method for Determination of Aggregate Impact Value.
6. BS 812, Part 113 (Draft), Method for Determination of Aggregate Abrasion Value.
7. BS 882-92, Specification for Aggregates from Natural Sources of Concrete.
8. Canadian Standards Association (CSA) A23.2-14, Test methods and standard practices for concrete.
9. Canadian Standards Association (CSA) A23.1-14 - Concrete materials and methods of concrete construction/Test methods and standard practices for concrete.
10. M.L. Gambhir, Concrete Technology, TATA McGraw Hill, India, 1989.
11. BS 3148-80, Water for Making Concrete (Including Notes on the Suitability of Water).

12. A.M. Neville and J.J. Brooks, Concrete Technology, E.L.B.S. Longman, Singapore, 1993.
13. Hubert Woods, Durability of Concrete Construction, ACI Monograph No.4, USA, 1984.
14. ACI 211.1-91 (Reapproved 2009), Standard Practice for Selecting Proportions for Normal, Heavy Weight and Mass Concrete.
15. ACI 212.3R-91 (Revised 2016), Chemical Admixtures for concrete.
16. CCAA- Cement Concrete & Aggregates Australia- Use of Recycled Water in Concrete Production- August 2007.
17. Lobo C and Mullings GM 'Recycled Water in Ready Mixed Concrete Operations' Concrete In FOCUS Vol. 2, No. 1, pp 17–26 Spring 2003- National Ready Mixed Concrete Association.

Chapter 7
Poor Construction Methods

7.1 Introduction

Good-quality reinforced cement concrete is made from cement, sand, aggregate, water, and reinforcement, and so is bad-quality concrete. The quality of concrete, in fact, mostly depends on operations like mixing, transporting, placing in forms, compacting, and curing. It is important that the constituent materials remain uniformly distributed within the concrete mass during the various stages of its handling, and that full compaction is achieved followed by proper curing. When any of these conditions is not satisfied the strength and durability of hardened concrete are adversely affected. To obtain good-quality concrete, the availability of experienced, knowledgeable, and trained personnel at all levels is also required. The designer should have the knowledge of construction operations as well.

This chapter mainly considers the poor construction practices generally adopted on sites, its adverse effects on concrete produced, and possible guidelines for controlling the same.

A wide variety of poor construction practices, resulting from bad workmanship and inadequate quality control and supervision, which lead to many troubles in production and performance of concrete, can be categorised as under.

7.2 Use of Excess Water Content

The strength and workability of concrete depend to a great extent on the amount of water used in mixing. Overwet mixes exhibit excessive bleeding and increased shrinkage. It is much more important to obtain a truly plastic mix through proper proportioning than to try to overcome deficiencies by adding more water.

© Springer Nature Switzerland AG 2019

A. Surahyo, *Concrete Construction*,

https://doi.org/10.1007/978-3-030-10510-5_7

7.2.1   Effects of Using Excess Water Content

More water than the optimum amount increases workability but reduces strength; an increase of 10% may reduce the strength by approximately 15%, whereas an increase of 50% may reduce the strength by one-half. With an excess of more than 50% the concrete becomes too wet and liable to separation [1].

Excess water not only produces low-strength concrete but also reduces density and durability, and increases drying shrinkage, which result in cracking and surface crazing of concrete. Excess water also leaves voids and cavities after evaporation, leading to increase in porosity and permeability of the hardened concrete. Porous and permeable concrete will allow ingress of moisture and air into it, which will result in the corrosion of steel. As corrosion leads to increase in the volume of steel, cracking and spalling of the concrete cover will follow.

When too much water is added to concrete, the excess water along with cement comes to the surface by capillary action and this cement-water mixture forms a scum or thin layer of chalky material known as laitance. This laitance prevents bond formation between the successive layers of concrete and forms a plane of weakness. Excess water may also leak through the joints of the formwork and make the concrete honeycombed.

7.2.2   Protective Measures

Using cement, water, and aggregates from a specific source, concrete mixed and cured under given conditions will have improved properties at lower water/cement (w/c) ratio. Therefore, the use of minimum amount of water necessary to produce the required slump and workability is extremely important. As a rule, the smaller the percentage of water, the stronger the concrete.

Generally the water/cement ratio required for full hydration of cement is approximately 0.36 by weight of cement [2]. But, with this water content, a normal concrete mix will be extremely dry and difficult to place and compact. Additional water is, therefore, required for absorption of aggregates and to lubricate the mix, so as to make it workable. Water in excess of that required for hydration and absorption must be kept to the minimum. Hence, the correct water-to-cement ratio for the mix should be strictly enforced.

In order to have a proper w/c ratio, it is suggested to select a w/c ratio which is consistent with a specified compressive strength. Based on this idea, the required water/cement ratio to produce given compressive strength of concrete with ordinary Portland cement in normal conditions, Table 4.7 proposed by ACI-211 could be followed. However, for special exposure conditions, ACI-318 suggests the values mentioned in Table 7.1.

The use of appropriate slump of concrete mixes will also help to control the unnecessary use of excess water content. A concrete mix with excess flowability,

Table 7.1 Requirements for special exposure conditions based on ACI-318 [3]

Exposure conditions	Maximum w/c ratio (by weight)	Maximum compressive strength N/mm² (psi)
Concrete intended to have low permeability when exposed to water	0.50	28 (4000)
Concrete exposed to freezing and thawing in a moist condition or to deicing chemicals	0.45	31 (4500)
For corrosion protection: reinforced concrete exposed to chlorides from deicing chemicals, salt or brackish water, or spray from these sources	0.40	34 (5000)

which is usually the result of too much water in the mix, will generally be unstable and will probably segregate during the compaction process. In order to have proper slump, the requirements for maximum and minimum slump for different types and stages of construction specified by ACI-211 are produced in Table 4.4.

Many admixtures, which act as water reducers or plasticisers, are available in the market and could also be used to reduce the water content with increased workability of concrete mix. Chapter 2 addresses the various types of plasticisers which are used for this purpose. The use of entrained air increases workability of the mix and helps in reducing the water content. An increase of 1% air is equivalent to an increase of 1% in fine aggregate or increase in the unit water content by 3% [4]. The characteristics of aggregate also affect the water requirements. It increases, as aggregates used are crushed, angular, and rough textured. However, to make concrete of equal workability, the required mixing water decreases when well-graded or uncrushed aggregates are used. Table 4.3 clearly indicates this variation when crushed and uncrushed aggregates of same size are used. However, it may be kept in mind that crushed aggregates result in a concrete of higher strength compared to a similar mix made with uncrushed aggregates, as discussed in Sect. 2.3. Excess water content utilised by the use of crushed aggregates, however, can be controlled by using plasticisers to produce mix of required workability.

Moreover, the smaller the maximum size of the aggregate, the higher water content needed. This is evident from Tables 4.3 and 4.8. The use of excess water content due to this reason can be controlled if coarse aggregate used in concrete mix follows the grading requirements of BS-882, ASTM-C33, and OPSS.PROV-1002, as provided in Chap. 2. Similarly, the grading of the fine aggregate also has a considerable effect on the water requirement of the concrete. Changing the grading of sand from a coarse one to a fine one can result in an increase in the water content of 25 kg/m³ in order to maintain the required workability of concrete [5]. However, the workability can be maintained at the same water content if finer sand is used having reduced fines content. Figure 4.4 shows how the fines content of the mix can be reduced as the sand becomes finer, i.e. as the percentage passing the 600 μm test sieve increases.

The water required for proper concrete consistency (slump) is affected also by such things as when rate of delivery is slow or irregular, haul distances are long, and weather is warm. Adding water in excess of the proportioned w/c ratio to compensate for slump loss resulting from hot weather, delays in delivery or placing should be avoided. Loss of water, which results in loss of workability during warm weather, can be minimised by expediting delivery and placement, and by controlling the concrete temperature. Use of retarder will be helpful to prolong the setting time of concrete in such climate. Usually all mixing water should be added at the central plant. However, ACI-304 [6] suggests that in hot weather it is desirable to withhold some of the mixing water until the mixer arrives at the site. Then the remaining required water may be added with an additional 30 revolutions at mixing speed to adequately incorporate the additional water into the mix. CSA-A23.1-14 [7] also suggests to add water on site, when the measured slump or slump flow of the concrete is less than that required, with a condition that the specified water-to-cementing material ratio is not exceeded and no more than 60 min has elapsed from the time of batching. It further suggests that the mixing water added should not be more than the lesser of 16 L/m^3 or 10%. In each case, the mixer drum shall be turned at mixing speed for at least 30 revolutions (or equivalent time limit) after the addition of water.

7.3 Segregation/Inadequate Placing

Segregation is the separation of the coarse aggregate from the rest of the mix. However, as per ACI-309 [8], it can be defined as a mixture's instability, caused by a weak matrix that cannot retain individual aggregate particles in a homogenous mass.

7.3.1 Effects and Reasons of Segregation/Inadequate Placing

Blemishes, sand streaks, porous layers, and honeycombing are direct results of segregation. It adversely affects the strength, durability, and other properties of the hardened concrete. In experiments conducted by the Missouri State Highway Department, a difference of 14% in the flexural strength of vibrated concrete was attributed mainly to the effect of segregation [9].

The result of poor placement techniques and horizontal movement of concrete also produces voids, segregation, honeycombing, scouring, form streaking, sand streaking, low in-place strengths, and generally poor-quality concrete.

Usually segregation is the result of:

- Using harsh, extremely wet, and dry mixes as well as those deficient in sand
- Over-vibration
- Dropping concrete from excessive heights during placing

When the mix is too dry, coarse particles tend to separate out during transportation to the site. While dumping this mix on pavements or floors, the lack of cohesiveness in the mix encourages the course particles to roll down the sides of heaps of concrete. Now, if the material is roughly spread, this segregation will persist throughout the compacting operation since the machine used for compaction cannot rearrange the particles in the concrete to make a uniform mass when the original material is not uniform. On the other hand, when the mix is too wet, the cement water paste rises to the surface and coarse particles settle at the bottom. The grout also leaks out through the formwork that results in honeycombing.

Over-vibration results in separation of coarse aggregate towards the bottom of the form and of the cement paste towards the top. Such concrete would obviously be weak, and the laitance on its surface would be too rich and too wet, so that it leads to crazing of concrete. Similarly, when concrete is dropped from excessive heights, it also results in separation of the cement paste from the aggregate and turns to honeycombed concrete.

7.3.2 Protective Measures

The likelihood of segregation can be greatly reduced by the choice of suitable grading and with a correct method of handling, transporting, and placing. Proper grading results in cohesive mix and a uniform mass that keeps the concrete ingredients intact. The grading limits for fine and coarse aggregates specified by BS-882 [10], ASTM-C33 [11], and OPSS.PROV-1002 [12] shown in Tables 2.8–2.13 in Part I can be referenced. The concrete should be placed in its final position rapidly so that it is not too stiff to work. The concrete should be placed as closely as possible to its final position and should never be moved by vibrating and allowing it to flow, as this will result in segregation. Over-vibration should also be avoided.

The quality of concrete along with other factors depends on the care with which it is placed in the final position. OPSS-1350 [13] specifies that, when concrete is transported to the site by means of agitating or mixing equipment, discharge of the concrete shall be completed within 1.5 h after introduction of the mixing water to the cement and aggregates; except when the air temperature exceeds 28 °C and the concrete temperature exceeds 25 °C, the concrete shall be discharged within 1 h after the introduction of the mixing water. When concrete is transported by means of non-agitating equipment, discharge shall be completed within 30 min after introduction of the mixing water to the cement and aggregates. Use of retarders does not change the specified concrete discharge time.

The temperature of formwork, steel reinforcement, or any other material on which the concrete is to be placed shall not exceed 30 °C. During cold weather, excavations prepared for concreting and any existing concrete, steel reinforcement, structural steel, forms, or other surfaces against which concrete shall be placed shall be at a minimum temperature of 5 °C for a period of 12 h prior to commencement of placing concrete [14]. When placing the concrete, care should be taken to drop the concrete vertically and from not too great a height. OPSS-904 [14] recommends that the concrete shall be deposited within 1.5 m of its final position. When concrete is to be dropped more than 1.5 m, fully enclosed vertical drop chutes extending to the point of deposit shall be used.

Chutes are frequently used for transferring concrete from upper to lower elevations. Chutes shall have sufficient slope to deliver concrete of the approved consistency and shall have a maximum length of 15 m [14]. They should have round corners, be constructed of steel or be metal lined, and should have sufficient capacity to avoid overflow. The slope should be constant and steep enough to permit concrete of the slump required to flow continuously down the chute without segregation. The flow of concrete at the end of a chute should be controlled to prevent segregation. Circular pipes are also used for placing concrete vertically from higher to lower elevations. However, care should be taken to have its diameter at least eight times the maximum aggregate size [15]. It should be plumb, secure, and positioned so that the concrete will drop vertically.

The concrete should be placed in uniform layers. Hand shovelling and placing in large heaps and in sloping layers should be avoided. On a slope, the concreting should begin at the lower end of the slope. To avoid cracking due to settlement, concrete in columns and walls should be allowed to stand at least 1 h before concrete is placed in slabs or beams which they are to support. The length of delay in starting the concrete for slabs or beams will depend upon the temperature and setting characteristics of the concrete used, but concreting should begin soon enough to permit a proper bonding of the new layer with the old layer by vibration.

For walls and columns, ACI-304 [15] recommends that all concrete should be placed in lifts not exceeding 24 in. (600 mm). While concreting thin sections like fins and walls, of appreciable heights, concrete should be placed in horizontal layers of about 6 in. (150 mm) to 8 in. (200 mm) in depth. The forms must also be checked for tightness to avoid any loss of mortar, which may result in honeycombing.

While placing concrete, the formation of cold joints in successive pours must be avoided. Cold joints are places of discontinuity within a member where concrete may not tightly bond to itself. If a part of concrete in one placement sets, and then the rest of the concrete is placed in it, a cold joint may occur as shown in Fig. 7.1. The result is a weak connection between placements because during the setting, laitance forms resulting in formation of a weak plane. Leakage occurs when the structure is put into service at a later stage.

Placing concrete under water, a long downpipe or tremie ensures accuracy of location of the concrete and minimum segregation. In the tremie method, concrete is fed by gravity through a vertical pipe, which is gradually raised. CSA-A23.1-14 [7] recommends that the tremie pipe shall have a diameter at least eight times the maximum size of aggregate, and w/c ratio should not exceed 0.45. Precautions shall be undertaken to prevent the loss of the cementing material paste by the washing action of water. The use of anti-washout admixtures may be used, provided that they should not adversely affect the overall quality, and place ability of the concrete. CSA further recommends that the concrete shall not be placed in water having a temperature below 5 °C except when the strength gain of the concrete is sufficient.

A cohesive mix with high cement content and high proportions of fines should be selected while concreting underwater. Tremie concrete shall not be placed above the

Fig. 7.1 Cold joint
(reprinted with permission
from Concrete Repair and
Maintenance [16])

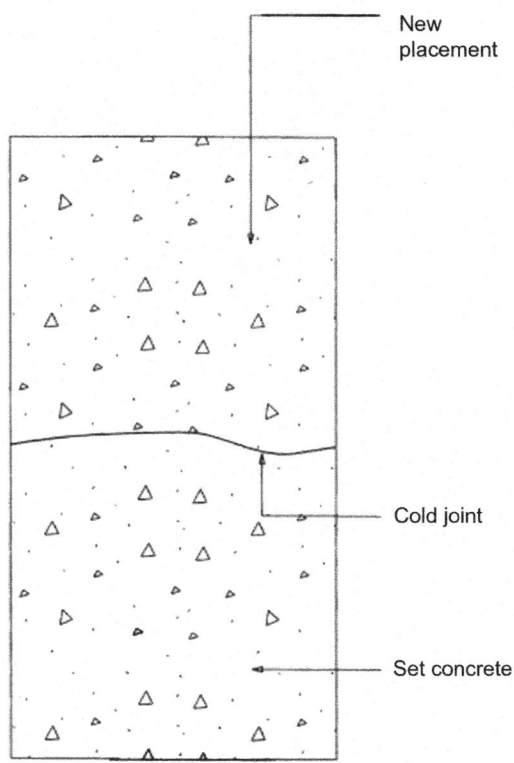

existing water level. OPSS.PROV 1350 [13] recommends that for tremie concrete, the minimum cementing material content shall be 415 kg/m³ of concrete.

When choosing equipment for placing, consideration must be given to the ability of the equipment to economically place the concrete in the correct location and without altering its quality. The selection of equipment is influenced by the method of concrete production. Certain types of equipment such as buckets, hoppers, and buggies/trollies will suit batch production while other equipment such as belt conveyors and pumps are more suitable for continuous production. In order to avoid inadequate placing which results in segregation, some methods of concrete placing, based on ACI-304 [15], are shown in Figs. 7.2, 7.3, 7.4, 7.5, 7.6, and 7.7.

7.4   Poor Compaction/Consolidation

The process of compacting is carried out almost simultaneously with placing of concrete. The purpose of compacting is to expel voids and air bubbles in the concrete mass entrapped during mixing or placing. The density, and consequently the strength and durability of concrete, depends upon the quality of this compaction. Compaction of concrete may be done either manually or mechanically. In case of

Fig. 7.2 Methods showing
filling of concrete hoppers,
based on ACI-304R-89
[15]

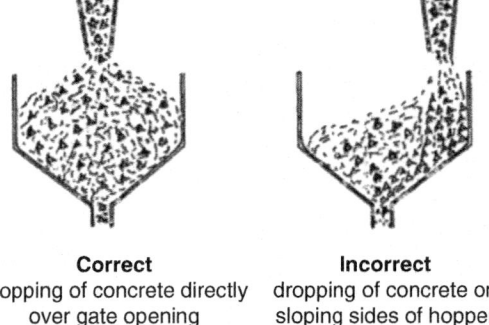

Correct
dropping of concrete directly
over gate opening

Incorrect
dropping of concrete on
sloping sides of hopper

Fig. 7.3 Methods showing
discharge of concrete from
a hopper, based on
ACI-304R-89 [15]

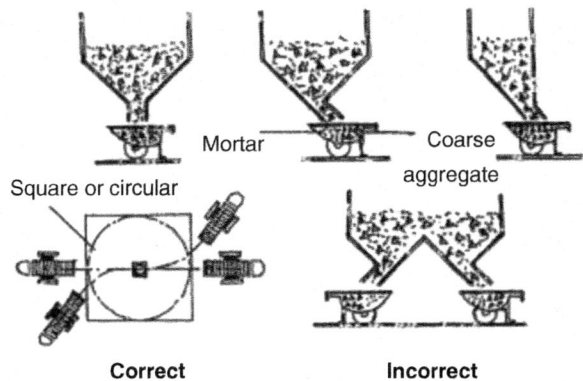

Correct   **Incorrect**

Sloping hopper gates without end control, causes segregation
in filling trollies/buggies, however discharge from center opening
to get vertical drop into centre of trolley as shown on left avoids
segregation.

Fig. 7.4 Methods showing
discharge of concrete at the
end of concrete chutes,
based on ACI-304R-89
[15]

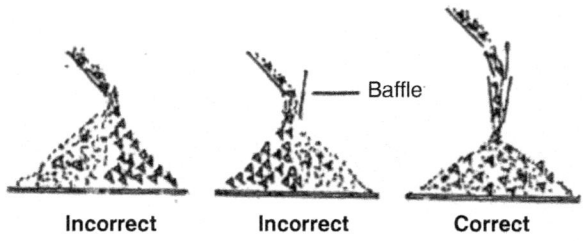

Incorrect   **Incorrect**   **Correct**

The sloping discharge at right from drum mixers, truck mixers,
etc., prevents separation of mortar and coarse aggregate, however
either of the arrangements at the left leads to segregation, which is
result of improper or lack of control at end of any concrete chute.

Fig. 7.5 Methods showing placing of concrete from trolley/buggy, based on ACI-304R-89 [15]

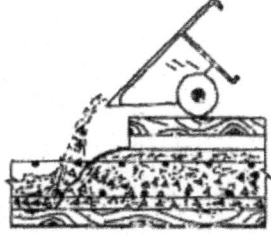
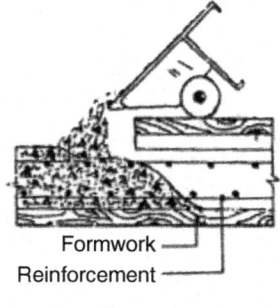

Incorrect
To dump concrete away
from concrete in place

Formwork
Reinforcement

Correct
Placing slab concrete into
face of concrete in place

Fig. 7.6 Methods showing placing of concrete in deep wall by concrete pump or hose, based on ACI-304R-89 [15]

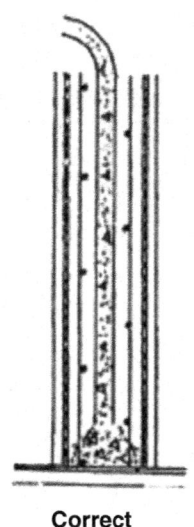
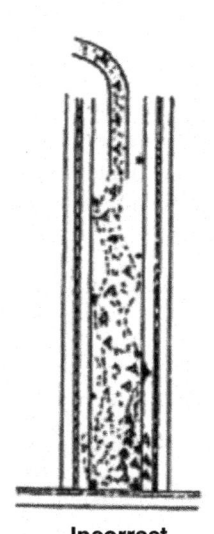

Correct   **Incorrect**

While placing concrete by concrete pump or hose, striking
against form and bars should be avoided to overcome
segregation by taking hose deep enough to bottom.

manual compaction, vertical sections are consolidated by rodding, whereas flat surfaces like slabs and floors are compacted by tamping or ramming by tamping rod. In case of mechanical consolidation or compaction, vibration is the most widely used system. For vibration, many types of internal and external vibrators are available in the market. Internal vibration is the most effective method of consolidating plastic concrete for most applications.

Fig. 7.7 Methods showing placing of concrete in top of narrow form, based on ACI-304R-89 [15]

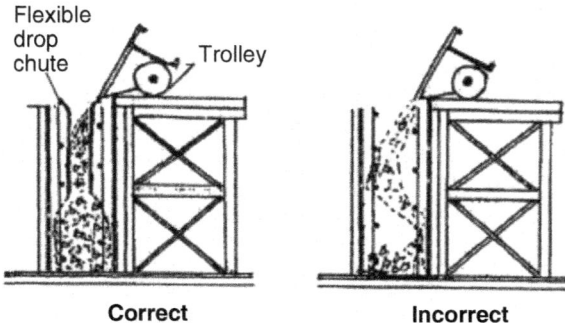

Correct   **Incorrect**

While placing concrete from buggy/trolley or chute, striking against form and bars should be avoided as it causes separation and honeycombing at bottom.

7.4.1   Reasons and Effects of Poor Compaction

In the case of hand-rammed concrete, inadequate compaction is the most common fault, whereas, in the case of vibration, non-uniform compaction can occur due to inadequate and insufficient vibration that leads to air pockets, or over-vibration, which leads to segregation. In case of internal vibration, poor compaction is also the result of:

- Inadequate penetration
- Excessive spacing between insertions
- Short immersion time
- Vibrator too small in diameter
- Frequency too low
- Amplitude too small

Insufficient or under-vibration can cause porous and non-homogeneous conditions as well as local defects. Generally, the volume of entrapped air in the porous or unconsolidated concrete is in the range of 5–20% [8]. The presence of air voids in concrete reduces the density and greatly reduces the strength. Only 5% of air voids reduces the strength of concrete by 30% [2].

Over-vibration can occur if vibration is continued for a prolonged time. Over-vibration is generally the result of using oversized equipment, improper procedures, and high slump. It usually results in segregation, excessive form deflection, sand streaking, form damage, and loss of almost all of the entrained air in the concrete.

Poor consolidation can also lead to frost damage as well as reinforcement corrosion and other chemical attacks, primarily as a consequence of penetration into the concrete by aggressive matters such as water, chlorides, carbon dioxide, and oxygen. A well-consolidated surface layer of concrete is thus especially important for the quality and service life of the concrete structure.

7.4.2   Protective Measures

The best way of removal of entrapped air or voids from the concrete, and to convert it in a homogeneous dense mass, is the appropriate method of compaction. The concrete mix is usually designed on the basis that after being placed in forms it may be thoroughly compacted with available equipment. Therefore, thorough compaction is necessary for a successful, durable, and dense concrete.

The most widely used compaction method is mechanical; however, manual compaction is also used sometimes on smaller jobs.

7.4.2.1   Mechanical Compaction

The most widely used method for compaction is vibration. Vibration consists of subjecting freshly placed concrete to rapid vibratory impulses, which liquefy the mortar, and reduces the internal friction between aggregate particles, thereby resulting in the proper settlement of concrete. Equipment of concrete vibrators can be divided into two main classes: internal and external. However, external vibrators may further be divided into three kinds: form vibrators, surface vibrators, and vibrating tables.

Internal Vibrator

This is the most common type of vibrator and consists of a steel tube named as poker. This poker is connected to an electric motor or a diesel engine through a flexible tube.

Operation Method

While using internal vibrators, the poker must be moved from place to place, with a uniform spacing so that the concrete is vibrated within the range of about every 2 ft. (0.6 m), or the distance between insertions should be about 1½ times the radius of action, or should be such that the area visibly affected by the vibrator overlaps the adjacent just-vibrated area by a few inches. The vibrator should be immersed quickly through the entire depth of the freshly deposited concrete and at least 6 in. (150 mm) into the preceding layer, as shown in Fig. 7.8. The vibrator should be moved in an up-and-down motion, generally for 5–15 s, to combine the two layers together. In this manner, monolithic concrete is obtained, thus avoiding a plane of weakness at the junction of the two layers, possible settlement cracks, and internal effects of bleeding. After completion of vibration, the poker must be withdrawn slowly so that the hole left by the poker closes fully by itself without any air being trapped. The actual completion of compaction can be judged by the appearance of the surface of the concrete, which should neither be honeycombed nor contain an excess of mortar.

Fig. 7.8 Placing of
vibrators [15]

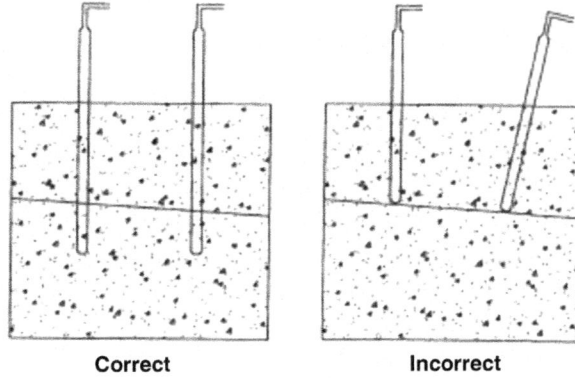

The poker should always be used in a vertical position in beams, columns, and walls. However, in slabs, as recommended by ACI-309R-87 [8], the poker may be sloped towards the horizontal as necessary to operate in a fully embedded position. During vibration in a sloped position, care should be taken to avoid the lifting of steel reinforcement. In order to have adequate vibration results, care should also be taken to deposit the concrete in appropriate layers. If concrete is deposited in thick layers of more than 300 mm (12 in.), more air may be trapped than if it is placed in thinner, even layers. These layers should be as level as possible so that the vibrator does not need to move the concrete laterally, since this will result in segregation. Where mixes of dry or stiff consistencies are required, the placement rate should be slower in order to avoid bug holes and honeycombing.

Mostly in congested reinforcement areas, the vibrator cannot reach the concrete. In such cases, it may be helpful to vibrate exposed portions of reinforcing bars. This vibration will help in the mobility of concrete and increase the concrete to steel bond through the removal of entrapped air and water from underneath the reinforcing bars.

Vibration Period

Generally, the vibration of a concrete mix can be considered to be sufficient when the air bubbles cease to appear and sufficient mortar appears to close the surface interstices and facilitate easy finishing operations. However, to reach this position, every mix needs an optimum period of vibration depending upon the characteristics of the mix. This optimum period can be estimated by conducting trials with different periods of vibration to obtain compaction without segregation, and then selecting the period which gives maximum strength of concrete cubes.

For proper vibration period in different casting conditions, based on research made by Swedish engineers [17], Table 7.2 may be followed. The vibration effort mentioned in the table is equal to the effective time of vibration per concrete volume as measured in seconds per cubic meter or cubic yard. This measure, originally introduced by the US Bureau of Reclamation, has later been used in Sweden in studies of concrete vibration.

Based on data provided in Table 7.2, it may be noted that for super-plasticised concrete with a slump of about 200 mm (8 in.), the vibration effort can be reduced

Table 7.2 Guide values for adequate vibration effort and practical vibrator capacity for different casting conditions [17]

Degree of difficulty	Adequate vibration effort		Practical vibrator capacity	
	s/m³	s/yd³	m³/h	yd³/h
Simple castings: large open forms easily accessible	200–300	150–230	6–10	8–13
Castings of medium difficulty: walls, columns, beams, slabs with normal reinforcement density	300–400	230–310	4–6	5–8
Difficult castings: narrow beams, and walls with dense reinforcement	400–600	310–460	3–4	4–5

by 50% and the corresponding practical vibrator capacity increased by 100%. The consequences of over-vibration can be minimised once the period of vibration is controlled; however, the use of well-proportioned mix with a proper slump and adequate equipment is also of great help in this regard.

Adequacy of Equipment

Experience indicates that the effectiveness of an internal vibrator, including operation method as above, also depends on the head diameter, frequency, and amplitude. Table 7.3 recommended by ACI-301 [18] provides the ordinary range of characteristics, performance, and applications of internal vibrators. Recommended frequencies are given along with suggested values of eccentric moment, average amplitude, and centrifugal force. Approximate ranges are also given for the radius of action and rate of concrete placement. These are empirical values based mainly on previous experience.

These vibrators usually work at frequencies of 6000–12,000 vibrations/min. The vibrator should have an adequate radius of action and it should be capable of flattening and de-aerating the concrete quickly. To obtain an adequate radius of action, an internal vibrator must operate at high vibration intensity. An increase in amplitude also results in an increased radius of action. However, it may be noted that the radius of action of an internal vibrator is substantially less in reinforced concrete than in non-reinforced concrete. A reduction of 50% is not uncommon according to ACI-309.1R [4].

In selecting the vibrator and vibration procedures, consideration should be given to the vibrator size relative to the form size. If a vibrator too large for the application is used, it will result in high concentration of cement paste at the surface level that will lead to crazing. However, if a vibrator of low frequency is used, the compaction achieved will not be satisfactory and that may result in formation of some pockets of honeycombed concrete.

External Vibrators

1. *Form Vibrators*

These vibrators are clamped to the outside of the form or mould. They vibrate the form, which in turn transmits the vibration to the concrete. While using this

Table 7.3 Range of characteristics, performance, and applications of internal vibrators, as per ACI-301-96 [18]

Group	Diameter of head (in.)	Frequency (vibrations/ min)	Eccentric moment (in.-pound)	Average amplitude (in.)	Centrifugal force (pounds)	Radius of action (in.)	Rate of concrete placement (yd^3/h/ vibration)	Application
1	¾ to 1½	10,000–15,000	0.03–0.10	0.015–0.03	100–400	3–6	1–5	Plastic and flowing concrete in very thin members and confined places
2	1¼ to 2½	9000–13,500	0.08–0.25	0.02–0.04	300–900	5–10	3–10	Plastic concrete in thin walls, columns, beams, precast piles, thin slabs, and along construction joints
3	2 to 3½	8000–12,000	0.20–0.70	0.025–0.05	700–2000	7–14	6–20	Stiff plastic concrete (<3 in. slump) in general construction such as walls, columns, beams, prestressed piles, and heavy slabs
4	3 to 6	7000–10,500	0.70–2.50	0.03–0.06	1500–4000	12–20	15–40	Mass and structural concrete of 0–2 in. slump deposited in quantity up to 4 yd^3 in heavy construction
5	5 to 7	5500–8500	2.25–3.50	0.04–0.08	2500–6000	16–24	25–50	Mass concrete in gravity dams, large piers, massive walls, etc.

type of vibrator, concrete should be placed in layers of suitable depth, as air cannot be expelled through too great a depth of concrete. It is better to deposit in layers of 10–15 in. (250–400 mm) thick. Each layer should be vibrated separately. In order to distribute the proper intensity of vibration over the desired area of form, the size and spacing of form vibrator need proper skills and efficiency. The amplitude should be fairly uniform over the entire surface. ACI-309R-87 [8] suggests that initially the spacing should be kept in the range of 4–8 ft. (1.2–2.4 m) and in case adequate and uniform vibration is not achieved the vibrators should be relocated till the proper results are obtained.

In some cases internal and form vibrators are combined to get better results. Generally, form vibrators tend to draw mortar to the form and when used in combination with internal vibrators have proved effective in reducing the size and number of air voids on the surface. As these vibrators are clamped to the formwork, the formwork has to be strong and tight so as to prevent distortion and leakage of grout.

For efficient form vibration, ACI-309.1R [4] recommends a minimum form acceleration of 1–3 g (9.8–30 m/s^2) for fluid to plastic mixes when the form is filled with concrete. Furthermore, depending on specific conditions, vibration frequencies between 3000 and 12,000 vibration/min (50 and 200 Hz) are suitable for form vibrator. High-frequency form vibration results in a better surface appearance than does form vibration at lower frequency.

2. *Surface Vibrators*

Surface vibrators are applied to the top surface of the concrete. They are used for compacting shallow elements and where internal vibrators cannot work properly, for example in road and runway pavements or grade slabs. Surface vibrators can compact effectively very dry mixes. The levelling effect of surface vibrators helps in the finishing operation of concrete. Surface vibrators are of three types:

(a) Vibrating screeds: It consists of a single or double wooden or steel beam of about 4–5 m length preferably spanning the slab width. One or two motors, depending on the screed length, are mounted on the beam.

They are mostly suitable for horizontal surfaces whose depth is not more than 200 mm. Vibrating screeds strike off and straight edge the concrete in addition to providing compaction. For stiff concrete mixes, large amplitude is required to attain a considerable depth of compaction. Frequencies of 3000–6000 vibrations/min have been found satisfactory [8]. Vibrating screeds should be moved forward as rapidly as proper consolidation allows; otherwise, too much mortar will be brought to the surface that will result in surface crazing.

(b) Plate or pan vibratory tampers: It consists of a flat steel plate of about 0.2 m^2 in area on which an electric motor is mounted. These vibrators are used for small jobs. They do not produce a finish level surface but are suitable for patching or for irregularly shaped slabs. The pan vibratory unit must be positioned behind the surface strike-off equipment. A surface should not be allowed to build up in front of the pan because it will dampen the motor.

(c) Vibratory roller screed: Like the vibrating screed, it is also a dual-purpose equipment and strikes off as well as compacts the concrete. This equipment is suitable for plastic mixtures, having slump of more than 50 mm (2 in.). It is generally used for irregular areas and hand placements. The frequency range mostly used is 100–400 vibrations/min. Usually two passes give better results, when vibratory roller screed is used. The first pass strikes off and compacts the concrete and the second provides the surface finish. On the first pass, maximum frequency should be used, and a lower frequency on the second pass.

3. *Vibrating Table*

It consists of a steel or reinforced concrete table with a supporting frame having steel springs. The external vibrators are rigidly clamped to the supporting frame to provide vibration. Vibration is transmitted from the table to the concrete. Vibrating tables are very efficient in compacting stiff and harsh concrete mixes required for the manufacture of precast elements in factories and test specimens in the laboratories. As per ACI-309R-87 [8], for stiff mixtures, frequency below 6000 vibrations/min and a high amplitude of over 0.005 in. (0.13 mm) are normally preferred, but for plastic mixtures the amplitude should not be less than 0.001 in. (0.025 mm). For vibrating concrete sections of different sizes, the table should have a variable amplitude and frequency.

The acceleration imparted to the concrete by the table mainly depends on the effectiveness of table vibration. Acceleration in the range 3–10 g is generally recommended. The vibrators are usually moved around to get uniform vibration. In order to get good compaction of very stiff mixtures, pressure is usually applied to the top surface during vibration.

7.4.2.2   Manual Compaction

In case of manual compaction, rodding for vertical sections should be done with a steel or a wooden rod by moving it up and down until the concrete is thoroughly worked into empty places and completely fills the corners and other awkward positions.

Flat surfaces like slabs should be tamped with a tamping rod, which serves the double purpose of compacting and finishing the concrete to the required level. Tamping should be done properly, and every inch of the surface should be attended; otherwise inefficient tamping will result in honeycombing. Where ramming is needed, usually in heavy masses of reinforced concrete, it should be continued till the slurry of the cement and water starts to appear on the upper face of the concrete. Ramming should not be heavy, as otherwise there is a tendency for concrete to rise against the side of the shutters, and also to exert excessive pressure on the shuttering face.

7.5 Inadequate Cover to Reinforcement

The thickness and quality of concrete cover over steel are of great importance as this cover protects steel from the factors that promote corrosion. The role of the correctly specified concrete cover and its attainment during construction is crucial to a concrete structure so as to remain durable for its design life. Concrete cover is the quality and thickness of concrete between the outer surface of the structure and the nearest embedded steel reinforcement. It is therefore of critical importance to ensure that this relatively thin layer must meet the specified requirements.

Attainment of correct cover must be verified and recorded prior to commencement of concrete placement. The cover of concrete should be well compacted, effectively cured, and essentially crack free. These properties will ensure that capillary discontinuity is achieved and would also limit the ingress over time of carbon dioxide, oxygen, water, and harmful ions of chloride and sulphate-based salts.

7.5.1 Effects of Inadequate Cover to Reinforcement

When the concrete cover that protects the reinforcing steel is damaged and the bond between the concrete and steel reinforcing bar is broken, the passive layer of steel will break down and active corrosion of the steel will start. Inadequate cover to reinforcement permits ingress of moisture, gases, and other substances that leads to corrosion of concrete reinforced bars resulting in cracking and spalling of concrete.

7.5.2 Protective Measures

To ensure adequate durability of concrete by providing proper protection to the reinforcement, and to employ a sufficient thickness of concrete around each bar to develop the necessary bond resistance between the steel and the concrete, it is necessary to provide an adequate cover of concrete over the bars. The cover required to protect the reinforcement against corrosion depends on the exposure conditions, and the quality of the concrete as placed, compacted, and cured. The requirements for minimum cover recommended by ACI-318-95 [3], BS-8110 [19], and CSA-A23.1-14 [7] based on various exposure conditions are given in Tables 7.4, 7.5, 7.6, and 7.7, respectively. However, BS-8110 further specifies that the nominal cover should always be greater than the size of aggregate and diameter of the main bar.

The position of reinforcement should be checked before and during concreting to make sure that the specified concrete cover is maintained. In case where concrete is cast against uneven surfaces, the nominal cover should generally be increased beyond the values given in above-mentioned tables to ensure that an adequate mini-

Table 7.4 Requirements for minimum concrete cover for the protection of reinforcement, based on ACI-318 [3]

Exposure conditions	Minimum cover in mm		
	Reinforced concrete cast in situ	Precast concrete	Prestressed concrete
Concrete cast against earth or permanently exposed to earth	75	–	75
Concrete exposed to earth or weather:			
Wall panels	40–50	20–40	30
Slabs and joints	40–50	–	30
Other members	40–50	30–50	40
Concrete not exposed to earth and weather:			
Slabs, walls, and joints	0–40	15–30	20
Beams, columns	40	10–40	20–40
Shells, folded plate members	15–20	10–25	10
Non-prestressed reinforcement	–	–	20
Concrete exposed to deicing salts, brackish water, seawater, or spray from these sources:			
Walls and slabs	50	40	–
Other members	60	50	–

Table 7.5 Requirements for nominal cover to reinforced and prestressed concrete of different grades under specified exposure conditions, based on BS-8110 [19]

Environment	Exposure conditions	Nominal cover of concrete in mm including links				
Mild	Concrete surfaces protected against weather or aggressive conditions	25	20	20	20	20
Moderate	Concrete surfaces continuously under water or in contact with non-aggressive soils	–	35	30	25	20
Severe	Concrete surfaces exposed to severe rain, alternate wetting and drying, or freezing while wet	–	–	40	30	25
Very severe	Concrete surfaces exposed to seawater spray, severe freezing while wet, deicing salts, corrosive fumes	–	–	50	40	30
Most severe	Concrete surfaces exposed to abrasion, e.g. seawater carrying solids or acidic water with pH 4.5 or machinery or vehicles	–	–	–	–	50
Minimum grade MPa (psi)		30 (4400)	35 (5100)	40 (5800)	45 (6500)	50 (7300)

Note: Maximum size of aggregate in this table is considered 20 mm (¾ in.)

Table 7.6 Requirements for minimum concrete cover for the protection of reinforcement, based on CSA-A23.1-14 [7]

Exposure condition	Exposure class (see Table 7.7)		
	N, NF	F-1, F-2, and sulphates	C-XL, A-XL, C-1, C-3, A-1, A-2, A-3
Cast against and permanently exposed to earth, including footings and piles	75 mm	75 mm	75 mm
Beams, girders, and columns	30 mm	40 mm	60 mm
Slabs, walls, joists, shells, and folded plates	20 mm	40 mm	60 mm
Ratio of cover to nominal bar diameter[a]	1.0[b]	1.5	2.0
Ratio of cover to nominal maximum aggregate size	1.0[b,c]	1.5	2.0

Notes:
1. Greater cover or protective coatings might be required for exposure to industrial chemicals, food processing, and other corrosive materials. See PCA IS001.08T
2. For information on the additional protective measures and requirements for parking structures, see CSA S413
3. For information on the additional protective measures and requirements for bridges, see CAN/CSA-S6
4. For complete details with respect to Tables 7.6 and 7.7 produced here, CSA-A23.1-14 can be referenced
[a]The cover for a bundle of bars shall be the same as that for a single bar with an equivalent area
[b]This refers only to concrete that will be continually dry within the conditioned space (i.e. members entirely within the vapour barrier of the building envelope)
[c]The specified cover from screeded surfaces shall be at least 1.5 times the nominal maximum aggregate size to reduce interference between aggregate and reinforcement where variations in bar placement result in a cover smaller than specified

mum cover is obtained. For this reason, the minimum nominal cover of 75 mm (3 in.) should generally be provided to the concrete casted directly against the earth. However, if blinding concrete is used then the minimum cover of 40 mm (1½ in.) excluding blinding concrete is advisable. Moreover, BS-5337 [20] specifies that the minimum cover to all reinforcement must not be less than 40 mm (including links), and that this value should be increased where the surface is liable to erosion, abrasion, or contact with particularly aggressive liquid.

As adequacy of cover also depends on the quality of concrete, the requirements of Table 4.1 for a durable concrete could be followed. The table specifies minimum cement contents and maximum w/c ratios based on various exposure conditions. The limits on w/c ratio and cement content will automatically be assured by specifying the minimum grades of concrete indicated in the table.

In addition to durable concrete mix, its compaction and curing play a very important role in achieving good-quality concrete cover. Usually when concrete is compacted, the centre portion of the concrete known as "heartcrete" becomes well compacted, whereas the concrete of cover portion known as "covercrete" is not compacted so well as it has to pass through reinforcement. Additionally, if the curing is poor this cover will suffer more and turn to a low-quality and permeable

Table 7.7 Requirements for concrete exposure classes as per CSA-A23.1-14 [7]

Exposure class	Maximum w/c ratio	Maximum specified compressive strength (MPa)	Air content category (ref: Table 8.4)	Definitions of exposure classes
C-XL	0.40	50 within 56 days	1 or 2	Structurally reinforced concrete exposed to chlorides or other severe environments with or without freezing and thawing conditions, with higher durability performance expectations than the C-1 classes
C-1	0.40	35 within 56 days	1 or 2	Structurally reinforced concrete exposed to chlorides with or without freezing and thawing conditions Examples: bridge decks, parking decks and ramps, portions of structures exposed to seawater located within the tidal and splash zones, concrete exposed to seawater spray, and saltwater pools. For seawater or seawater spray exposures the requirements for S-3 exposure also have to be met
C-2	0.45	32 at 28 days	1	Non-structurally reinforced (i.e. plain) concrete exposed to chlorides and freezing and thawing Examples: garage floors, porches, steps, pavements, sidewalks, curbs, and gutters
C-3	0.50	30 at 28 days	2	Continuously submerged concrete exposed to chlorides, but not to freezing and thawing Examples: underwater portions of structures exposed to seawater. For seawater or seawater spray exposures the requirements for S-3 exposure also have to be met
C-4	0.55	25 at 28 days	2	Non-structurally reinforced concrete exposed to chlorides, but not to freezing and thawing Examples: underground parking slabs on grade
F-1	0.50	30 at 28 days	1	Concrete exposed to freezing and thawing in a saturated condition, but not to chlorides Examples: pool decks, patios, tennis courts, freshwater pools, and freshwater control structures

(continued)

Table 7.7 (continued)

Exposure class	Maximum w/c ratio	Maximum specified compressive strength (MPa)	Air content category (ref: Table 8.4)	Definitions of exposure classes
F-2	0.55	25 at 28 days	2	Concrete in an unsaturated condition exposed to freezing and thawing, but not to chlorides Examples: exterior walls and columns
N	As per mix design	For structural design	None	Concrete that when in service is exposed to neither chlorides nor freezing and thawing nor sulphates, either in a wet or a dry environment Examples: footings and interior slabs, walls, and columns
N-CF	0/55	25 at 28 days	None	Interior concrete floors with a steel-trowel finish that are not exposed to chlorides, nor to sulphates either in a wet or a dry environment Examples: interior floors, surface-covered applications (carpet, vinyl tile) and surface-exposed applications (with or without floor hardener), ice-hockey rinks, and freezer warehouse floors
A-XL	0.40	50 within 56 days	1 or 2	Structurally reinforced concrete exposed to severe manure and/or silage gases, with or without freeze/thaw exposure. Concrete exposed to the vapour above municipal sewage or industrial effluent, where hydrogen sulphide gas might be generated, with higher durability performance expectations than A-1 class
A-1	0.40	35 within 56 days	1 or 2	Structurally reinforced concrete exposed to severe manure and/or silage gases, with or without freeze/thaw exposure. Concrete exposed to the vapour above municipal sewage or industrial effluent, where hydrogen sulphide gas might be generated Examples: reinforced beams, slabs, and columns over manure pits and silos, canals, and pig slats; and access holes, enclosed chambers, and pipes that are partially filled with effluents

(continued)

Table 7.7 (continued)

Exposure class	Maximum w/c ratio	Maximum specified compressive strength (MPa)	Air content category (ref: Table 8.4)	Definitions of exposure classes
A-2				Structurally reinforced concrete exposed to moderate-to-severe manure and/or silage gases and liquids, with or without freeze/thaw exposure Examples: reinforced walls in exterior manure tanks, silos and feed bunkers, and exterior slabs
A-3	0.50	30 at 28 days	2	Structurally reinforced concrete exposed to moderate-to-severe manure and/or silage gases and liquids, with or without freeze/thaw exposure in a continuously submerged condition. Concrete continuously submerged in municipal or industrial effluents Examples: interior gutter walls, beams, slabs, and columns; sewage pipes that are continuously full (e.g. force mains); and submerged portions of sewage treatment structures
A-4	0.55	25 at 28 days	2	Non-structurally reinforced concrete exposed to moderate manure and/or silage gases and liquids, without freeze/thaw exposure Examples: interior slabs on grade

Notes:
1. "C" classes pertain to chloride exposure
2. "F" classes pertain to freezing and thawing exposure without chlorides
3. "N" class is exposed to neither chlorides nor freezing and thawing
4. All classes of concrete exposed to sulphates shall comply with the minimum requirements of S class noted in Tables 2 and 3. In particular, Classes A-1 to A-4 and A-XL in municipal sewage elements could be subjected to sulphate exposure
5. No hydraulic cement concrete will be entirely resistant in severe acid exposures. The resistance of hydraulic cement concrete in such exposures is largely dependent on its resistance to penetration of fluids
6. Decision of exposure class should be based upon the service conditions of the structure or structural element, and not upon the conditions during construction
7. The minimum specified compressive strength may be adjusted to reflect proven relationships between strength and water-to-cementing materials ratio provided that freezing and thawing and deicer scaling resistance have been demonstrated to be satisfactory. The water-to-cementing materials ratio shall not be exceeded for a given class of exposure
8. For further details, on concrete requirements for sulphate attack Tables 1, 2 and 3 mentioned within these notes reference can be made to CSA A23.1-14/A23.2-14—Concrete materials and methods of concrete construction/test methods and standard practices for concrete. © 2014 Canadian Standards Association

concrete. Hence extra care and efforts should be used to improve the compaction in covercrete and thereafter its curing. It is also advisable to design sections that will allow concrete to be placed and compacted easily and quickly.

A research conducted at Dundee University, UK [21], reveals that the quality of concrete cover is also affected adversely due to continued use of oiled, plywood, grp (glass fibre-reinforced plastics), and steel form face contact materials. Test results indicated that the use of various cement types and admixtures helps to improve the durability of the heartcrete, but had little effect on the quality of the covercrete. However, when traditional face contact materials were replaced with a controlled permeability formwork (CPF) liner, it provided a well-cured, low-porosity, high-strength surface with improved durability irrespective of mix strength, cement type, and admixture addition [22].

Moreover, as per Concrete Society Technical Report No. 31 [23], the concretes cast against CPF are rated as having almost no or low absorption, whereas those cast against impermeable (traditional) formwork ranges from very high for grade 20 concrete to low for grade 50 concrete.

Since much of the deterioration of reinforced concrete is due to the provision of insufficient cover to the bars, a designer should not hesitate to increase the minimum cover, if it is thought desirable to do so. However, excessive thickness of cover is to be avoided, since any increase will also increase the surface crack width.

Good workmanship is required to ensure that the reinforcement is properly placed and that the specified cover is obtained. In order to maintain correct concrete cover, spacers, chairs, and other supports should be used to get the specified cover to the steel reinforcement. Spacers used should be durable and should not cause spalling of the concrete cover. Usually concrete spacer blocks, which are machine pressed, are recommended. If manufactured at site, they shall be made from a mix of one part cement and two parts of sand and small-size aggregate. Site manufactured blocks shall be well compacted and water cured for a minimum of 7 days after casting. Concrete spacers shall be comparable in strength, durability, porosity, and appearance to the surrounding concrete. Any wire cast into the spacer blocks shall be positioned well away from the exposed surface and shall be galvanised.

However, if plastic-made spacers are used, they shall be of such design that when the concrete is placed, a thin layer of grout shall seep between the spacers and the formwork so that they are hidden when the formwork is struck off. In no case small pebbles or brickbats shall be used for this purpose.

7.6 Incorrect Placement of Steel

The improper placement of reinforcing steel can greatly reduce the strength and life of a structure and possibly lead to a structural failure. For example, lowering the top bars or raising the bottom bars by ½ inch more than that specified in a 6-inch-deep slab could reduce its load-carrying capacity by 20% [24]. The failure of a structure does not necessarily mean the collapse of a structure. A structure fails when it can no longer be used in the manner in which it was intended. Costly repairs and early replacement of structures, however, are too common and are often caused by improper reinforcing steel placement.

Plain concrete has low tensile and flexural strengths and a high compressive strength. Steel reinforcement is provided to overcome the deficiencies in the tensile and bending strengths. Steel reinforcement is available in the form of plain steel bars, deformed steel bars, cold-drawn wire, welded wire fabric, and deformed welded wire fabric. The common types, grades, and sizes of steel reinforcement in use based on ASTM are provided in Table 7.8.

CAN/CSA-G30.18-M92 (R2007) Standard [26] specifies two types of hot-rolled deformed billet-steel bars, designated regular (R) and weldable (W). (R) Grades are intended for general applications. (W) Grades are intended for applications where welding, bending, or ductility is of special concern. Bars that are supplied to CSA G30.18 grade W, and to ASTM A706, are characterised as weldable and require minimal preparation to produce quality welds.

CAN/CSA-G30.18-M92 Standard specifies three minimum yield strength levels, namely 300 MPa, 400 MPa, and 500 MPa designated as grades 300R, 400R and 400 W, and 500R and 500 W, respectively. Properties of deformed reinforcing bars are presented in Table 7.9. Rebar sizes are commonly referred to as the bar designation number combined with the letter "M". Thus, 10 M rebar has a designation number of 10 and a diameter of 11.3 mm. Similarly, bar mark with letter or prefix "S" denotes stainless steel bar, and letter "C" denotes coated steel bar.

Table 7.8 ASTM standard reinforcing bars [25]

ASTM designation	Type	Metric grades	Metric sizes (mm)	US customary grades	US customary sizes (in.)
A 615M/A615	Billet	300	10–19	40	3–6
		420	10–57	60	3–11, 14, 18
		520	19–57	75	6–11, 14, 18
A 706M/A706	Low alloy	420	10–57	60	3–11, 14, 18
A 996M/A996	Rail	350	10–25	50	3–8
		420	10–25	60	3–8
A 996M/A996	Axle	300	10–25	40	3–8
		420	10–25	60	3–8

Notes: The four metric strength grades of steel, 300, 350, 420, and 520, have minimum yield strengths of 300, 350, 420, and 520 MPa, respectively. However US customary grades 40, 50, 60, and 75 have minimum yield strengths of 40,000, 50,000, 60,000, and 75,000 psi

Rail-steel bars are manufactured by rolling used railroad rails, whereas axle-steel bars are manufactured by rolling used railroad car axles

The deformed bars carry symbols such as S- for billet steel, W- for low-alloy steel, R- for rail steel, and A- for axle steel

The bar number for metric bar sizes denotes the approximate diameter of the bar in millimetres. For example, a No. 13 bar is about 13 mm in diameter (actually 12.7 mm). US customary bar sizes No. 3 through No. 8 have similar designations, the bar number denoting the approximate diameter in eighths of an inch (for example, a No. 5 bar is about $\frac{5}{8}$ in. in diameter)

Table 7.9 Properties of deformed reinforcing bars as per CSA

Bar designation	Nominal dimensions			Mass per unit Length (kg/m)
	Diameter (mm)	Area (mm²)	Perimeter (mm)	
10 M	11.3	100	35.5	0.785
15 M	16.0	200	50.1	1.570
20 M	19.5	300	61.3	2.355
25 M	25.2	500	79.2	3.925
30 M	29.9	700	93.9	5.495
35 M	35.7	1000	112.2	7.850
45 M	43.7	1500	137.3	11.775
55 M	56.4	2500	177.2	19.625

Table 7.10 Deformed reinforcement bar grades as per BS-4449

Standard	Grade	Yield strength (MPa)
BS 4449: 1997	460A	460
BS 4449: 1997	460B	460
BS 4449: 2005	B500A	500
BS 4449: 2005	B500B	500
BS 4449: 2005	B500C	500

However, reinforcing bar grades as per BS-4449 [27] are provided in Table 7.10. BS-4449 has modified the grades in 2005 from 460 to 500 as shown in the table.

Other Reinforcing steel: Some of the other types of reinforcing bars used in concrete for various applications are epoxy-coated reinforcing steel, stainless steel, hot-dipped galvanised reinforcing steel, fibre-reinforced polymer (FRP) bars, and welded wire fabric (WWF). WWF is a prefabricated reinforcement consisting of parallel series of high-strength, cold-drawn, or cold-rolled wire welded together in square or rectangular grids.

7.6.1 Reasons and Effects of Incorrect Placement of Steel

Properly placed reinforcement in concrete improves its compressive and tensile strength. Use of steel reinforcement needs proper knowledge as it may have an adverse effect on durability and performance of concrete, if not used with care and understanding. Incorrect placement of steel can result in insufficient cover, leading to corrosion of the reinforcement. If the bars are placed grossly out of position or in the wrong position, collapse can occur, when the element is fully loaded.

There are many reasons to control the proper placement of reinforcing steel in structures. Incorrect reinforcing steel placement can and has led to serious concrete structural failures. The common cases are cantilevered slabs or beams (Fig. 7.9)

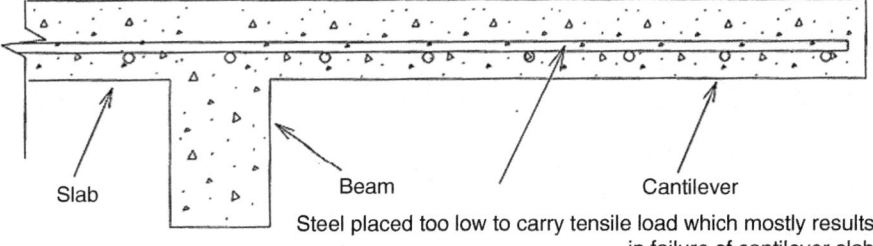

Slab   Beam   Cantilever

Steel placed too low to carry tensile load which mostly results
in failure of cantilever slab

Fig. 7.9 Improper steel placement in cantilever (reprinted with permission from Concrete Repair and Maintenance)

Fig. 7.10 Improper steel placement in column (reprinted with permission from Concrete Repair and Maintenance)

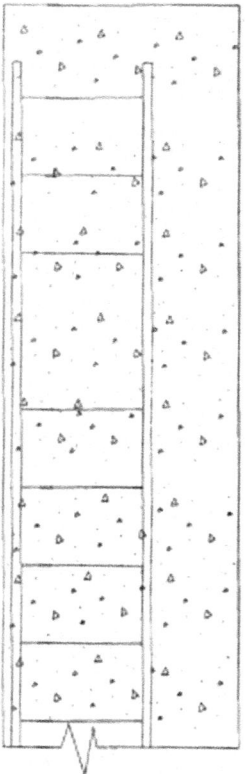

[16]. In such cases, top of the slab or beam is in tension and will pull apart or crack if there is no reinforcing steel near the top of the beam.

The second reason for proper placement of steel is to have adequate concrete cover to steel so as to protect it from corrosion. If the concrete cover is inadequate, it will not provide the necessary long-term protection. Shifted reinforcing bar cages in columns, walls, or beams (Fig. 7.10) are the common examples in this case. Good workmanship is required to ensure that the reinforcement is properly placed and that the specified cover is obtained. For adequate concrete cover, Tables 7.4, 7.5, and 7.6 can be referenced.

Additionally, concrete bonds to steel, and both expand and contract to about the same degree with temperature changes because the coefficients of linear expansion of concrete and steel are nearly same. The good bond between concrete and steel allows an effective transfer of stress or load between the steel and concrete so both materials act together in resisting beam action. For these reasons, steel is the most common material used to reinforce concrete. Moreover, reinforcement is also used to resist shear in beams mostly in the form of stirrups.

7.6.2 Protective Measures

The steel should be inspected before concrete is placed. Bars and wire fabric (mesh) should not be kinked or have unspecified curvatures when positioned. Kinked and curved bars, including those displaced by workers walking on them, may cause the hardened concrete to crack, when the bars are tensioned by service loads. The inspector must ensure that reinforcing steel is placed in its specified location with a proper spacing and is of the correct grade, type, size, shape, and length.

The congested reinforcement should always be avoided. Congestion causes problems when the clear spacing between reinforcing bars or between a bar and the form is less than $1\frac{1}{3}$ times the maximum size of coarse aggregate used in the concrete mix [28]. This condition usually occurs at links and bends in reinforcement and at beam-column connection. Sometimes, congestion is caused by multiple layers of reinforcement. To minimise congestion issues, spacing between reinforcements, embedment, and form must be at least $1\frac{1}{3}$ times the nominal maximum size of the coarse aggregate. Moreover, the clear distance between multiple layers of reinforcements, particularly in beams, should be at least 1 in. (25 mm). The upper layer bars should be positioned directly above the corresponding bars below. The intention of doing this is to provide enough space for all the concrete mix to pass between the bars so that it could be properly compacted.

Additionally, flowing concrete can be used in congested areas where the member itself is unusual in shape or size or a large amount of reinforcement is present. The use of flowing concrete is discussed in Chap. 1.

The top reinforcement in slabs shall be rigidly supported by mild steel chairs, from the bottom reinforcement. Plastic coated or galvanised steel chairs shall be used where the reinforcement is in contact with exposed concrete surfaces. Chair spacing shall be at 1.0 m centres in both directions or less if required for working loads. At all intersections the bars should be securely tied together with 16-gauge flexible iron wire (galvanised). The ends of the wire ties should not project towards the face of the concrete.

In order to control further the proper placement of steel reinforcement, the construction engineer must also be familiar with the anchorage bond, hooks and bends, and laps and joints.

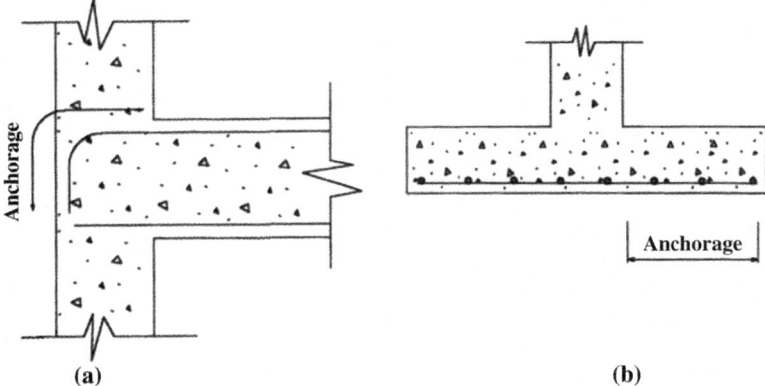

Fig. 7.11 (**a**) Beam to external column; (**b**) column base

7.6.3 Anchorage Bond

Anchorage is the embedment of a bar in concrete so that it can carry load through bond between the steel and concrete. In order that the bar is not pulled out when subjected to tensile stress, the bar has to have sufficient surface area in contact with the concrete. Therefore, a certain minimum length must be embedded to develop a force in the bar. Based on BS-8110 [19], the calculated anchorage lengths in tension and compression for a grade 460 deformed bar in grade 30 concrete are 37 × dia. of bar and 29 × dia. of bar, respectively. Anchorages for bars in a beam to external column joints, and in a column base past the face of column, are shown in Fig. 7.11.

However, for the anchorage of bars at supports, it is required that each tension bar should extend a length equal to 12 times the bar size beyond the centre line of the support or 12 times the bar size plus half of effective depth from the face of the support.

7.6.4 Hooks and Bends

Hooks and bends are used to shorten the length required for anchorage, or where it is not possible to use straight bars due to limitations of space. They increase the grip and provide good bond against slipping in the concrete. These effective anchorage lengths in two common forms of hooks and bends mostly used under Canadian and US standards are as follows (see Fig. 7.12):

1. 180 hook or a semicircular bend plus an extension of at least four bar diameters but not less than 60 mm at the free end of the bar
2. 90 bend plus an extension of at least 12 bar diameters at the free end of the bar

The diameter (D) of bar for 90° and 180° hooks can be taken as follows [29]:

Fig. 7.12 Standard hooks
and bends

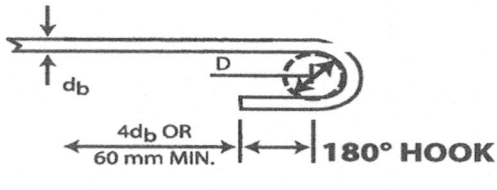

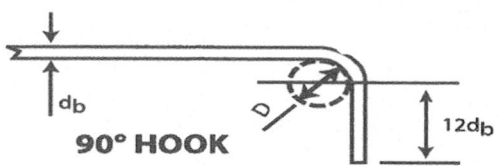

1. Grade 400 uncoated bars:

   (a) $D = 6\ d_b$ for 10 M to 25 M
   (b) $D = 8\ d_b$ for 30 M to 35 M
   (c) $D = 10\ d_b$ for 45 M to 55 M

2. Epoxy-coated bars:

   (a) $D = 8\ d_b$ for 10 M to 30 M
   (b) $D = 10\ d_b$ for 35 M to 55 M

In simply supported beams, all bars running through in the lower flange must extend to the centre of the support before hooks or bends begin. In case of continuous beams, these bars must extend for 0.10 L (effective length) beyond the centre of support, before hook begins.

The hooks will not be effective unless the concrete is thoroughly consolidated all around it, with a minimum cover of about 25 mm (1 in.) on sides and 40 mm (1½ in.) on top. Hooks should not be too close to the free surface of the concrete, as they have a tendency to straighten out under stress and may burst off a thin concrete cover.

The congestion of reinforcement at ends or junctions of heavy beams should be avoided by providing bends, instead of hooks, or by avoiding hook or bend for a bar in compression, except on outer face of concrete. Moreover, distribution bars should not be hooked or bent, as hooks will tend to localise the cracks and make the distribution steel non-effective [30].

7.6.5   Rebar Lap Splicing

The lap splice is created by overlapping two lengths of rebar, and then wiring them together. In order to transfer stresses from one bar to another, the lengths of reinforcing bars are spliced. The other reason for bar splicing is that the limitations on the length of reinforcing steel bars, due to fabrication, transportation, and constructability restraints, make it impossible to place continuous bars in one piece throughout the structure. A sufficient lap length is needed to adequately transfer loads between the bars. Lap lengths can be longer than specified, but never shorter. Inadequate lap length can cause severe cracking in the concrete around the lap.

The lengths of reinforcing bars can be joined by lapping, mechanical couplers, or butt or lap welded joints. Only lapping is discussed here. The minimum lap length should not be less than 15 times the bar diameter or 300 mm. However, adequate requirements for tension laps as per BS-8110 [19] are the following:

(a) The lap length should not be less than the tension anchorage length, i.e. 37 × dia. of bar.
(b) If the lap is at the top of the section and the cover is less than two bar diameters, the top length is to be increased by 1.4.
(c) If the lap is at the corner of a section and the cover is less than two bar diameters, the lap length is to be increased by 1.4.
(d) If conditions "b" and "c" both apply, the lap length is to be doubled.

The length of compression laps as per BS-8110 should be 1.25 times the length of compression anchorage, that is, 1.25 (29 × dia. of bar). All laps should, wherever possible, be located at points where the tensile or compressive forces in the bars are low, and they should preferably be staggered.

CSA-A23.3-04 [31] and ACI 318 [3] specify that the minimum length of lap for tension lap splices for Class A or B splices shall be as follows (but not less than 300 mm):

1. Class A splice length = 1.0 λ_d
2. Class B splice length = 1.3 λ_d

Where λ_d is the tensile development length for the specified yield strength. For values of λ_d Table 3.1 under Part II of CSA A23.3-04 can be referenced.

Since Class B laps are 30% longer than Class A splices, mostly Class B splices are considered except in certain instances.

The compression and tension splices (based on concrete strength) recommended by OCCDC (Ontario Cast-in-Place Concrete Development Council) [29] are provided in Table 7.11.

7.7 Poor Curing

The setting of cement in concrete takes place on account of hydration. Hydration is the heat concrete generates as it hardens because of the chemical reaction of the cement and water. The process of hydration continues for a long period but its rate decreases later on. The mixing water is usually sufficient for initial hydration but most of the quantity of water is lost by evaporation due to high concrete temperatures, high air temperature, high wind velocity, and low humidity, alone or in

Table 7.11 Compression and tension lap lengths [29]

Bar size	Compression splice mm (in.)	Tension splice (Class B) mm (in.) Concrete strength (MPa)					
		20	25	30	35	40	45
10 M	330 (13)	473 (19)	423 (17)	390 (15)	390 (15)	390 (15)	390 (15)
15 M	470 (19)	670 (26)	599 (24)	547 (22)	506 (20)	474 (19)	446 (18)
20 M	570 (23)	816 (32)	730 (29)	666 (26)	617 (24)	577 (23)	544 (21)
25 M	740 (29)	1319 (52)	1179 (46)	1077 (42)	997 (39)	932 (37)	879 (35)
30 M	870 (34)	1564 (62)	1399 (55)	1277 (50)	1183 (47)	1106 (41)	1042 (41)
35 M	1050 (41)	1868 (74)	1671 (66)	1525 (60)	1412 (56)	1321 (52)	1245 (49)

Note: Multiply above values by 1.5 for epoxy-coated bars, 1.3 for top reinforcement, 1.7 for epoxy top reinforcement

combination, at the surface and the heat generated during setting. Hence, to prevent this loss of water, curing is required. Concrete must be kept moist till it gains sufficient strength.

Well-cured concrete has better surface hardness and will better withstand surface wear and abrasion. Concrete also becomes more watertight with better curing preventing moisture and chemicals from entering into the concrete, thereby increasing durability and service life of the concrete.

The most common methods of curing are as follows:

1. *By Continuous Ponding or Frequent Applications of Water to Concrete Surface*
   This method includes continuous ponding of finished concrete surface in water. The water curing can also be done by frequent spraying of water or covering the concrete surface constantly with wetted gunny bags (burlap), sand, earth, straw, or sawdust.
2. *Membrane Curing of the Concrete*
   Liquid membrane-forming curing compounds are provided to prevent excessive loss of water from the concrete by evaporation. These compounds are usually made from waxes, natural and synthetic resin, and solvents of high volatility at atmospheric temperature. White or grey pigments are mostly added to these compounds so as to provide heat reflection and to make the compound visible. For application purposes, manufacturer's instructions must be followed.
3. *Covering the Surface with an Impermeable Material*
   The concept of covering the fresh concrete surface with impermeable materials is the same as that of liquid membrane, i.e. to prevent loss of water by evaporation. These impermeable materials usually consist of polyethylene sheets or fibre-reinforced paper. After applying water to concrete, the polyethylene sheet is placed to cover the fresh concrete as soon as possible.
4. *Steam Curing*
   In this method, steam is used as a source of curing to increase the curing temperature of concrete that increases its rate of development of strength. The basic

motto of steam curing is to get a sufficiently high-early strength so that the concrete products, particularly the precast products, may be handled soon after casting.

All these methods are further explained in the following sections. However, curing in special conditions like hot and cold weather is also addressed in Chap. 11.

7.7.1 Effects of Poor Curing

Like batching, and mixing and placing, curing operation is also of great importance and requires close supervision, because it is extremely difficult to prove that proper curing had not been applied except laboratory testing. A good concrete after placing becomes good in service, if curing is properly achieved, whereas the same good concrete can result in low strength, if curing is inadequate.

In general, it has been observed that less importance is given to concrete curing and usually it is terminated within 3 or 4 days, which is not a good practice. Lack of curing will increase the degree of cracking within a concrete structure. The early termination of curing will allow for increased shrinkage at a time when the concrete has low strength. The lack of hydration of cement due to drying will result not only in decreased long-term strength, but also in the reduced durability of the structure.

The hardened cement paste has a porous structure, the pore sizes varying from very small to very large, and is called gel pores and capillary pores. The necessity for curing arises from the fact that the hydration of cement can take place only and when these capillary pores remain saturated. This is why loss of water by evaporation from the capillaries must be prevented. Furthermore, water lost internally by self-desiccation has to be replaced by water from outside. Thus, when concrete is continuously wet-cured for the required time, the capillary system is always full of water and hydration proceeds uninterrupted.

Figure 7.13, taken from ACI, demonstrates the effects of wet curing on the strength development of concrete specimens removed from moist curing at various ages and subsequently exposed to laboratory air [32]. It can be seen that early transfer from a moist atmosphere to a dry one detrimentally affects the strength of concrete. Moreover, as the specimens dry out, further strength gain is almost nil.

7.7.2 Protective Measures

In order to obtain good-quality and durable concrete, the placing of an appropriate mix must be followed by curing in a suitable environment, during the early stages of hardening. Curing and protection should start immediately after the compaction of the concrete to protect it from freezing, abnormally high temperatures or temperature differentials, premature drying, excessive moisture, and moisture loss for the period of time necessary to develop the desired properties of the concrete.

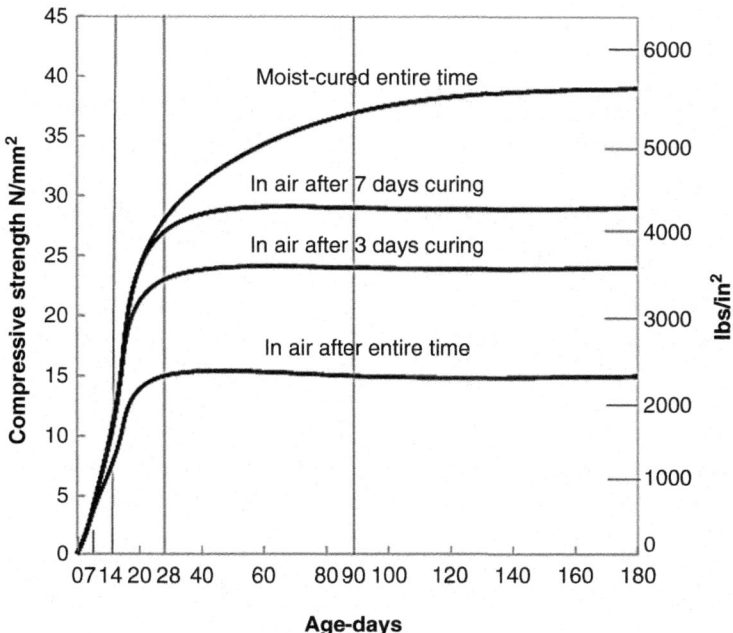

Fig. 7.13 Effects of continuous and partial moist curing on the strength of concrete with a water/ cement ratio of 0.5, taken from ACI-306R-88 [32] (Price 1950)

The protective measures based on common curing procedures are briefed as follows:

Ponding and sprinkling with water: Under hot weather conditions, the high temperatures are likely to result in excessive moisture loss. Maintaining mixing water in the concrete is the major concern. Soon after concrete is placed, it should be protected from direct rays of sun and wind velocity by covering with hanging curtains on vertical surfaces, and on flat surfaces, by covering with polythene sheets or gunny bags (burlap). After the concrete has begun to harden, it should be protected from quick drying with moist gunny bags. Once the concrete has set, the concrete surface should be cured by continuous spraying or ponding with water or covering the concrete with wetted burlaps. Columns and other small members can be cured by wrapping wet burlap around them. Water can be sprayed easily to keep these bags continuously wet.

Curing in cold weather is different as in that case the biggest concern will be the maintaining of an adequate and conducive temperature for hydration. For massive members, the heat generated by the concrete during hydration will be adequate to provide a satisfactory curing temperature. For non-massive members, a good alternative is steam curing, which provides both moisture and heat. In any case, the concrete temperature shall be maintained at no less than 10 °C throughout the curing period as recommended by CSA-A23.1-14 [7] because the rate of compressive strength gain in concrete reduces significantly below 10 °C. Where moist curing is not done, very low temperatures may be avoided by insulating the member appropriately. Heaters are also used under the insulated members (tarps) after ponding the concrete surface.

Delay form removal: Evaporation of water from vertical members like beams and columns is largely prevented by the forms, which should be left in place as long as possible by periodically spraying them with water. However, recently the use of controlled permeability formwork (CPF) liner has proved very helpful in enhancing concrete curing. CPF aids concrete curing in two ways. In the critical period between pouring and stripping of the formwork, the water-rich CPF liner acts like a curing membrane. Upon stripping, the reduced porosity of the surface means that there is no rapid moisture loss, as can occur with conventionally cast surfaces [22]. The benefits of CPF are also addressed in Sect. 7.5.2.

Curing Compound: While using membrane curing, the concrete is coated with a sealing compound, which is usually done by spraying. The curing membrane serves as a physical barrier to prevent loss of moisture from the concrete to be cured. Membrane curing may not assure full hydration like moist curing, but is adequate and particularly suitable for concrete members in contact with soil. The coating should be applied at a uniform rate. Usually two applications at right angles to each other are suggested for complete coverage. The coating must be protected against damage.

The curing compound shall preferably be water-based membrane forming and of a type approved by the engineer and be applied as directed by the manufacturer. When curing compound is used, the exposed concrete shall be thoroughly sealed immediately after the free water has left the surface. Formed surfaces shall be sealed immediately after the forms are removed and necessary finishing has been completed. The curing compound shall be applied by power-operated atomising spray equipment in one or two separate applications. Hand-operated sprayers may be used for coating small areas.

Curing compound solutions containing pigment shall be thoroughly mixed prior to use and agitated during application. If the solution is applied in two applications, the second application shall follow the first within 30 min. Satisfactory equipment shall be provided, together with means to properly control and assure the direct application of the curing solution on the concrete surface so as to result in a uniform coverage rate of at least 0.27 L/m^2. If rain falls on the newly coated concrete before the film has dried sufficiently to resist damage, or if the film is damaged in any other manner during the curing period, a new coat of solution shall be applied to the affected portions equal in curing value to that specified above.

It has been observed [33] that if the water/cement ratio is greater than about 0.42, membrane curing is adequate. However, if water/cement ratio is smaller than 0.42, autogenous shrinkage will develop rapidly even if the curing membrane has been applied. In fact, a membrane will prevent evaporation of water from the concrete but will not allow ingress of water to cover the water loss by self-desiccation. When the membrane ceases to be effective, drying shrinkage will also develop. It is only at water/cement ratio in excess of about 0.5 that membrane curing is fully satisfactory.

Covering sheets: While using coverings like waterproof/reinforced paper or plastic sheeting, as a means of curing with the aim to reduce the loss of water from the concrete by evaporation, care must be taken that the sheets are sealed airtight, and corners and edges are adequately protected against loss of moisture. Coverings can be placed as soon as the concrete has been finished. In order to avoid tearing of plastic sheeting, a minimum thickness is required to ensure adequate strength in the sheet; ASTM C 171 [34] recommends to use at least 4.0 mm thick sheet for curing concrete. The plastic sheet reinforced with glass or other fibres and waterproof paper may also be used, as they are more durable. Waterproof paper generally consists of two layers of kraft paper cemented together and reinforced with fibre.

Various types of insulated tarps are also available in the form as concrete curing blankets that are built to protect concrete moisture loss in very cold weather. Using the toughest woven polyethylene fabric for correct insulation and extended longevity, these concrete curing blankets are manufactured with flexible foam that provides the highest possible R-value. Fresh concrete has its own heat: heat of hydration. In many circumstances, if properly trapped and insulated, this heat will be generated in sufficient quantity to be the only source of heat necessary during a 3-day curing cycle while insulating it from the environment and ambient air temperature.

Steam curing: While using steam curing method, a rapid rise in temperature at the time of setting of concrete must not be permitted as it would be detrimental to the concrete because the green/fresh concrete is too weak to resist the air pressure set up in the pores by the increased temperature. A delay in the application of steam curing by 3–5 h (in which normal moist curing could be done) is thus advisable, and then the temperature can be raised at the rate of 22–33 °C (72–91 °F) per hour [2]. The adverse effect is more with higher w/c ratio of the concrete mix and with rapid hardening than with ordinary Portland cement. Usually, mixes with low w/c ratio respond more favourably to steam curing than mixes with higher w/c ratio.

7.7.3 Curing Period

As regards period of curing, ACI-308 [35] suggests that if the temperature is above 10 °C (50 °F), then curing should be carried on minimum for 3 days for rapid-hardening Portland (type III) cement, a minimum of 7 days for ordinary Portland (type I) cement, and a minimum of 14 days for low-heat Portland (type IV) or modified (type II) cement.

For basic curing (curing type 1), CSA-A23.1-14 [7] suggests 3-day curing at ≥10 °C or for the time necessary to attain 40% of the specified strength and for additional curing (curing type 2) for abrasion and air pollution resistance and extended wet curing (curing type 3); CSA suggests 7-day curing at ≥10 °C or for the time necessary to attain 70% of the specified strength. If silica fume concrete is being used, then additional curing procedures should be adopted. CSA further suggests that concrete be allowed to air-dry for a period of at least 1 month after the end of the curing period, before exposure to deicing chemicals.

Since adequate curing is rarely achieved on construction sites, most of the professionals/experts suggest that item of curing should be detached from the general item of concrete placing, and paid separately as a line item. This change no doubt will automatically improve the curing implementation meeting required standards.

7.8 Inadequate Formwork

The formwork generally forms a part of concrete construction practice, as it influences the performance of hardened concrete appreciably. Formwork may be built from a number of materials including steel, aluminium, and fiberglass; however the most common formwork material is plywood sheathing supported by wood studs, walers, joists, and stringers. OPSS-919 [36] specifies that plywood for formwork shall be 7 ply, 17 mm minimum thickness exterior grade Douglas Fir plywood according to CSA 0121-08 (R-2013).

The formwork must be designed and built accurately, so that the desired shape, size, position, and finish of cast concrete are obtained. It should also be designed to withstand the pressure resulting from placement and vibration of concrete, and to maintain specified tolerance.

7.8.1 Effects of Inadequate Formwork

Inadequate formwork supports result in settlement and cracking of the concrete before it has developed sufficient strength to support its own weight. Grout leakage occurs where formwork joints do not fit together properly. The result is a porous area of concrete that has little or no cement or fine aggregate.

Non-rigid joints in the formwork will result in bulging, twisting, or sagging due to dead or live loads. Excessive deformations may disfigure the surface of the concrete. Too much release oil causes concrete surface to be oily and splotchy, whereas lack of release oil causes fast deterioration of formwork surfaces. Too dry timber causes adhesion to the concrete surface.

7.8.2 Protective Measures

Forms must be supported on false work of adequate strength and sufficient rigidity to keep deflections within acceptable limits. The vertical supports or shores/props should be installed plumb and with adequate bearing and bracing. However, inclined shores must be braced securely against slipping or sliding. OPSS-919 [36] provides that studs and joists shall be spaced not more than 400 mm on centres. Edges of abutting sheets shall be nailed to the same stud or joist with 50 mm nails at not more than 200 mm centres. The bearing ends of shores should be square. All formwork joints must be properly sealed so that mortar/cement slurry should not leak out during concreting and vibration.

The formwork should be capable of being dismantled and removed from the cast concrete without shock, disturbance, or damage, as the sudden removal of wedges is equivalent to an impact load on the partially hardened concrete. Preferably, the formwork should be chamfered at the junctions to facilitate its free and easy withdrawal. If nails are used, they should be driven until they are in the concrete surface and their heads should be slightly projected outside for easy removal. The nails should preferably be driven at an angle both ways.

Form surfaces should be cleaned before concreting. They should be treated with suitable non-staining form release oil or other coating that will prevent the concrete from sticking to them. The oil or coating should be brushed or sprayed evenly over the forms. It should not be permitted to get on construction joint surfaces or reinforcing bars, because it will interfere with bond. Where form ties have to pass through the concrete, they should be as small in cross section as possible, because the holes they form sometimes have to be plugged to stop leaks. End of form ties should be removed without spalling adjacent concrete.

Supporting forms and shores must not be removed from beams, slabs, and walls until these structural units are strong enough to carry their own weight and any approved superimposed load. In no case should supporting forms and shores be removed from horizontal members before concrete strength is at least 70% of design strength. OPSS-919 [36] requires that formwork shall not be removed until the concrete has attained a minimum strength of 20 MPa; however, formwork for cast-in-place barrier walls and parapet walls may be removed 24 h after completion of the placement.

The minimum period for striking formwork given in Table 7.12 shall be followed when ordinary or sulphate-resisting cements are used. When the forms are removed before the specified curing is completed, measures should be taken to continue the curing and provide adequate thermal protection for the concrete. However, it may be noted that in case of porches, sunshades, or cantilevers, removal of formwork is not based only on the requirement of time period, but before removal of props it may be

Table 7.12 Minimum period requirements for striking formwork

Type of formwork	Minimum period in days (24 h)	
	Cold weather (number of days)	Normal weather (number of days)
Sides of beams, columns, walls, and foundations	4	2
Soffits and props/supports of slabs	14	10
Soffits and props/supports of beams	18	14

ensured that load taken by these elements (due to their own weight) is counter-balanced at the support by casting proposed slab or erecting wall. Many incidents of collapse have happened due to this carelessness.

For fair finish concrete, cardboard or steel cylinder forms are widely used. Usually, columns are cast prior to the placement of the slab/beam formwork. If the column is cast little higher than the required height, it will penetrate the slab/beam concrete, due to which critical shear stresses may occur because of inadequate shear capacity area between the column and the slab/beam. The smooth form surface may not provide adequate shear transfer, and may result in potential punching shear failure. Hence, exact height of column or little less should be cast to avoid such a problem. The balance height of column is always cast with beam later on. The problem is illustrated in Fig. 7.14.

Where a slab load is supported on one side of the beam only (see Fig. 7.15), edge beam forms should be carefully planned to prevent tipping of the beam due to unequal loading.

In multi-storied construction, supports/shores should be located in the same position on each floor so that they will be continuous in their support from floor to floor. Improper positioning of shores from floor to floor may create bending stresses for which the slab was not designed (see Fig. 7.16).

Building materials including concrete must not be dropped or piled on the form-work in such a manner as to damage or overload it. During and after concreting, but before the initial set of the concrete, the elevation, camber, and plumpness of form-work systems should be checked with the help of proper instruments. Formwork must be continuously watched during concreting so that any corrective measures found necessary may be promptly taken.

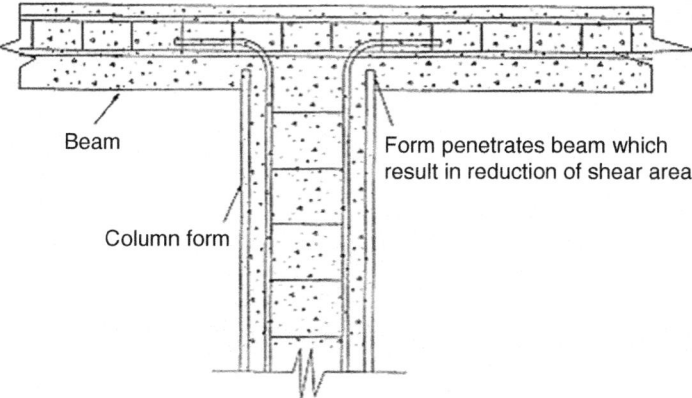

Fig. 7.14 Improper column form placement (reprinted with permission from Concrete Repair and Maintenance) [16]

Fig. 7.15 Bracing
arrangement for edge beam
(based on ACI-347R-94)
[37]

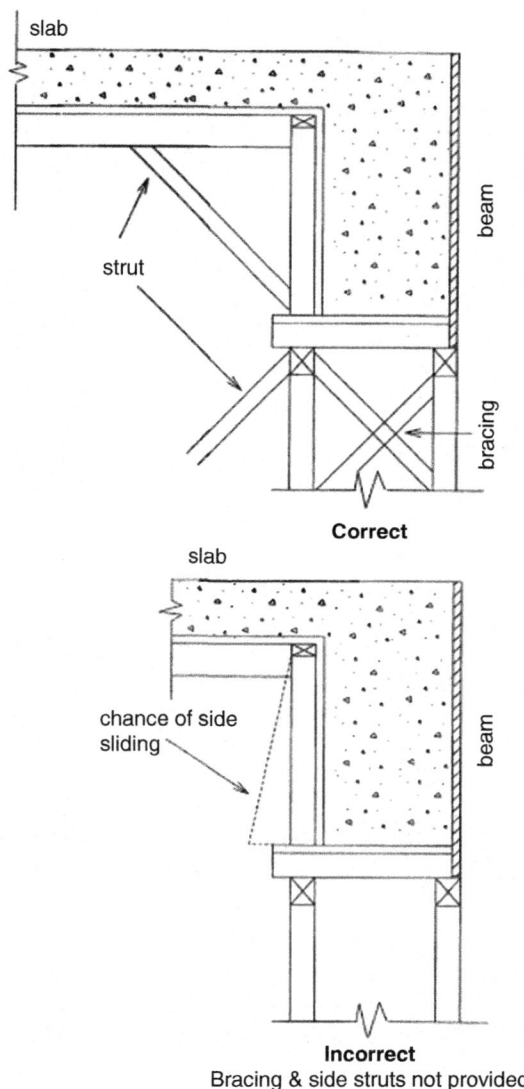

7.9 Incorrect Placement of Construction Joints

Joints are necessary in concrete structures for a variety of reasons. Larger concrete pours are usually planned to be placed continuously; hence construction joints are needed to have a break in concreting for some time. Since concrete undergoes volume changes, principally related to shrinkage and temperature changes, it can be desirable to provide joints and thus relieve tensile or compressive stresses that would be induced in the structure.

Fig. 7.16 Improper arrangement of re-shores (based on ACI-347R-94) [37]

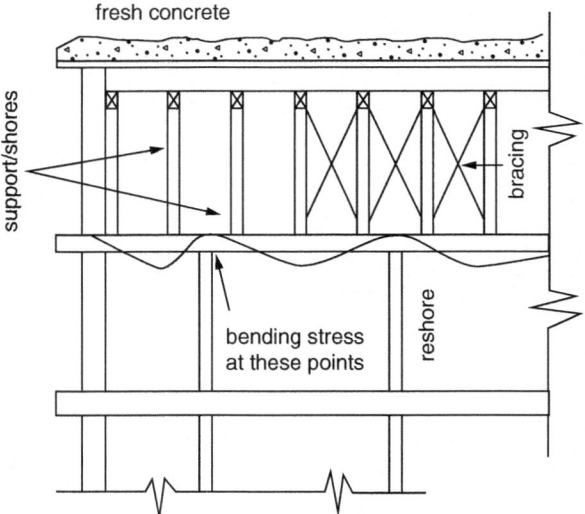

Correctly located and properly executed construction joints provide limits for successive concrete problems, without affecting the structure. The location of construction joints must be carefully considered and agreed before concrete is placed. They should normally be at right angles to the general direction of the member, and their number should be kept to the minimum for execution of the work.

7.9.1 Effects of Incorrect Placement of Construction Joints

The placement of construction joints at improper locations like points of high stress can result in the joints opening at these points, which lead to entry of water, resulting in the corrosion of steel reinforcement, and further deterioration of concrete. Particular care should be taken in placing of new concrete close to the joint to ensure that it is fully compacted; otherwise this will also lead to a permeable and porous concrete.

7.9.2 Protective Measures

The position of construction joints should be such that the strength of the member is not affected. Joints in beams and girders should be located at points of minimum shear, i.e. midway between supports, with a vertical plane at right angles to the direction of the main reinforcement. Where a beam intersects a girder, ACI-318 [3] requires that the construction joint in the girder should be offset a distance equal to twice the width of the incident beam. However, in L-beams and T-beams, the ribs should normally be concreted together with the floor slab of which they form a part. In case of T-beams with continuous slabs, having stirrups for shear reinforcement, concreting may be left after completing the rib portion, provided that the stirrups project from the rib almost to the top of the slab. The slab must be cast over the rib within 24 h.

In slabs, which are not continuous, the joint may be left at the middle of span making it vertical and at right angles to the direction of span. However, in case of continuous slabs, concreting can be stopped directly over the centres of the beams making a vertical joint and allowing for the future adjoining slab. In slabs having reinforcement in two directions (two-way slab), concreting may be left somewhere near the middle of the slab like noncontinuous slab, or within the one-third span on either side.

Construction joints in columns and bearing walls should be located at the underside of floor slabs, beams, or girders and at the top of footings and floor slabs. Beams, girders, or slabs supported by columns or walls shall not be cast or erected until concrete in the vertical support members is no longer plastic and has been in place at least 1 h [18]. Delay in placing concrete in members supported by columns and walls is necessary to prevent cracking at the interface of the slab and supporting member, caused by bleeding and settlement of plastic concrete in the supporting member. In no case work be terminated in beams or slabs where shearing action will be great, e.g. near the ends or directly under a concentrated load.

The vertical construction joints in walls are often provided near re-entrant corners of walls, beside columns or other suitable locations. Vertical construction joints at or near the corner of the building should be avoided so that the corner should be tied together adequately.

It is always preferable to have construction joints either all horizontal or all vertical. In some instances, long lengths of walls or slab are constructed in alternate lengths, so that when intermediate pours are made later, the older concrete in earlier pours will have already taken up some of the shrinkage movements.

In road slabs it is always planned that the works should be broken off only at an expansion joint or at a contraction or warping joint. However, if due to certain reasons it is necessary to form a construction joint, a cross-form should be placed in position at the appropriate point. If possible, construction joints should be located not less than 5 ft. (1.5 m) from any other joint to which they are parallel [38]. The joint should be tied together either by continuing the reinforcement across the joint or by installing tie bars; the latter one is the simplest method, and on a large job a suitable cross-form and tie bars should be kept in readiness.

Construction joints should be provided with a groove for sealing against the ingress of water or grit. This is always advisable as a slight opening of the joint may occur, even with tie bars.

High-quality workmanship is necessary when forming the joints to ensure that the load-bearing capacity of the concrete in the area of the joint is not impaired. Various types of joints are used for different situations. In the most common types, grooves are formed in the first pour so that the next pour is keyed into it. The grooves are formed by timbers, which are either set in the exposed surface of the concrete or attached to the vertical formwork. Keyed joints are not recommended for construction joints subjected to hard-wheel traffic. In reinforced concrete, the aim must be to ensure reinforcement continuity with good bonding between the new concrete and old. Laitance must always be removed from old surface to expose coarse aggregate and a sound irregular concrete surface. The old surface may need to be slightly wetted prior to the new concrete being placed, to prevent excessive loss of mix water into it by absorption. To increase bond, a thin layer of cement grout or bonding agent may also be applied. If well constructed, such joints are completely watertight.

However, in water-retaining structures, water stops/water bars are often installed across construction joints. The water stops should be fixed so as to ensure that they are not displaced from their intended position during compaction of the concrete. Joints in water stop material should be as watertight as the continuous material. Great care is needed when placing concrete around water bars because the space is often congested. If the concrete is not properly compacted and is honeycombed, water can pass round the water bar and the required results will not be achieved. Insufficient care in placing concrete may even displace the water bars. CSA-A23.1-14 [7] suggests that where construction joints are specified in watertight construction, all specified layers of reinforcement shall be continuous across the joint unless otherwise detailed on the drawings.

7.10 Inadequate Mixing

Concrete is usually mixed by mechanical mixers; however, sometimes hand mixing on small jobs is also adopted. The object of mixing the required quantities of cement, aggregate, and water is to coat the surface of all the aggregate particles with cement paste, and blend all of them into a uniform mass. A proper and accurate measurement (batching) of all the materials used in the mixer to produce concrete is essential to ensure uniformity of proportions and aggregate grading in successive batches. In general, batching by weight is more reliable than batching by volume.

7.10.1 Mechanical Mixing

The main types of mechanical mixers are batch mixers and continuous mixers. The batch mixers are commonly used, in which one batch of concrete is mixed and discharged before any more materials are put into the mixer. However, continuous mixers are used on major works and produce a continuous flow of concrete. Satisfactorily designed mixers have a blade arrangement and drum shape which ensure an end-to-end exchange of materials parallel to the axis of rotation, or a rolling, folding, and spreading movement of the batch over itself as it is being mixed.

The common mechanical mixers are as follows:

7.10.1.1   Batch Mixers

Batch mixers are usually of two types based on the orientation of the axis of rotation: horizontal or inclined drum mixers or vertical pan mixers. The drum mixers have a drum, with fixed blades, rotating around its axis, whereas the pan mixers have either the blades or the pan rotating around the axis.

Drum Mixers

All the drum mixers have a drum with blades attached to the inside of the drum. The main purpose of the drum mixer is to mix the ingredients while rotating. Parameters that need control are the rotating speed of the drum and the angle of inclination of the rotating axis. There are three types of the drum mixers: tilting drum mixer, non-tilting drum mixer, and reversing drum mixer.

1. *Tilting Drum Mixer*
   In this mixer, the drum is kept horizontal while loading the ingredients, and then tilted upwards during mixing and for discharging it is tilted diagonally downwards. This is the most common type of mixer used on most of the sites.
2. *Non-tilting Drum Mixer*
   In this mixer, the orientation of drum is fixed and the discharge takes place by inserting materials at one end and discharging at the other end of the drum.
3. *Reversing Drum Mixer*
   This is similar to non-tilting mixer, except that the same opening is used to add the materials and to discharge the concrete. The drum rotates in one direction for mixing and in the opposite direction for discharge.

Pan Mixers

1. *Pan Mixer*
   All pan mixers either fixed or rotating work on the same principle consisting of a circular pan, which is fitted internally with blades. Mixing is effected either by rotation of the blades and the pan or by rotation of the blades in a stationary pan. The concrete is discharged in the bottom of the pan, which is sealed by a door during the mixing process.

7.10.1.2   Continuous Mixers

Continuous mixers batch by volume or by weight, and when once they have been set the batching is done more or less automatically. Materials are continuously fed into the mixer at the same rate as the concrete is discharged. They are usually non-tilting drums with screw-type blades rotating in the middle of the drum. The drum is tilted downward towards the discharge opening.

In these types of mixers which use feeding screws to supply aggregate and cement to the mixer, the screws must be examined carefully for wear, which may tend to alter the mix proportions. Also, a watch must be kept to ensure that the cement and sand do not arch over the screws, and that there is a constant supply and a free flow of material.

7.10.1.3   Ready-Mix Concrete

Ready-mix concrete (RMC) is a type of concrete which is manufactured in a batching plant of a factory, based on a required set of proportions. It is then delivered to a construction site, by truck mounted with mixers. Ready-mix concrete is an extremely versatile and cost-effective product that has made it one of the most common building materials in the world. By adjusting the basic ingredients, and with the use of chemical additives, the characteristics and performance of concrete can be customised to meet a wide range of applications and construction requirements. Using a predetermined concrete mixture reduces flexibility, both in the supply chain and in the actual components of the concrete.

In the batch plant, the quantities of dry ingredients are measured by weight and added to a central mixer inside the plant or charged directly into the truck drum from overhead silos or conveyors. Water and liquid admixtures are measured by volume or by weight. Ingredients are added to the drum through a metal chute or hopper at the upper rear of the truck. Truck mixer drum capacities vary depending on the expected use of the vehicle and the production capacity of the plant.

Ready-mix concrete is mostly preferred over on-site concrete mixing because of the volume it can produce with precision of proportion of mixtures. Ready-mix concrete is bought and sold by volume such as cubic meters or cubic yards. Ready-mix concrete is manufactured under controlled operations and transported and placed at site using sophisticated equipment and methods.

There are three principal categories of ready-mix concrete:

1. *Transit-Mixed Concrete*

    In transit-mixed concrete, also called truck-mixed or dry-batched concrete, materials are batched at a central plant and are completely mixed in the truck in transit. Frequently, the concrete is partially mixed in transit and mixing is completed at the jobsite. Transit mixing keeps the water separate from the cement and aggregates and allows the concrete to be mixed immediately before placement at the construction site. This method avoids the problems of premature hardening and slump loss that result from potential delays in transportation or placement of central-mixed concrete. Additionally, transit mixing allows concrete to be hauled to construction sites further away from the plant. A disadvantage of transit-mixed concrete, however, is that the truck capacity is smaller than that of the same truck containing central-mixed concrete.

2. *Shrink-Mixed Concrete*

   In shrink-mixed concrete, concrete is partially mixed at the plant to reduce or shrink the volume of the mixture and mixing is completed during transit in truck or at the jobsite. Central-mixing plants that include a stationary, plant-mounted mixer are often actually used to shrink mix, or partially mix the concrete.

3. *Central-Mixed Concrete*

   Central-mixed concrete is completely mixed in a stationary mixer within the plant and then discharged from the plant mixer. It may be hauled in a ready-mix truck, operating as an agitator, or it may be an open-top truck body with or without an agitator. Each batch of ready-mix concrete is tailor-made according to the requirement of the buyers and is delivered to them in a plastic condition, usually in the ready-mix truck.

   The tendency of concrete to segregate limits the distance it may be hauled in vehicle not equipped with an agitator. As per ASTM-C94 [39], the total volume that a truck can handle is limited to 63% of the drum volume. It is especially important to keep the inside of the drum thoroughly clean by washing out after each load has been discharged.

7.10.2 Hand Mixing

Hand mixing should always be avoided because in this case it is very difficult to achieve uniformity. However, if it is permitted than hand mixing should be done on a clean, hard, and watertight platform with some upstand at all edges to prevent the materials being washed or shovelled off during mixing. First of all, aggregate (fine and coarse) should be spread in a uniform layer on the platform, then cement is spread over the aggregate, and dry materials are mixed by turning over properly about three times with a shovel until the mix appears uniform. In hand mixing, it is always suggested to use 10% more cement than the required quantity. Now water is gradually added so that neither water nor cement should escape. The mix is turned over again, for about three times, until it appears uniform in colour and consistency.

7.10.3 Adverse Effects of Inadequate Mixing and Possible Protective Measures

The good-quality concrete is not just a matter of mixing the constituent materials to obtain some kind of mass, but it is a scientific process which is based on some well-established principles and governs the properties of concrete mixes in fresh as well as in hardened state. The mixing should be carried out in such a way that the ingredients of concrete are mixed thoroughly so that the ingredients are uniformly distributed in the concrete mass. The uniformity must be maintained while discharging the concrete from the mixer.

Inadequate mixing results in production of non-uniform concrete and larger variations in strength and workability of concrete will occur. The concrete achieved in this way will be of poor quality, permeable, and low strength. In the operation of mixing, speed of revolution of the drum, size of the batch, mixing time, type and size of mixer, and method of loading the dry materials affect the uniformity of concrete.

7.10.3.1 Effect of Speed of Revolution of Mixer Drum

Tests have proved that if the speed of revolution of mixer drum is reduced to say 11 rev/min against the rated value of say 17 rev/min, the reduction in the cube strength of batch is around 5%. However, if the speed of the revolution of the mixer drum is allowed to fall up to say 7 rev/min, then the reduction in cube strength of batch is around 40% [9].

It is, therefore, very important that mixers should be properly maintained so that the speed of the drum is always correct. The rated speed of the mixer, specified by the manufacturers, should be strictly followed.

7.10.3.2 Effects of Size of Batch

If the size of batch is increased or decreased than the rated capacity of the drum, the strength of concrete decreases. The size of batch is mostly affected by poor discharge particularly in non-tilting mixers. The poor discharge has more effect when the batch is small than when it is large. Due to this poor discharge, while discharging, the material retained in the drum is usually mortar and the discharge concrete has, therefore, a high aggregate/cement ratio; hence it affects the uniformity of concrete. The variations in the cube strength that can occur as a result of overfilling or under-filling are found to be 25%, in comparison with those for the correct size of charge.

Overloading or under-loading the mixer should, therefore, be avoided as this will result in production of poor-quality and non-uniform mix. However, variation by ±10% is found harmless. The uniformity is also affected by the mix proportions: lean dry mixes with large size of coarse aggregate are more difficult to mix uniformly than rich and more workable concretes using ¾ in. (20 mm) aggregate.

7.10.3.3 Effect of Mixing Time

Mixing time should be sufficient to produce uniformly mixed concrete with the required slump. The minimum mixing time required to produce uniform concrete with reliable strength depends on the type and size of mixer, design of drum and

Table 7.13 Mixing time of three types of concrete mixer

Concrete quality	Type of mixer	Minimum time required			
		Charging	Mixing	Discharging	Total
Ordinary	Tilting	40	50–55	20	110–115
	Non-tilting	25–35	35–65	40	100–140
	Open-pan	30	15–35	20	65–85
High	Non-tilting	25–35	50–75	40	115–150
	Open-pan	30	30–45	20	80–95

(Crown © 1955; HM Stationary Office) [9]

blades, speed of revolution of mixing drum, and quality of blending of ingredients during charging of the mixer. Large mixers generally require longer mixing time than smaller ones. Tests have also proved that the mixer which needed 5 min for proper mixing could be adjusted to take half the time, i.e. 2½ min by removing one-third of the blades from the mixer.

Based on various tests made by Transport Research Laboratory, UK [9], the time required for mixing in various three common types of mixers is produced in Table 7.13. The data indicates that the open-pan mixers are likely to have the shortest mixing cycles (about 1½ min), tilting and non-tilting mixers both generally requiring about 30 s longer. However, ACI-304 [15] recommends that site mixers of less than 1 yd^3 (¾ m^3) capacity should mix for not less than 1 min, and 15 s should be added for each additional 1 yd^3 (¾ m^3) of capacity.

For central mixers, manufacturers generally recommend 1 min for 1 yd^3 (¾ m^3) plus ¼ min for each additional cubic yard (cubic meter) of capacity, which could be used for establishing initial mixing time. However, this could be adjusted later on, based on the results of the mixer performance.

In general, a mixing time of less than 1 min produces non-uniform concrete and mixing beyond 2 min causes no significant improvement in these properties.

7.10.3.4   Effect of Type and Size of Mixer

Tilting and Non-tilting Drum Mixers: It was observed by conducting various tests at the Transport Research Laboratory, UK [9], that the concrete mixed in a tilting mixer is less uniform than that produced in a non-tilting mixer irrespective of the mix proportions. This difference happens as the fines have the tendency to remain near the base of the drum in tilting mixer; hence they are not properly mixed with the other constituent materials. In operating tilting-drum mixers, the angle of drum is of great importance and should not exceed 30° to the horizontal. If the angle is too steep, the tendency for the fines to remain near the base of the drum is increased, and an alteration of as little as 5° may cause serious segregation.

While using tilting mixer, however, it is observed that the average mix proportions are more accurate than non-tilting mixers. This happens because while discharging the batch from non-tilting mixers, considerable amount of mortar is left sticking to the drum and blades resulting in the increased aggregate/cement ratio. Poor discharge, particularly when mix is dry, must be avoided.

In general, design of the mixer is more important than the type or size of the machine, as a poor mix will be produced if the drum is too short or the blades are badly designed.

Pan Mixers: Pan mixers produce concrete of more uniform strength than any of the rotating-drum mixers. They are particularly efficient with stiff and cohesive mixes and are, therefore, often used for precast concrete, as well as for mixing small quantities of concrete or mortar in the laboratory.

However, this type of mixer has a slight tendency to throw coarse aggregate to the top and sides of the pan and never to be properly mixed with the mortar. This segregation increases when mixing speed is increased in order to reduce mixing time. When the concrete is discharged through the bottom of the pan, there is probably some remixing during discharge. The segregation may not be then so noticeable, particularly when a special blade is introduced into the pan to guide the concrete towards the discharge door.

Continuous Mixers: If proper attention is given, continuous mixers produce concrete of similar quality and uniformity to that from paver mixers supplied from batching plants. However, some variations in the crushing strength of cubes made from concrete mixed in continuous mixers have been reported [9], which was attributed to the fluctuations in the water supply. For efficient operations of continuous mixers, attention must be given to the following points:

1. The quantities of materials being fed to the mixer must be checked frequently.
2. Preferably all the cement should be obtained from the same source.
3. The use of finely graded sand should be avoided as sometimes it results in the interruption of flow by accumulating in the bottom of the hopper.
4. Concrete consistency should be monitored continuously and if necessary the flow of water could be adjusted accordingly.
5. The mixer should be set horizontally so that concrete should not pass too quickly through the drum.
6. The speed of mixer should be controlled so that it should not deviate from that specified by the manufacturer.

7.10.3.5 Effect of Method of Loading

The order of feeding the ingredients into a mixer is also a very important factor for producing uniform and good-quality concrete. By conducting various tests [9], it has been found that mixing the solid materials, i.e. dry premixing for one-third of the total mixing time before adding the water, produces much lower strength concrete. The reduction in strength can be attributed to loss of cement during the dry mixing. The other problem of dry premixing is sticking of fines in the drum, which also has a serious effect on the quality of concrete.

Hence, in order to obtain cohesive concrete mix, it is generally preferable to introduce one-third of the total mix water first, followed by the coarse aggregate into the mixer. Mix this for 30 s and add more one-third water during mixing. Now add all the fine aggregates and cement and again mix for 30 s. Finally, the remaining one-third of the mix water may be added and the ingredients shall be mixed for 60 s continuously or as per time mentioned in Table 7.13 depending on the type and capacity of the mixer. However, it may be noted that if water or cement is fed too fast or is too hot, cement balls are produced, which should be avoided.

In truck mixers, all loading procedures must be designed to avoid packing of the material, particularly sand and cement, in the head of the drum during charging. The probability of packing is decreased by placing about 10% of the coarse aggregate and water in the mixer drum before the sand and cement. Generally, 70–100 revolutions at mixing speed are specified for truck mixing. However, ASTM-C94 [39] limits the total number of revolutions to a maximum of 300. This is to limit grinding soft aggregates, loss of slump, wear on the mixer, and other undesirable effects on concrete in hot weather. Prior to discharge of concrete transported in truck mixers, the drum should again be rotated at mixing speed for about 30 revolutions to re-blend into the batch possible stagnant spots near the discharge end.

Chemical admixtures should be charged to the mixer at the same point in the mixing sequence batch after batch. Liquid admixtures should be charged with the water or on damp sand, and powdered admixtures should be fed into the mixer with other dry ingredients. When more than one admixture is used, each should be batched separately unless premixing is shown to be permissible and they should be properly diluted before they enter the mixer.

Bibliography

1. Charles E. Reynolds and James C. Steedman, Reinforced Concrete Designer's Hand Book, E & FN Spon, London, 1994.
2. A.M. Neville and J.J. Brooks, Concrete Technology, E.L.B.S. Longman, Singapore, 1993.
3. ACI 318-08 (Revised 2014), Building Code Requirements for Structural Concrete.
4. ACI 309.1 R-93 (Revised 2008), Behavior of Fresh Concrete During Vibration.
5. D.C. Teychenne, J.C. Nicholls, R.E. Franklin, D.W. Hobbs, second edition amended by B.K. Marsh, Design of Normal Concrete Mixes. Building Research Establishment Ltd., UK, 1997
6. ACI 304.2R-91 (Revised 2017), Placing Concrete by Pumping Methods.
7. Canadian Standards Association (CSA) A23.1-14 - Concrete materials and methods of concrete construction/Test methods and standard practices for concrete.
8. ACI 309R-87 (Revised 2005), Guide for Consolidation of Concrete.
9. Road Research Laboratory, Ministry of Transport, Concrete Roads Design and Construction, published by Her Majesty's Stationary Office, printed by Lowe and Brydone Ltd, London, 1966.
10. BS 882-92, Specification for Aggregates from Natural Sources of Concrete.
11. ASTM (American Society for Testing and Materials) C33-86 (Revised 2016), Specifications for Concrete Aggregates.

12. Ontario Provincial Standard Specification- OPSS.PROV 1002, Material Specifications for aggregate – Concrete, April 2013.
13. Ontario Provincial Standard Specification-OPSS.PROV1350, Material Specifications for Concrete – Materials and Production- November 2013.
14. Ontario Provincial Standard Specification- OPSS.PROV 904, Construction Specifications for Concrete Structures- November 2014.
15. ACI 304R-89 (Revised 2000), Guide for Measuring, Mixing, Transporting and Placing Concrete.
16. Peter H. Emmons, Concrete Repair and Maintenance Illustrated, R. S Means Company Inc. USA 1993.
17. Lars Forssblad & Stig Sallstrom, Concrete Vibration- What's Adequate ACI Journal, Concrete International, Vol. 17, No. 9, September 1995.
18. American Concrete Institute (ACI), Publication SP-15(95), Field Reference Manual, ACI 301-96 (Revised 2016).
19. British Standard Institution, BS 8110- Part l:1985, Structural Use of Concrete: Code of Practice for Design and Construction.
20. BS 5337-76 (Amended-82), Code of Practice for the Structural Use of Concrete for Retaining Aqueous Liquids.
21. R.K. Dhir, M.J. McCarthy, and P.A. Mc Kenna, Total Design Using CPF, University of Dundee, 1999, 167 pp., CTU Report 1399.
22. David Wilson, Controlled Permeability Formwork (CPF) Concrete, Concrete Society Journal CONCRETE, Vol. 34, Nov. 6, June 2000.
23. Concrete Society (1987), Permeability Testing of Site Concrete; A Review of Methods and Experience, Technical Report 31. The Concrete Society, Slough, p. 96.
24. Concrete Construction Staff- Placing Reinforcing Steel- Concrete Construction, March 2005.
25. ACI Education Bulletin E2-00 (Reapproved 2006)- "Reinforcement For Concrete Materials and Applications" - Developed by Committee E-701, Materials for Concrete Construction.
26. CAN/CSA-G30.18–M92 "Billet-Steel Bars for Concrete Reinforcement" metallurgy.
27. BS 4449:2005+A3:2016- Steel for the reinforcement of concrete. Weldable reinforcing steel. Bar, coil and de-coiled product. Specification.
28. ACI 309.3R-92 (Reapproved 1997), Guide to Consolidation of Concrete in Congested Areas.
29. OCCDC (Ontario Cast-in-Place Concrete Development Council) Reinforced Concrete Reference Guide. www.occdc.org.
30. P.N. Khanna, Civil Engineers Hand Book, Engineers Publishers New Delhi, India, 1994.
31. CAN/CSA-A23.3-04 (R2010) - Design of Concrete Structures.
32. ACI 306R-88 (Revised 2016), Cold Weather Concreting.
33. P.C. Aitcin, A.M. Neville & P. Acker, Integrated View of Shrinkage Deformation. ACI Journal, Concrete International, Vol. 19, No. 9, September 1997.
34. ASTM C171- 16, Standard Specification for Sheet Materials for Curing Concrete.
35. ACI 308-92 (Reapproved 1997), Standard Practice for Curing Concrete.
36. Ontario Provincial Standard Specification-OPSS 919, Construction Specifications for Formwork and Falsework- November 2011.
37. ACI 347R-94 (Revised 2014), Guide to Formwork for Concrete.
38. ACI 302. 1R-89 (Revised 2015), Guide for Concrete Floor and Slab Construction.
39. ASTM C94-94 (Revised 2017), Central Mixed Concrete Specification for Ready-mixed Concrete.

Chapter 8
External Factors

8.1 Introduction

Besides problems due to internal causes like effects of salts in the constituent materials, interaction between the constituent materials such as alkali-aggregate reaction, volume changes, absorption, and permeability, the concrete is also influenced by a lot of external factors, which create their own problems. The main external factors causing concrete structures to fail can be summarised as:

1. Restraint against movement
2. Exposure conditions

   (a) Weather conditions
   (b) Effects of salts from soil, ground, or seawater
   (c) Freezing and thawing
   (d) Wetting and drying
   (e) Leaching
   (f) Abrasion

3. Overloading
4. Settlement
5. Fire resistance

8.2 Restraint Against Movement

Restraint against movement causes cracking that permits ingress of moisture and leads to corrosion of steel. Movement in concrete is due to elastic deformation and creep under constant load, shrinkage on drying and setting, temperature changes, changes in moisture content, and settlement of the foundations. The design should

© Springer Nature Switzerland AG 2019
A. Surahyo, *Concrete Construction*,
https://doi.org/10.1007/978-3-030-10510-5_8

include sufficient movement joints, and sometimes steel reinforcement, to prevent serious cracking.

The provision for movement joints depends to a considerable extent on the nature of the structure and the usage. For instance, an elevated structure may be subjected to few restraints, while an underground structure may be massive and restrained. On the other hand, temperature and moisture variations may be greater in exposed structures than those which are buried. If warm liquids are involved, then this must be reflected in the provision of adequate joints.

The type of member, and construction sequence, is also an important consideration. In some instances of buildings, roofs may be separated from the walls by sliding joints. If the roof is to be designed as unrestrained then great care must be taken to minimise the restraints to thermal movement during construction. If significant restraints cannot be avoided, reinforcement must be designed to limit the likely cracking. Where roof and wall are monolithic, joints in the roof should correspond to those in the wall, which in turn may be related to those in the floor slab.

8.2.1   Types of Movement Joints

The joints should be clearly indicated for both member and structure as a whole. The joints are to permit relative movement to occur without impairing structural integrity. The main types of movement joints used are as follows:

8.2.1.1   Expansion Joints

Expansion joints provide gaps, i.e. complete discontinuity of concrete and steel reinforcement, to allow for expansion and contraction, both due to climatic changes such as temperature, moisture, etc. Such joints are also called as isolation joints. However, in some cases of pavements, steel is continued for transfer of load across the joint. Expansion joints are provided vertically in building and wall structures, and horizontally in concrete pavement and floor structures as shown in Fig. 8.1.

Fig. 8.1 Expansion joint in floor slab

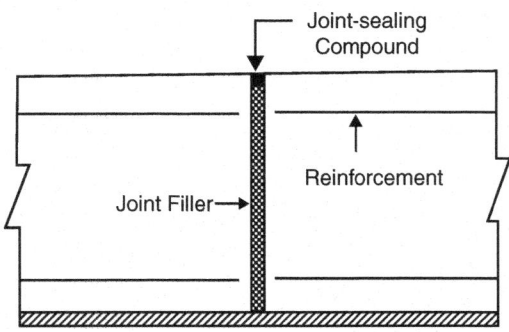

They are usually provided at much greater intervals depending on various factors. In road pavements, based on the slab thickness and weather conditions, Table 8.1 provides the information for maximum spacing of expansion joints for both type of slabs, i.e. reinforced and un-reinforced, required steel reinforcement, and width of expansion joints [1].

Under normal conditions, expansion joints may be provided at intervals of approximately 30 m (100 ft.). Moreover, in order to avoid further chances of cracks in un-reinforced slabs, the thickness of slab is usually increased so as to reduce loading stresses or by providing warping or contraction joints in addition to expansion joints at the spacing shown in Table 8.2.

Table 8.1 Spacing of joints

Thickness of slab not less than[a] (in.)	Weight of reinforcement not less than (lb/yd²)	Weather	Minimum spacing of expansion joints (ft.)	Width of expansion joints (in.)
(a) Reinforced concrete slabs without contraction joints				
10	14	Warm	150	¾
		Cold	150	1
8	10	Warm	120	¾
		Cold	120	1
6	7	Warm	80	¾
		Cold	80	1
4	5	Warm	40	½
		Cold	40	¾
(b) Un-reinforced concrete slabs				
8	–	Warm	120	1
		Cold	90	1
6	–	Warm	90	1
		Cold	60	1

[a]The thickness of slab and amount of reinforcement are determined by the amount of traffic and the subgrade conditions (Crown © 1955, reproduced with the permission of the Controller of HM stationary office [1])

Table 8.2 Suggested spacing of contraction/control joints in floors on ground [2]

Slab thickness in (mm)	Slump greater than 4 in. (100 mm)		Slump less than 4 in. (100 mm) Spacing ft. (m)
	Aggregate less than ¾ in. (20 mm) Spacing ft. (m)	Aggregate greater than ¾ in. (20 mm) Spacing ft. (m)	
4 (100)	8 (2.4)	10 (3.3)	12 (3.9)
5 (125)	10 (3.3)	13 (4.3)	15 (4.9)
6 (150)	12 (3.9)	15 (4.9)	18 (5.9)
7 (175)	14 (4.6)	18 (5.9)	21 (6.9)
8 (200)	16 (5.2)	20 (6.6)	24 (7.9)
9 (225)	18 (5.9)	23 (7.5)	27 (8.8)
10 (250)	20 (6.6)	25 (8.2)	30 (9.8)

In order to prevent the opening of joints on runways and major highways and to ensure transfer of load across the joint, steel reinforcement is provided. The load is transferred across the joint by a series of steel dowel bars, each bar being bonded into one slab and sliding in the other. For a typical section through a dowel bar expansion joint see Fig. 8.2. The bar length of 12 in. (300 mm) on metal cap side is usually coated with bitumen or grease or painted to prevent bonding, whereas the other side is cast in concrete without any coating so as to have proper bonding. It is essential that the dowel bars in a joint shall be able to slide freely, thus permitting the joint to open and close without undue restriction; otherwise the main purpose of the joint will not be achieved.

In buildings, the expansion joint is usually provided when its length is more than 200 ft. (60 m). In order to be effective, expansion joints should extend entirely through the building, completely separating it into independent units. Column footings that are located at expansion joints need not be cut through unless differential settlements or other foundation movements are anticipated; and in that case, separate smaller buildings should be built on separate foundations to allow for those movements. Expansion joints should be carried down through foundation walls; otherwise the restraining influence of the wall below grade, without a joint, may cause the wall above to crack in spite of its joint; reinforcement must never pass through such an expansion joint. The flow of wind and water through the joint must be stopped by weatherproofing. Many gasket and sealant products are available to maintain a complete seal while accommodating movement of the expansion joint.

The actual width of an expansion joint must be greater than the computed maximum joint to provide for construction tolerances, and for the width and compressibility or expandability of the joint sealant and compression seal.

Expansion joints in structural floors or suspended slabs must be designed to prevent water from leaking to the floor below. The roof at an expansion joint is always a critical place because of the possibility of leakage. A proper detail is shown in Fig. 8.3.

Wall and roof joints must be continuous over parapets. The water seal or flashing must be continuous so that there should be no place for leakage.

Expansion joints in bridge deck are provided to accommodate horizontal movements, generally caused by temperature variations and those caused by end

Fig. 8.2 Dowel-bar expansion joint

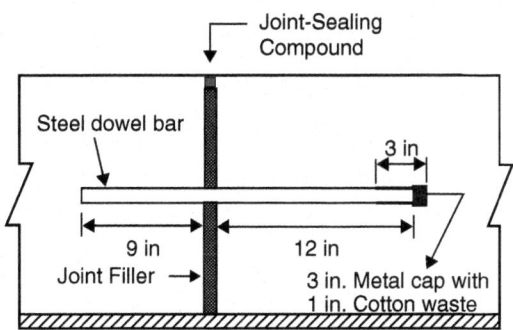

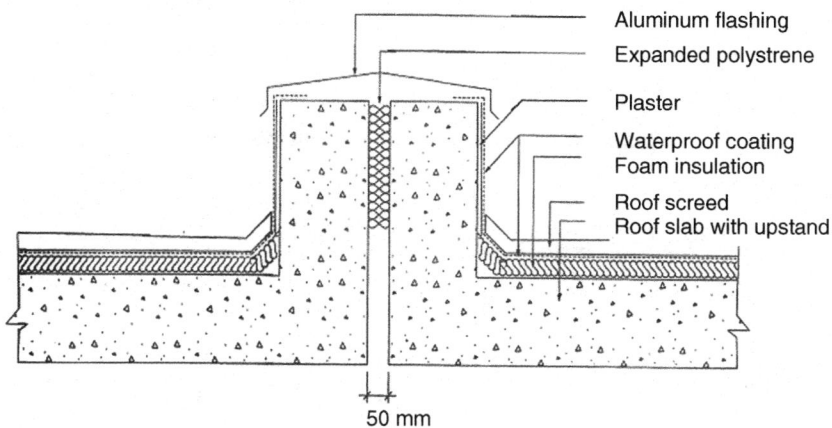

Aluminum flashing

Expanded polystrene

Plaster

Waterproof coating
Foam insulation

Roof screed
Roof slab with upstand

50 mm

Fig. 8.3 Expansion joint detail in roof

rotations at simple supports. It is preferable to provide sealed deck joints in bridges to avoid contamination with water, dirt, and deicing salts.

8.2.1.2 Contraction/Control Joints

These joints are generally used in the construction of concrete floors, road pavement, and walls. Contraction joints are essentially breaks in the structural continuity of the concrete, permitting it to contract when the temperature falls below the temperature of laying. They also help in controlling cracks due to initial drying shrinkage and warping. Because concrete is very much weaker in tension than in compression, contraction joints normally have to be spaced at closer intervals than expansion joints. It is also a usual practice to cast alternate concrete panels to avoid cracks due to contraction. Usual size for floor and pavements adopted on most of the sites is panel of 4 m × 4 m (13 ft. × 13 ft.).

However, as suggested by Portland Cement Association [2], a rule of thumb for non-reinforced slabs is that the joint spacing should not exceed 24 slab thickness for concrete made with less than 3/4 in. (20 mm) coarse aggregate; 30 slab thickness for concrete with greater than 3/4 in. (20 mm) coarse aggregate; or 36 slab thickness for low-slump concrete. Suggested joint spacing is given in Table 8.2.

For reinforced slab on ground, joint spacing normally varies from 30 to 80 ft. (9 to 24 m). The percentage of reinforcement increases with an increase in joint spacing. To be effective in controlling cracks, the steel must be placed as near the surface as possible by providing adequate cover on top.

The contraction joints in non-reinforced concrete roads usually suffer from the disadvantage that they are liable to open due to continuous process of expansion and contraction of panels between expansion joints. When the temperature first rises above that at which the concrete was laid, the whole section of road between the

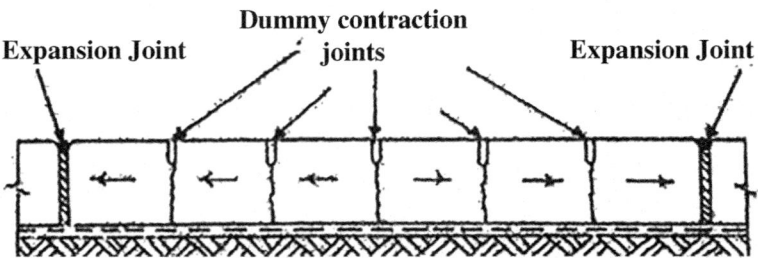

Fig. 8.4 Movement of slabs during a rise in temperature shortly after construction

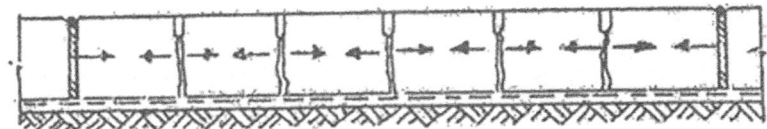

Fig. 8.5 Movement during a subsequent fall in temperature (Crown © 1955, reproduced by permission of the Controller of HM stationary office [1])

expansion joints will expand as single unit, and if it behaves symmetrically will move outwards from the midpoint between the expansion joints. This behaviour of concrete is shown in Fig. 8.4.

When the temperature falls, the compressive forces in the expansion joints will not be sufficient to push the bays back and each bay will contract about the midpoint of its length, i.e. halfway between each pair of contraction joints. After several such operations, the contraction joints will be open, as shown in Fig. 8.5 [1]. However, where heavy traffic is expected, the opening of joints can be avoided by providing warping-type joints. Usually, the width provided for contraction joints is 3/4 in. (20 mm), whereas for depth Portland Cement Association suggests to take 1/4 × depth of slab and BS-5337 recommends 1/3 × depth of slab, and for walls 1/4 × thickness of wall. These grooves are formed by sawing or by use of a premoulded strip inserted in the fresh concrete.

In exterior walls the contraction joints should not be more than 20 ft. (6.0 m) apart with frequent openings. In walls without openings the joint spacing may be a little greater but should never be more than 25 ft. (7.5 m), and it is desirable to have a joint within 10 or 15 ft. (3 or 4 m) of a corner if possible [2]. Joint spacing in any exposed cast-in-place interior walls should be identical to joint spacing in outside walls. In general, the walls should have contraction joints spaced from one to three times the wall height. Contraction joints should be provided from the top of the wall footing and should be continued on the outside of the wall to the top of the parapet, then over the top of the parapet, and down the back of the parapet. However, on the inside face the joint should extend from floor to ceiling.

In multistorey buildings with a setback at one or more floors, the joints need not be continuous from one level to another, but may be offset at each roof line in order to locate the joints at the best sections in the respective walls.

8.2.1.3   Warping Joints

In addition to horizontal movement caused by a change in moisture and temperature in concrete slabs on ground, warping or curling of the slab can be caused by differences in moisture and temperature between the top and bottom of the slab. The edges at the joints tend to curl up when the surface of the slab is dryer or cooler than the bottom (Fig. 8.6). The slab will assume a reverse curl when the surface is wetter or warmer than the bottom.

The joints provided for this purpose are termed as warping joints. Warping joints are also simply breaks in the continuity of the concrete; however, opening of the joints is prevented by tie bars, the bars being bonded into concrete at each end. The behaviour of these bars should not be confused with that of dowel bars mentioned in expansion joints. These tie/dowel bars are used to withstand the temperature warping stresses and not for transfer of loads. The amount of curling (vertical movement at edges) may also be small with a short, thick slab.

8.2.1.4   Settlement Joints

Such joints permit adjacent members to settle or displace vertically as a result of foundation or other movements relative to each other. Entire parts of the building can be separated into blocks to permit relative settlement, in which case the joint must run through the full height of the structure. Figure 8.11 shows such an arrangement.

8.2.1.5   Sliding/Isolation Joints

Isolation joint totally separates a concrete element from another concrete element, or a fixed object such as a wall or column, so that each can move and not affect the other. For example, in a circular tank on a flat base, the walls may be designed as independent of the base by providing sliding joint. The essential requirement is that the two concrete surfaces are absolutely plane and smooth, and that the bond is broken between the surfaces by painting or using building paper, or using a suitable flexible rubber pad or expansion joint filler. Figure 8.7 shows a typical detail for such a joint, which must always be effectively sealed.

Fig. 8.6 Curling of grade slab

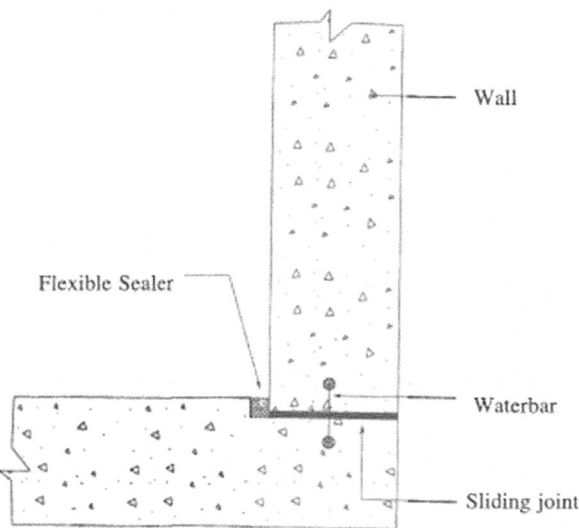

Fig. 8.7 Typical sliding joint between slab and wall

Sliding/isolation joints are provided to accommodate differential, horizontal, and vertical movements, and can be used at junctions with walls, columns, machine foundations, and footings. For a vertical section, joint filling should be of full depth and soft. It can be made of cork, foam rubber, or some other flexible material.

8.2.1.6 Uncommon Joints

1. Joints in Curbing

Where precast or stone curbs are laid on the surface of the concrete slab, a gap for expansion must be provided in the curbing, in line with the expansion joint in the slab. The gap in the curb should be filled with a preformed joint filler of the same width as that in the slab. At contraction joints, no gap need to be left, but the joint in the curb should coincide with the joint in the slab; if the two joints do not coincide, the curb often cracks at the slab joint as shown in Fig. 8.8. Curbs cast in situ should also have joints of similar type and at the same position as those in the slab.

8.2.1.7 Fillet Joints

These are used in instances where the gap to be sealed is too narrow to provide sufficient movement accommodation and sealing is achieved by taking the sealant up and out of the joint and on to the surface of the substrate.

As a rule, this type of joint should be restricted to internal applications where anticipated movement is limited.

Fig. 8.8 Cracking of curb induced by placing it across contraction joint of road pavement [1]

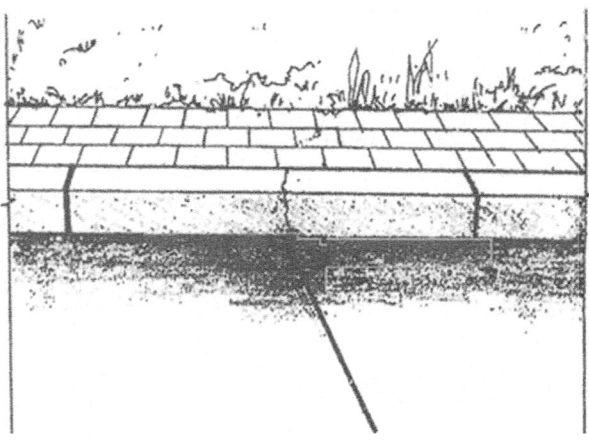

8.2.1.8 Other Building Joints

In addition to above, a building requires some of other minor but essential joints such as around window and door openings in concrete walls. Consider an aluminium window frame built into a concrete wall; the aluminium will tend to move more than the concrete for a given change of temperature because of the different coefficients of linear expansion. This difference needs to be accommodated by providing a joint; otherwise joint failure will occur leading to cracking of concrete.

8.2.2 Sealing of Joints

Mostly three types of materials are used in joints:

1. Joint fillers
2. Joint-sealing compounds
3. Water stops

The jointing material should be capable of accommodating repeated movement without permanent distortion or extrusion. The materials should remain effective over the whole range of temperature variation.

1. *Joint Fillers*

   Joint fillers are compressible sheet or strip materials used in expansion joints to provide the initial separation between the faces of the joint. They also provide a support for the sealing compound.

   In order to work properly, the joint filler must be compressible without extruding, elastic, durable, and sufficiently rigid to facilitate its support during construction. The preformed joint filler is mostly made from softwood, impregnated fibre board, cork, and polyethylene.

2. *Joint-Sealing Compounds/Sealants*

Joint-sealing compounds are impermeable ductile materials, which are required to provide a watertight seal by adhesion to the concrete, and resist the entry of grit and water.

All joints form breaks in the continuity of structures, and water and grit are liable to enter them. Entry of water into joint may cause deterioration of the concrete leading to costly repairs, and if grit or stones become packed into the joint, they may prevent free expansion of the concrete and cause spalling; in the extreme case "blow-ups" may occur. It is therefore a practice to fill the top 25 mm (1 in.) or so of the joint with a sealing compound in order to prevent, or minimise, the entry of water and grit into the joint.

Generally, sealants are needed in the weather proofing of buildings and other structures including houses, high-rise constructions, bridges, roads, and pavements. They are also used in water-retaining and water-exclusion applications in dams, canals, basements, etc. The exterior of a building must be weatherproofed to eliminate drafts and to prevent wind-driven rain from entering the building.

In order to work efficiently, the joint-sealing compound should have the following properties:

(a) Good adhesion to the concrete
(b) Resistance to the ingress of grit
(c) Extensibility without fracture
(d) Resistance to flow in hot weather
(e) Durability

The sealing compounds are usually made from bitumen and rubber-bitumen compounds, polyurethane, polysulphide polymer, and pale-coloured resinous compounds.

3. *Water Stops*

Water stops are generally preformed strips of durable impermeable material, which are wholly or partially embedded in the concrete during construction so as to span across the joint and provide a permanent watertight seal during the joint movement.

In order to work properly, it is important to ensure that proper compaction of the concrete against the water stops is achieved. They should be sufficiently wide and the distance of the water stop from the nearest exposed concrete face should not be less than half the width of the water stop [3]. The full concrete cover to all reinforcement should be maintained. While concreting the movement of water stops, edges should be prevented by tying the edges to adjacent reinforcement. They are usually made of rubber, PVC, or a flexible plastic.

8.2.3 Joint Failures

If joints are properly designed, carefully constructed, and adequately maintained, defects are unlikely to occur. However, as a result of poor construction methods, it has been observed that instead of providing structural stability, joints become vulnerable to deterioration for several reasons:

1. As they are usually difficult to construct, poor compaction may result at a joint.
2. If poorly constructed, they may provide access for the entry of aggressive waters or harmful gases.
3. If alignment of filler material is disturbed, they might be sealed wrongly, and the concrete may crack at an adjacent plane.
4. The inactive joints like construction joints may turn to be active without having any provision for sealing.

The most common faults, which result in failure of joints, along with possible protective measures can be summarised as below:

(a) Poor Joint Preparation

Poor joint preparation is the most common cause of joint failure. Joints to the correct dimensions can be either built in or cut using saws or grinders. Ensure that the joint is prepared wide enough to allow sufficient sealant to be applied to cope with the anticipated movement.

Components, such as doors and windows, should be shimmed into place leaving sufficient clearances for the application of the sealant. Cracks and excessively narrow joints may need to be opened out mechanically. If unsure, the joint dimensions should be measured.

(b) Spalling of Joint Edges

This defect is very common in joints and is usually due to the ingress of stones into the sealing compound. The risk of spalling may be reduced by keeping the road clean and free from stones, and by using proper sealant, which should have sufficient resistance to the entry of stones.

Use of mortar (may be rich mix, say 1:1) or weak concrete for making edges of joints may also be a cause of spalling. Such type of joint is liable to scale off during frost and will not be suitably resistant to wear and impact of vehicles. The edges thus must be prepared from the same mix, used for pavements. It is, however, more preferable to make joints mechanically by saw cutting into hardened concrete.

Another cause of spalling is non-alignment of the filling material and sealing groove. If preformed joint filler is not rigidly supported during construction, the filler may tilt vertically as shown in Fig. 8.9, or bow horizontally under the thrust of the compactors. If the groove for the sealing compound is then formed according to marking made on the side forms, a narrow wedge of concrete will be left, between the groove and the filler, which is particularly liable to spall.

(c) Inaccurate Alignment of Dowel Bars

Misaligned dowel bars are also a common cause of joint failure in concrete structures. If the dowel bars are not properly aligned, spalling of the concrete at the joint face and bending of the dowel bars may occur. Hence, for doweled joints the accuracy of bar alignment is very important. The bars must be securely fixed so that their alignment is not disturbed during concreting.

(d) Inadequate Expansion Space

If the space for expansion of slabs is inadequate, compression failures may occur. Insufficient expansion space may result due to construction mistake or by the filling of joint with grit. This could be avoided by good workmanship and blowing the grit off the joint before sealing the joint.

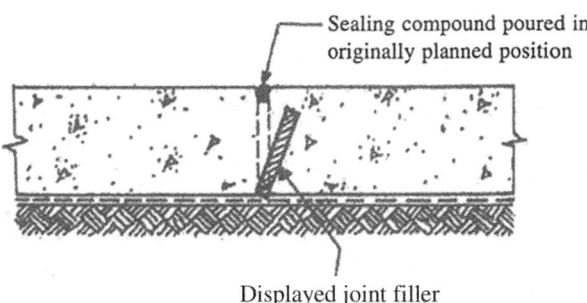

Sealing compound poured in
originally planned position

Displayed joint filler

Fig. 8.9 Joint-filling material displaced (Crown © 1955, reproduced by permission of the controller of HM stationary office [1])

(e) Level Difference

Sometimes the difference in floor levels adjacent to the joints also becomes the cause of joint failure. It is strongly recommended that the profile across the joints should be checked with a straight edge before the concrete has set, so that corrections can be made if necessary.

(f) Faults that Restrict the Opening and Closing of a Joint

The opening and closing of a joint are restricted mostly due to the following reasons:

1. When dowel bars become bonded into the concrete at both ends, particularly in load-transferring joints
2. When cap provided on the sliding end of the dowel bar (see Fig. 8.2) is displaced or filled with concrete during construction
3. When dowel bars are not in alignment

All the above faults occur due to poor construction practices. However, proper supervision with good construction practices can overcome such faults.

8.2.4 Sealant Failures

Majority of sealant failures are caused by poor workmanship, but the other factors like improper selection of sealant are also of importance. There are generally three basic modes of failure: adhesive failure, cohesive failure, and substrate failure [4].

(a) Adhesive Failure

This is a loss of adhesion along the bond line between the sealant and the surface of the substrate. Possible causes are:

• Joint movement exceeding the sealant design capacity
• Joint movement during sealant cure
• Poor surface preparation
• Non-use of primer where needed

- Wrong joint profile

(b) Cohesive Failure

Cohesive failure occurs when the sealant itself fails to hold together and splits and tears in the body of the joint. Possible causes may be:

- Improper sealant selection
- Inadequate mixing of multicomponent systems
- Entraining air, during either mixing or application
- Contamination from the substrate sides

(c) Substrate Failure

This occurs when the substrate fails before the sealant and is readily identifiable from evidence of the substrate adhering to the sealant. Possible causes are:

- Wrong choice of sealant
- Poor surface preparation
- Shoddy materials

In order to avoid failure of sealants, the points to be considered with possible solutions could be summarised as below:

1. *Surface Preparation*

   The joint surface should be cleaned and made free of contamination, particularly free from old sealant, coatings, laitance, etc. When getting rid of dust and dirt, brushing followed by a solvent wipe is usually sufficient.

   If the substrate has not been properly prepared then a loosely adhering or friable surface will be very prone to spalling or adhesive failure. Unremoved laitance from precast concrete is a familiar example of this problem. Contamination may also be present from previous, incompatible sealant, such as bituminous materials when refurbishment work has been carried out using elastomeric. If contamination is heavy, then grit blasting is ideal, with any dust blown or brushed out of the joint afterwards.

2. *Mixing*

   When multi-components are involved, there is always a chance of poor mixing. This is usually easy to spot where the components of the unmixed sealants are of different colours or if there are still lumps of pasty material readily visible.

   Before mixing, the data sheets must be read and understood properly. All the contents of any tins of activator need to be added and properly mixed, with times being measured, using the correct mixing blade. Entrained air should be avoided. The mixing, at too high a rate, also needs to be avoided so as to minimise the effect of mechanical heating.

3. *Priming*

   A primer forms a layer between the sealant and the underlying surface and improves the adhesion of one to the other. It can provide a barrier to prevent water from reaching the bond line of the sealant and reducing the adhesion.

Joints should be primed properly. It is possible to over-prime and this is evidenced by pools of hardened patches of primer, particularly at the base of horizontal joints. The under-prime will result in dis-bonding. Proper primer should be used based on required conditions. Primers available are smooth surface primers and porous surface primers, and for wet condition and for special situations.

The primer should be applied to the sides of the joint with any pooling being corrected. The open time is critical and if uncertain should be timed so as to ensure that the sealant is applied after allowing the primer to dry within its open period.

The primer often contains solvent and may be flammable. Appropriate safety precautions are necessary. Frequently primers are moisture sensitive and must be stored in airtight containers and used up quite quickly after opening.

4. *Application*

It is pointless to select a high-performance, technically advanced sealant and then degrade its performance by using the wrong ancillaries and/or the wrong application techniques.

As a general rule, sealants should only be applied at temperatures higher than 5 °C [4], unless specified by the manufacturer. Adhesion failure will occur if a sealant is applied to a joint containing moisture or frost. The use of sealant should also be avoided when humidity levels may cause surface condensation (max. 85%).

The sealant should not be applied in very hot environmental conditions, as the expansion of the substrate will minimise joint dimensions. In summer season, it is better to apply the sealant either in the early morning or in the late afternoon.

The sealant should be poured evenly, steadily, and directly into the joint. The width of nozzle tip, used for filling the joint, should not be so wide as to give a messy joint with material smearing onto the surface of the substrate or so narrow as to require more than a single pass to fill the joint effectively. This tends to lead to air being entrained into the joint.

Awkward joints can often be tackled using extra-long nozzles, or at a pinch, screwing one nozzle onto another. Joints exposed to traffic should always be recessed.

5. *Tools*

Inadequate use of tools while filling sealant in joints could lead to poor adhesion at the joint sides. A poorly tooled joint shows rounded flow marks as though simply poured in.

For neat clean joints, mask the joint edges with tape and remove immediately after sealant application. Use a suitable sized putty knife, spatula, or similar tool to press the sealant into the joint, to ensure full and firm contact with the joint surfaces and to avoid entrapped air.

Though tooling is used to create an aesthetically neat joint with a slightly concave surface, it is also essential to ensure that the side faces of the joint have been thoroughly "wetted". Tooling can be done dry or with slightly wetted tools provided that the guidelines in the manufacturer's data sheet are followed.

6. *Defective Materials*

Use of defective materials or time bond products should be avoided. Sometimes poor-quality fillers can also cause problems such as outgassing, leading to voids in the sealants.

7. *Wrong Choice of Sealant*

Proper sealants should be selected based on all known requirements. It is essential that each sealant and various ancillaries be used correctly according to the directions provided on labels and product data sheets. Un-reinforced concrete elements have low tensile strength and as a result high modulus sealants can have a greater tensile strength than the substrate, which can lead to substrate failure. A lower modulus material should be specified in such cases.

8. *Joint Design*

In order to cope with anticipated movement, the joints should be correctly designed in respect of the required number and width. The information of daily and annual maximum and minimum temperature ranges could be helpful in calculating the number of joints. The other factors, which could be considered while calculating the number of joints, are the width of joint, aesthetic considerations, and possibility of combining construction joints, expansion joints, or contraction joints.

For calculation of joint width, depth, and spacing, the topics on expansion and contraction joints addressed already in this chapter can be followed. Moreover, in order to accommodate the calculated movement within the joint, the specified sealant must have sufficient movement accommodation factor (MAF).

MAF [4] measures the ability of the sealant to cope with movement during its service life, both in compression and in tension.

MAF = joint movement/joint width × 100

In the example below (Fig. 8.10), the MAF = (20−16)/16 × 100 = 25%

Thus a sealant with an MAF of ±25% could cope with repeated cycles of stress, which changed the width of the joint from 75 and 125% of its initial width.

For free movement, three-sided adhesion of a sealant in a joint should be avoided. This could be achieved as follows:

Fig. 8.10 Movement accommodation factor

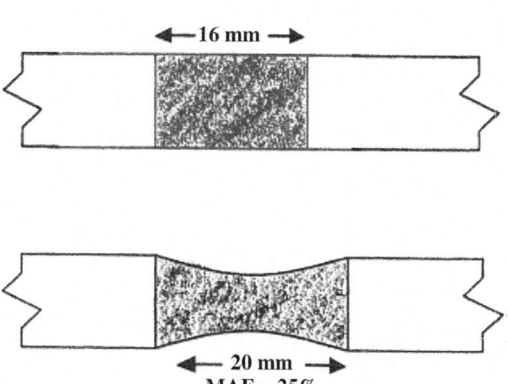

(a) Shallow joints: Place a polyethylene strip at the bottom of the joint.

(b) Deep joints: To control the depth of the applied sealant, use a closed-cell polyethylene foam backing strip at the bottom of a joint. Use a size that would compress 25% when inserted into the joint.

8.3 Exposure Conditions

Exposure conditions are one of the main causes of a number of defects in concrete structures. The degree of exposure anticipated for the concrete during its service life together with other relevant factors relating to mix composition, workmanship, and design must be considered.

The aggressiveness of the environment is a very important factor to consider when examining concrete that shows signs of possible distress. In mild exposure conditions, minor cracking or some minor defect in concrete quality (but not chloride contamination) may not lead to serious trouble. However, in severe exposure like freezing, heavy rains, sea splash, or alternate wetting and drying especially in hot arid climates, only the highest quality of undamaged concrete with proper cover to reinforcement should be expected to give satisfactory performance in the long run.

In general, the description about these exposure conditions, along with recommendations for the cover to steel and for concrete quality subjected to these exposure conditions as per BS-8110 [5] and BS-5328 [6], is given in Tables 7.5 and 4.1, respectively. However, the recommendations of ACI-318 [7] for the cover to steel and for concrete quality are specified in Tables 7.4 and 7.1, respectively, and the recommendations by CSA-A23.1-14 [8] are reproduced in Tables 7.6 and 7.7.

Now let us discuss the exposure conditions individually, to which the concrete will be exposed during its service life. The exposure conditions can be grouped as:

1. Weather conditions
2. Effects of salts from soil, ground, or seawater
3. Freezing and thawing
4. Wetting and drying
5. Leaching
6. Abrasion

The first two topics are discussed in Chaps. 9 and 11, respectively. The rest of the topics are addressed in this chapter.

8.3.1 Freezing and Thawing

Concrete always contains water in its capillaries, which expands on freezing as it is converted into ice. When water freezes there is an increase in volume of approximately 9% [9]. The water in the larger capillaries freezes at only a few degrees below the usual freezing point of water, whereas in smaller capillaries it freezes at lower temperatures. As the temperature of concrete drops, freezing occurs gradually so that the still unfrozen water in the capillary pores is subjected to hydraulic pressure by the expanding volume of ice. This pressure causes localised tension forces that fracture the surrounding concrete matrix. During warm seasons, the ice so formed will melt into water. This successive freezing (during cold weather) and thawing (during warm weather) of the concrete will lead to a cumulative expansion on the concrete volume. Additionally, on subsequent thawing, the expansion caused by ice is maintained so that there is now new space for additional water. On refreezing, further expansion occurs. Thus, this repeated cycles of freezing and thawing cause loss of strength, cracking, scaling, unusual whiteness, and disintegration of the concrete in the form of spalling.

Freezing and thawing deterioration generally occurs on horizontal surfaces that are exposed to water, or on vertical surfaces that are at the water line in submerged portions of structures.

8.3.1.1 Protective Measures

Resistance to freeze and thaw is improved by use of lowest practical water/cement ratio and total water content, durable aggregate, and adequate air entrainment. Adequate curing prior to exposure to freezing conditions is also important. Allowing the structure to dry after curing will enhance its freezing and thawing durability. Concrete placed 300 mm or more below ground level has proved unlikely to suffer any deterioration from frost. The requirements of w/c ratio, durable aggregate, and air entrainment are briefed as under:

W/C Ratio: As recommended by ACI-201 [10], concrete which is continuously or frequently wet and exposed to freezing and thawing should have a maximum w/c ratio:

1. Thin sections like bridge decks, railings, and curbs and concrete exposed to deicing salts: 0.45
2. All other structures: 0.50

However, for lightweight concrete, instead of w/c ratio, a 28-day strength of at least 4000 psi (28 MPa) is recommended by ACI-201 [10]. Moreover, the requirement of maximum w/c ratio based on BS-8110 varies from 0.45 to 0.55 based on the concrete cover thickness in severe and very severe exposure having problems of freeze and thaw. CSA [8] also recommends maximum w/c ratios varying from 0.40 to 0.55 for concretes exposed to freeze and thaw conditions.

Entrained Air: By use of air entrainment, the resistance of hardened concrete to freeze and thaw is increased and workability of fresh concrete is improved. However, too little entrained air will not protect cement paste against freezing and thawing. However, too much air will unduly penalise the strength. Recommended air contents of concrete based on different aggregate sizes under various exposure conditions by ACI-301 [11] are given in Table 8.3. However, BS-5328 [6] (Part I) recommends

Table 8.3 Recommended air content of concretes containing aggregates of different sizes, according to ACI-301-96

Maximum size of aggregate (mm)	Recommended total air content of concrete (%)		
	Mild exposure	Moderate exposure	Severe exposure
10	4.5	6.0	7.5
12	4.0	5.5	7.0
14	–	–	–
20	3.5	5.0	6.0
25	3.0	4.5	6.0
40	2.5	4.5	5.5
50	2.0	4.0	5.0
75	1.5	3.5	4.5
150	1.0	3.0	4.0

that average air content by volume of the fresh concrete at the time of placing, when concrete lower than grade 50 is used, should be:

7.5% for 10 mm maximum aggregate size
6.5% for 14 mm maximum aggregate size
5.5% for 20 mm maximum aggregate size
4.5% for 40 mm maximum aggregate size

Similarly, air contents as recommended by CSA-A23.1-14 [8] are produced in Table 8.4, which can be read in conjunction with Table 7.7 based on various exposure conditions.

Air-entrained concrete is produced through the use of an air-entraining admixture added at the concrete mixer, or air-entraining cement, or both if necessary. The preferred procedure is to use an air-entraining admixture, as the dosage can be adjusted to give the desired air content. For less severe conditions of freezing, good-quality concrete without air entrainment may be sufficient. Moreover, recently polypropylene fibres have been used successfully as a substitute for air entrainment for freeze/thaw protection [12].

Deicing chemicals used for snow and ice removal, such as sodium chloride, may accelerate damage caused by freezing and thawing and may lead to pitting and scaling. Additionally since the salt absorbs moisture, it keeps the concrete more saturated, increasing the potential for freeze/thaw deterioration. CSA-A23.1-14 [8] provides that air contents less than those shown in Table 8.4 might not give the required resistance to freezing and thawing or deicing salts, which is the primary purpose of air entrainment. Air contents higher than the levels shown might reduce strength without contributing further improvement to durability. However, properly designed and placed air-entrained concrete can withstand deicers for many years [13].

Aggregate: Some aggregates may absorb too much and cannot accommodate the expansion and hydraulic pressure that occur during the freezing of water. The result is expansion of the aggregate and possible disintegration of the concrete if enough of such particles are present. Hence aggregates liable to suffer from the action of

Table 8.4 Requirements for air content categories as per CSA-A23.1-14 [8]

Air content category	Range in air content[a] for concretes with indicated nominal maximum sizes of coarse aggregate (%)		
	10 mm	14–20 mm	28–40 mm
1[b]	6–9	5–8	4–7
2	5–8	4–7	3–6

Notes:

1. The above difference in air contents has been established based upon the difference in mortar fraction volume required for specific coarse aggregate sizes
2. Air contents measured after pumping or slip forming may be significantly lower than those measured at the end of the chute

For further reference See Table 4 of CSA23.1-14/CSA23.2-14—Concrete materials and methods of concrete construction/test methods and standard practices for concrete. © 2014 Canadian Standards Association

[a]At the point of discharge from the delivery equipment, unless otherwise specified
[b]For hardened concrete

frost should not be used for concrete that may be exposed to freezing and thawing conditions.

The use of aggregate with a large maximum size or a large proportion of flat particles is inadvisable as these require a high water content to provide adequate workability, due to which pockets of water may collect on the underside of the coarse aggregate.

Increased freeze/thaw resistance can also be achieved by using concrete sealers. Commonly, the materials used are low-viscosity epoxy, methacrylate, and polyester [14].

Additional precautionary measures required against freezing like temperature control of concrete during mixing and placing, covering by insulating materials at early stages, and accelerating the setting and strength development of concrete by using such cements and admixtures are discussed in Chap. 11 under the topic "Concreting in Cold Weather".

8.3.2 Wetting and Drying

Concrete changes length depending upon its moisture content. Moist concrete that dries out will shrink, while dry concrete that becomes moist will expand. As a rule, when concrete dries and becomes re-saturated, not more than two-thirds of the initial drying shrinkage will be recovered [15]. Such change in volume which occurs due to alternate cycles of wetting and drying of concrete will result in pattern cracking.

Curling of concrete slabs/pavements is a common example of change in moisture and temperature gradients across the thickness of a slab. Slab surfaces are usually dry on top, where they are exposed to air, and moist on the bottom, where they are exposed to soil. The drier surface has a tendency to contract in length relative to the moist bottom surface. The contraction of the top surface can only be relieved by the slab curling upward as already shown in Fig. 8.6.

While curing concrete with water, the situation of wetting and drying can take place if curing is done intermittently. Water curing should always be carried out continuously for the required period to avoid volume changes due to action of wetting and drying.

Alternate wetting and drying of structural members in seawater increase the deleterious effect of salt due to the phenomena of accumulation or crystallisation of salts in pores of concrete. However, permanently immersed concrete is attacked least, provided that the permeability of the concrete is low. The attack by seawater is also slowed down by the blocking of pores in the concrete due to the deposition of magnesium hydroxide, which is formed together with gypsum, by the reaction of magnesium sulphate with calcium hydroxide, present in concrete.

Wetting and drying leach lime out of concrete and make it more porous, which increases the risk of corrosion to the reinforcement. Wetting and drying also cause movement of the concrete that can cause cracking if restraint exists.

8.3.2.1 Protective Measures

A dense and low-permeable concrete, preferably with shrinkage-compensating cement, including properly designed movement joints, reduces the troubles caused by the action of wetting and drying. A well-compacted concrete and good workmanship, especially in the construction or movement joints, are of vital importance.

However, in seawater, use of sulphate-resisting cement with a minimum cover of 50–75 mm (2–3 in.) reduces the problems caused by the action of wetting and drying. For the requirements of minimum cement content and maximum w/c ratio based on exposure and sulphate concentration in seawater, Tables 9.4, 9.5, and 9.6 proposed by BS-5328 [6] and ACI-318 [7] are advisable. The permeability of concrete can also be decreased, by using concrete made with appropriate amounts of ground blast-furnace slag or pozzolans.

Additionally, water-repellent coatings, such as silane coatings (see Table 9.7), can provide excellent protection. However, coatings which restrict evaporation of free water from the interior of concrete should be avoided, as it will reduce the resistance of concrete to freeze and thaw.

8.3.3 *Leaching*

Calcium hydroxide, $Ca(OH)_2$, in hardened cement paste dissolves readily in water, particularly if the water is lime free and contains dissolved carbon dioxide. Thus, if concrete in service absorbs water through cracks or faulty joints, or through the areas of poorly compacted porous concrete, the $Ca(OH)_2$ in hardened cement dissolves and after evaporation leaves on the surface the calcium carbonate as white deposits. These deposits on the surface of the concrete resulting from the leaching of calcium hydroxide and subsequent carbonation and evaporation are termed efflorescence.

Snow water in mountain streams and reservoirs is often particularly aggressive because it is usually cold (calcium hydroxide is more soluble in cold than in warm water), pure, and consequently lime hungry, and like most surface water contains carbon dioxide. This produces a mild carbonic acid solution that has a higher capacity for dissolving lime than does pure water without it.

8.3.3.1 Effects of Leaching

Leaching can seriously impair the durability of concrete. Inside surfaces of concrete conduits, flumes, and canal linings develop a sandy appearance from having the cement matrix leached and weakened by contact with these lime-hungry waters.

Objectionable results of leaching are not confined to surfaces in contact with pure mountain water. Exposed surfaces of tunnel linings, retaining walls, abutments, and other structures, where groundwater has access to the opposite side, are often disfigured by lime deposits. These are formed by water that has come through the concrete, either along cracks or joints or through porous areas, taking lime into solution and becoming saturated with it. At the surface, the solution absorbs carbon dioxide, which reacts with the calcium hydroxide and causes precipitation of a white deposit of calcium carbonate.

8.3.3.2 Protective Measures

In order to reduce the effectiveness of the leaching action, the following ways are recommended by ASTM [25]:

1. Use of cements like:

    (a) Aluminous cements
    (b) Portland-blast-furnace slag cements
    (c) Portland-pozzolana cement with 20–30% of a good pozzolanic material that is strongly active in combining with lime to form insoluble lime silicates
    (d) "Low-lime" Portland cement with less tricalcium silicate than dicalcium silicate

2. Use of sufficient cement and a low water/cement ratio to ensure good dense concrete together with air entrainment, to minimise permeability and capillarity
3. Design and use of contraction and construction joints that are watertight and frequent enough to prevent intermediate cracking
4. Proper control of concrete mixes, placing procedures, and curing so that hardened concrete free from permeable imperfections is insured

5. When practical, provision for drainage facilities necessary to prevent water from standing behind structure walls
6. Consideration of a provision for a protective coating of durable and effective surface-sealing material

8.3.4 Abrasion

Abrasion is the wearing away of the surface by rubbing, friction, wave action, etc. Generally, the surface is uniformly worn away, including the cement matrix and aggregates. Abrasion reduces cover to reinforcement, which leads to corrosion.

Wear on concrete floors and concrete road surfaces is essentially a rubbing action due to traffic and is greatly increased by the introduction of foreign particles, such as sand, metal scraps, or similar materials. Many industrial floors are subjected to abrasion by steel or hard rubber-wheeled traffic, which can cause significant rutting. A concrete, which has a high water/cement ratio at or near the surface or has been cured inadequately, of course, would wear down readily.

8.3.4.1 Protective Measures

The resistance to abrasion can be improved by increasing the compressive strength of concrete. The finishing of concrete top surface plays a very important role in this context. If the slab carries too much free water on the surface, compressive strength at the surface may be seriously reduced. Therefore, the use of minimum amount of water necessary to produce the required slump and workability is extremely important. The high-range water-reducing admixtures (super-plasticisers) can be used to greatly increase slump without the need to increase the water content of the original mixtures. Improved finishing of concrete can also be achieved with use of toppings or other methods.

The other factor that helps in increasing the abrasion resistance is hardness of aggregates. Hard aggregates such as quartz, emery, and trap rock, as well as metallic aggregates, are frequently used in making wear-resist heavy-duty floors. It has also been found that concrete cured with steam and followed by fog curing to 7 days also increases the abrasion resistance. Recently the use of microsilica and fibre-reinforced concretes has proved very successful in achieving highly wear- or abrasion-resistant pavement surfaces. Urethane and epoxy floor coatings have traditionally been used to seal concrete surfaces and can also provide specialised chemical resistance. Penetrating liquid silicate hardeners are also commonly used to seal concrete surfaces through chemical densification.

Surface-applied dry shake-on hardeners is also commonly used to increase surface wear resistance from 150% to 400% over plain concrete [8]. These shake-on aggregate hardeners incorporate proportions of cement binder and special hard aggregates that are applied to the surface of fresh concrete in order to form a monolithic hardened surface. The application rate and aggregate selection vary depending upon the desired degree of protection required for an intended usage.

CSA [8] provides that the application of dry shake-on surface hardeners reduces the attainable floor tolerance. Floors with specified tolerances of Class D and higher commonly do not employ dry shake-on hardeners but use lower water-to-cement ratio concrete mixes (0.45), liquid hardeners, and/or abrasion-resistant toppings, depending upon the desired degree of abrasion resistance desired. Class D finish is considered as "extremely Flat grade slabs with advanced mechanical screeding, large pan float, highway straightedged, and steel trowel finished".

8.3.4.2 Cavitation

In hydraulic structures such as dams, spillways, tunnels, and other water-carrying systems, the wear occurs due to wave action and is generally known as cavitation. It is an impact-type abrasion and is caused by the abrupt change in direction and velocity of a liquid to such a degree that the pressure at some point is reduced to the vapour pressure of the liquid. The vapour pockets so created, upon entering areas of high pressure, collapse with a great impact, which eventually causes pits or holes in the concrete surface. Cavities are formed near curves and offsets, or at the centre of vortices. At higher velocities, the forces of cavitation may be great enough to wear away large quantities of concrete.

Protective Measures

This type of erosion can be minimised, to some extent, by careful attention to form alignment while designing the structure and avoidance of rough uneven surfaces. In general, dense concrete with hardwearing aggregate with smooth surfaces and extra covers allowing for wear is required. Introducing air into the water at a proper location near a cavitation area is helpful in reducing damage. Heavy rubber or rubber-like coatings, well bonded to the concrete, are also effective.

8.4 Overloading

A structural component is designed based on loads according to various standard code requirements. If the design value of the load effect exceeds the design value of the capacity with regard to safety factors, overloading occurs leading to failure, either in shear, tension, fatigue, or flexure. Unsuspected events by either human impacts or environmental influence can result in excessive load effect on a structure that was not considered in the design from the beginning. Environmental actions can be external forces inducing overloading either locally or globally, for example earthquake, strong wind, and heavy snow.

Extreme overloading will cause cracking and eventual collapse. Factors of safety in the original design allow for possible overloads, but vigilance is always required to ensure that the structure is never grossly overloaded. A change in function of the building or room can lead to overloading; for example if a classroom is changed to a library, the imposed load can be greatly increased. As the imposed or live load varies, deflection increases or decreases, and as deflection varies movements in partitions vary, which easily causes cracks in partitions.

Another example of overloading can be the screed/topping on roof. Mostly, the designers consider the dead load of 50–75 mm thick screed. However, while maintaining grades on roof, the thickness of screed generally increases to almost double. This added thickness increases the bending stresses.

Other items which usually result in overloading of structures could be heavy snows on roofs and bridges, auditorium crowds, water levels in tanks and pools, and ponding of rain on sagging roofs. An example of miscalculation is the damage to a prestressed bridge by overloading. The bridge was built in 1964 situated in northern Sweden near the polar circle. It spans a river where the ice in winter is 1 m thick. When the ice broke up during the spring of 1973, an ice barrier built up at the piers of the bridge. This caused such decreased water action that erosion took place around the actual pier followed by settlement of pier foundation [16].

Similarly, storage materials like dumping of concrete blocks and heavy bundles of reinforcing bars and the operation of equipment can easily result in loading conditions during construction far more severe than any load for which the structure was designed. The additional reasons for structural loading could be:

- Increased use of structure than designed for example bridges
- Unlimited visitor to the reinforced concrete structures or huge crowding such as new-year parties and related gatherings
- Early removal of formwork before the concrete reaches its design strength

8.4.1 Protective Measures

Tight control must be maintained to avoid overloading conditions. Damage from unintentional construction overloads can be prevented if the designers provide information on load limitations for the structure, and if the construction personnel care to comply with these limitations.

Changes of a structure function may lead to changes in applied loads. It is essential to determine whether the building can resist the new design loads and to estimate the remaining service life of the whole structure. Such assessment of residual service life includes investigation of current structure condition, which may be performed by visual inspection followed by non-destructive and destructive methods.

Engineers usually follow the design codes; however it is required to consider area and location-specific issues, such as maximum credible earthquake and wind speed and maximum snowfalls in that area. Designers must take these requirements into account in combination with other more common forms of loading (for example from vehicles and or due to temperature).

8.5 Settlement

Primarily, there are two types of settlement: uniform settlement and differential settlement.

Uniform settlement: Uniform settlement occurs when a building foundation settles by the same amount or same rate throughout its entire footprint area, effectively lowering the structure in place. If the whole of the foundation area of a structure settles to the same extent, there is no detrimental effect on the superstructure. For example, several structures in Mexico City have suffered from very heavy settlements but they are still functioning due to uniform settlement. However, if the uniform settlement is very excessive, the structure utility services such as water supply and sewage lines and electric and telephone supply lines would be affected even when the structure remains stable.

Differential settlement: When only part of the foundation is affected by ground failure, it is considered differential settlement and can cause much more severe damage to a building than uniform settlement. The differential or relative settlement between one part of a structure and another is of greater significance to the stability of the superstructure than the magnitude of the total settlement.

However, if there is relative movement between various parts of the foundations, stresses are set up in the structure. Serious cracking and even collapse of the structure may occur if the differential movements are excessive. That is why in a large number of cases the maximum differential settlement is the determining factor for the design of the foundation. Even slight amount of differential movement can cause substantial damage between two adjoining columns of a structure.

8.5.1 Causes and Possible Remedies for Differential Settlement

1. *Variations in Soil Strata*

    This case occurs when one part of a structure is founded on a compressible soil and the other part on an incompressible material. For example, in areas of irregular bedrock surface, one part of a structure may be founded on shallow rock and the other on soil or compressible weathered rock. In such cases, settlement of compressible soil will be more and there will be differential settlement of the structure. Such settlement can be minimised by replacing compressible soil with approved gravelly material and compacting it in layers of maximum 200 mm (8 in.) with vibratory rollers.

2. *Variations in Foundation Loading*

    If a building has a high tower at its centre with low projecting wings, differential settlement between the tower and the wings would be expected because

Fig. 8.11 Variation in
foundation loading [2]

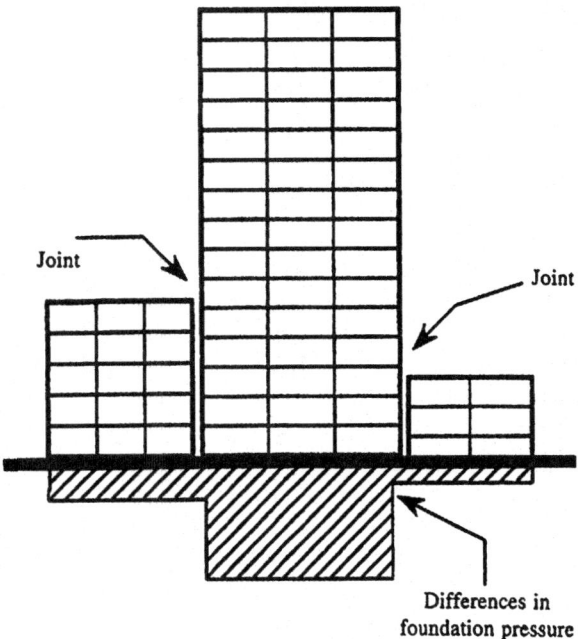

Fig. 8.11 Variation in
foundation loading [2]

Joint

Joint

Differences in
foundation pressure

various parts will transmit different intensities of stresses to the soil. Provision of
an expansion joint between various blocks of the building, so that each block can
move as a separate unit, will be of great help in this case (Fig. 8.11).

3. *Compressible Foundations*

When the large flexible raft foundations or large loaded areas having a number
of independent column foundations are constructed directly on compressible
soil, the maximum settlement will take place at the centre of the area and the
minimum at the corners. Hence, there will be a differential settlement. Such
settlement could be minimised if the large loaded area is founded on a relatively
incompressible stratum, like dense gravel, overlying compressible soil.

4. *Delay in Time of Construction of Adjacent Parts of a Structure*

Sometimes extensions to a structure are made many years after completion of
the original structure. In such cases, consolidation settlements of the original
structure might be virtually complete, whereas the new structure will settle an
equal amount, causing differential settlement between the old and new structures. Distortion and cracking between the old and new structures could be prevented by providing vertical joints.

8.5.1.1 Protective Measures in General

In order to reduce the total and differential settlements, one or a combination of the
following methods may be adopted [26]:

(a) By providing raft foundation
(b) By providing deep basements that will reduce the net bearing pressure on the soil
(c) By providing piers or piles, so as to transfer the foundation loading to deep and firm/less compressible soil by means of basements
(d) By providing additional loading on lightly loaded areas in the form of embankments.

Many other factors can also cause settlement and ground movement problems. Some of these factors may be shrinkage of clays from ground dewatering or drying out in droughts, tree roots causing disruption, ground movement from nearby excavations, etc.

8.6 Fire Resistance

Concrete is a porous substance bound together by water-containing crystals. The binding material can decompose if heated to too high temperature, with consequent loss of strength. The rise in temperature causes a decrease in the strength and modulus of elasticity for both concrete and steel reinforcement. However, the rate at which the strength and modulus decrease depends on the rate of increase in the temperature of the fire and the insulating properties of concrete. The loss of moisture causes shrinkage and the temperature rise causes the aggregates to expand, leading to cracking and spalling of the concrete. The cement mortar converts to quick lime at temperature of 840 °F (450 °C) and causes surface scaling and cracking.

Reinforcing steel is much more sensitive to high temperatures than concrete. Once the reinforcing steel is exposed by the spalling action, the steel expands more rapidly than the surrounding concrete, causing buckling and loss of bond to adjacent concrete, where the reinforcement is fully encased. High temperature also causes reinforcement to lose strength. Hot-rolled steels (reinforcing bars) retain much of their yield strength up to about 800 °F (426 °C), while cold-drawn steels (prestressing strands) begin to lose strength at about 500 °F (260 °C) [17]. It may be noted that duration of fire (until the reinforcing steel reaches the critical strength) depends on the protection to the reinforcement provided by the concrete cover.

8.6.1 Protective Measures

Concrete is a material with very good fire resistance and protects the reinforcing steel; however, concrete structures must still be designed for fire effects. The resistance to fire of reinforced concrete construction depends primarily on the type of aggregate, the thickness of the various parts comprising the member, and the cover of concrete over the reinforcement, and is rated/expressed by the number of hours of effective fire resistance. The fire resistance rating, in fact, is the time in hours that the assembly is able to withstand exposure to the standard fire before the criterion of failure is reached. Figure 8.12 shows the relation between time rating and temperature based on ASTM time-temperature curve [18] for fire test. In Canada, fire testing as per CAN/ULC-S101 [19] is used to establish fire resisting ratings.

8.6.1.1 Concrete Cover and Member Thickness

Meeting the requirements for fire protection with respect to adequate thickness of member and concrete cover over reinforcement, the recommendations of BS-8110 [5] are produced in Fig. 8.13 and Table 8.5, respectively.

The 2015 International Building Code (IBC) [20] contains prescriptive requirements for building elements in Section 720. This section contains tables describing various assemblies of building materials and finishes that meet specific fire endurance ratings. The tables in the IBC are compatible with the tables in ACI 216.1 except for the provisions for the use of high-strength concrete columns found in ACI 216.1-07 [17]. Hence fire ratings of various reinforced concrete members including concrete cover requirements proposed by IBC [20] (2015 edition), based on type of aggregates, are reproduced in Tables 8.6, 8.7, 8.8, 8.9, and 8.10.

The minimum thickness of members and concrete cover based on recommended fire resistance rating specified under Supplementary Guidelines to the Ontario Building Code (OBC) [21] is also reproduced in Tables 8.11, 8.12, and 8.13. OBC has developed fire resistance ratings based on the type of concretes prepared with using various types of aggregates, such as the following:

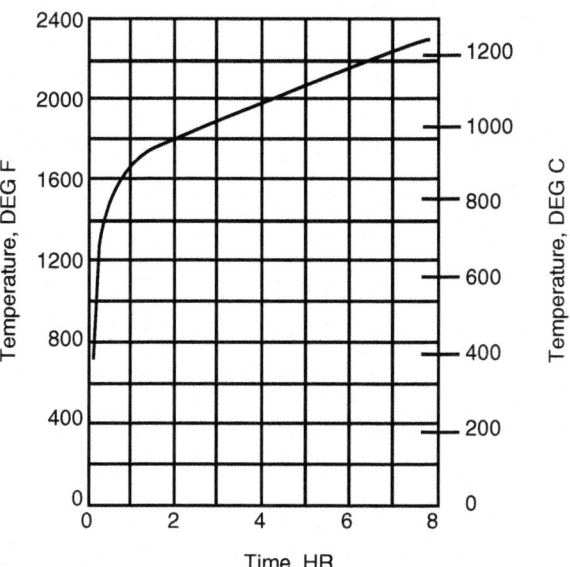

Fig. 8.12 ASTM time-temperature curve for fire test [18] (Copyright ASTM- Reprinted with Permission)

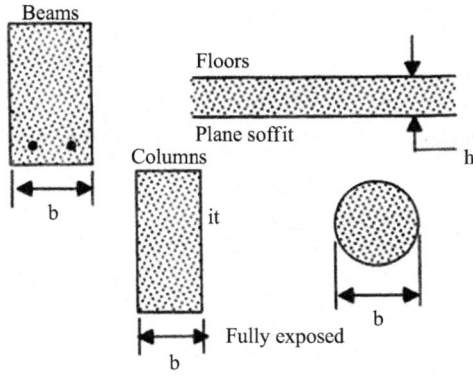

Fig. 8.13 Minimum dimensions of reinforced concrete members for fire resistance based on BS-8110 [5] (*p = area of steel relative to concrete)

Fire resistance	Minimum Dimensions (mm)			
(Hours)	Beam width (b)	Floor thickness (h)	Fully exposed column width (b)	Wall thickness $0.4\%< p^{*} <1\%$
0.5	200	75	150	100
1.0	200	95	200	120
1.5	200	110	250	140
2.0	200	125	300	160
3.0	240	150	400	200
4.0	280	170	450	240

* **p** = area of steel relative to concrete.

1. Type S concrete: In this concrete the coarse aggregate used is granite, quartzite, siliceous gravel, or other dense materials containing at least 30% quartz, chert, or flint.
2. Type N concrete: In this type the coarse aggregate used is cinders, broken brick, blast-furnace slag, limestone, calcareous gravel, trap rock, sandstone, or similar dense material containing not more than 30% of quartz, chert, or flint.

Table 8.5 Nominal cover to all reinforcement to meet specified periods of fire resistance based on BS-8110 [5]

Fire resistance (h)	Nominal cover (mm)				
	Beams		Floors		
	Simply supported (SS)	Continuous (C)	Simply supported (SS)	Continuous (C)	Columns
0.5	20	20	20	20	20
1.0	20	20	20	20	20
1.5	20	20	25	20	20
2.0	40	30	35	25	25
3.0	60	40	45	35	25
4.0	70	50	55	45	25

Table 8.6 Minimum equivalent thickness (in.) for reinforced concrete walls, floor, and roof slabs [20]

Aggregate type	Minimum thickness for fire rating (h)				
	1	1½	2	3	4
Siliceous	3.5	4.3	5.0	6.2	7.0
Carbonate	3.2	4.0	4.6	5.7	6.6
Sand-lightweight	2.7	3.3	3.8	4.6	5.4
Lightweight	2.5	3.1	3.6	4.4	5.1

Source: IBC Tables 722.2.1.1, 722.2.2.1

Table 8.7 Minimum concrete cover for reinforced concrete floor and roof slabs [20]

Aggregate type	Minimum concrete cover (in.) for fire rating of					
	Restrained	Unrestrained				
	4 h or less	1 h	1½ h	2 h	3 h	4 h
Siliceous	¾	¾	¾	1	1¼	$1\frac{5}{8}$
Carbonate	¾	¾	¾	¾	1¼	1¼
Sand-lightweight or lightweight	¾	¾	¾	¾	1¼	1¼

Source: IBC Table 722.2.3(1)

3. Type L concrete: Type of concrete in which all the aggregate is expanded slag, expanded clay, expanded shale, or pumice.
4. Type L_1 concrete: Type of concrete in which all the aggregate is expanded shale.
5. Type L_2 concrete: Type of concrete in which all the aggregate is expanded slag, expanded clay, or pumice.
6. Type L40S concrete: Type of concrete in which the fine portion of the aggregate is sand and low-density aggregate in which the sand does not exceed 40% of the total volume of all aggregates in the concrete.
7. Type $L_1$20S and Type $L_2$20S concretes: Types of concrete in which the fine portion of the aggregate is sand and low-density aggregate in which the sand does not exceed 20% of the total volume of all aggregates in the concrete.

Table 8.8 Minimum concrete cover for reinforced concrete beams[a] [20]

Restraint[b]	Beam Width[c] (in.)	Minimum concrete cover (in.) for fire rating of				
		1 h	1½ h	2 h	3 h	4 h
Restrained	5	¾	¾	¾	1	1¼
	7	¾	¾	¾	¾	¾
	≥10	¾	¾	¾	¾	¾
Unrestrained	5	¾	1	1¼	NP	NP
	7	¾	¾	¾	1¾	3
	≥10	¾	¾	¾	1	1¾

Source: IBC Table 722.2.3(3)
Notes:
[a]The concrete cover for an individual reinforcing bar is the minimum thickness of concrete between the surface of the bar and the fire-exposed surface of the beam. For beams in which several bars are used, the concrete cover for corner bars used in the calculation shall be reduced to one-half of the actual value. The concrete cover for an individual bar must not be less than one-half of the value given in the table, nor less than ¾ in
[b]Tabulated values for restrained assemblies apply to beams spaced more than 4 ft. on centre. For restrained beams spaced 4 ft. or less on centre, a minimum concrete cover of ¾ in. is adequate for a fire rating of 4 h or less
[c]For a beam width between the tabulated values, the minimum concrete cover can be determined by interpolation
"NP" = Not permitted

Table 8.9 Minimum size of reinforced concrete columns [20]

Aggregate type	Minimum thickness for fire rating (in.)				
	1 h	1½ h	2 h	3 h	4 h
Siliceous	8	9	10	12	14
Carbonate	8	9	10	11	12
Sand-lightweight	8	8½	9	10½	12

Source: IBC Tables 722.2.4
Notes:
1. For 2- and 3-h fire ratings, the minimum dimension can be reduced to 8 in. for rectangular columns with two parallel sides at least 36 in. in length
2. For a 4-h fire rating, the minimum dimension can be reduced to 10 in. for rectangular columns with two parallel sides at least 36 in. in length

Table 8.10 Minimum concrete cover for reinforced concrete columns [20]

Aggregate type	Minimum thickness for fire rating (in.)				
	1 h	1½ h	2 h	3 h	4 h
Siliceous	1	1½	2	2	2
Carbonate	1	1½	2	2	2
Sand-lightweight	1	1½	2	2	2

Source: IBC Section 720.2.4.2
Note: 1. Concrete cover is measured to the main longitudinal reinforcement in the column

Table 8.11 Minimum thickness of reinforced and prestressed concrete floor or roof slab (mm) [21]

Type of concrete	Fire resistance rating						
	30 min	45 min	1 h	1.5 h	2 h	3 h	4 h
Type S concrete	60	77	90	112	130	158	180
Type N concrete	59	74	87	108	124	150	171
Type L40S or Type L concrete	49	62	72	89	103	124	140

Source: OBC Table 2.2.1.A

It may be noted that the aggregates in type S and type N concretes apply to the coarse aggregates only. Coarse aggregate for this purpose is considered that retained on a 5 mm sieve as per grading requirements under Canadian Standards Association CSA-A23.1/A23.2.

For complete details standards specified above can be referenced.

For minimum thickness of concrete cover to concrete columns, OBC [21] specifies that:

1. Where the fire resistance requirement is 3 h or less, the minimum thickness of concrete cover (in mm) to vertical steel reinforcement shall be = 25 × number of required fire resistance hours or 50 mm, whichever is less.
2. Where the fire resistance requirement is more than 3 h, the minimum thickness of concrete cover (in mm) to vertical steel reinforcement shall be = 50 + 12.5 times the number of required fire resistance in excess of 3 h.
3. Where the concrete cover in sentence 2 above exceeds 62.5 mm, wire mesh reinforcement with 1.57 mm diameter wire and 100 mm openings shall be incorporated midway in the concrete cover to retain the concrete in position.

8.6.1.2 Aggregates

As regards to type of aggregates, aggregate used in concrete can be classified into three types: carbonate, siliceous, and lightweight. Carbonate aggregates include limestone and dolomite. Siliceous aggregate includes materials consisting of silica and includes granite and sandstone. Lightweight aggregates are usually manufactured by heating shale, slate, or clay.

Generally calcareous aggregates or limestone are of some advantage in concrete structures exposed to fire. During fire, carbonates are converted to oxides, which absorb heat and this heat absorption delays the temperature rise beyond the decomposition temperature. Thus, the use of calcareous aggregates improves the resistance of a concrete member in a fire. However, aggregates containing quartz such as granite, sandstone, and quartzite are more susceptible to fire damage.

In addition to this, the lightweight aggregates have also greater fire resistance, because they have a lesser tendency to spall. This lighter material reduces the thermal conductivity of the concrete and thus insulates the concrete better from

Table 8.12 Minimum concrete cover over reinforcement in concrete slabs (mm) [21]

Type of concrete	Fire resistance rating						
	30 min	45 min	1 h	1.5 h	2 h	3 h	4 h
Type S, N, L40S, or L concrete	20	20	20	20	25	32	39
Prestressed concrete slabs type S, N, L40S, or L concrete	20	25	25	32	39	50	64

Source: OBC Table 2.2.1.B

Table 8.13 Minimum cover to principal steel reinforcement in reinforced concrete beams (mm) [21]

Type of concrete	Fire resistance rating						
	30 min	45 min	1 h	1.5 h	2 h	3 h	4 h
Type S, N, or L	20	20	20	25	25	39	50

Source: OBC Table 2.9.1
Notes:
1. The minimum thickness of cover shown in Table 8.13 is where width of the beam or joist is at least 100 mm
2. Where the upper extension or top flange of a joist or T-beam in a floor assembly contributes wholly or partly to the thickness of the slab above, the total thickness at any point shall be not less than the minimum thickness of slab described in Table 8.11 for the fire resistance required

the heat source. Also, blast-furnace slag is more fire resistant than are other normal-weight aggregates because of its lightness and mineral stability at high temperature [22].

High alumina cements are also being used satisfactorily with suitable aggregates in making castable refractory concrete. Such concretes, at temperatures of 800 °C (1500 °F) or higher, result in formation of ceramic bond, which aids in maintaining a useful strength after cooling, and are found not to crack or expand [15].

The use of polypropylene fibres in the concrete has also been found to provide better results in minimising/preventing concrete spalling in a fire. Many researchers consider that concrete with polypropylene fibres is less sensitive to spalling, as during fire exposure the fibres melt and form an "escape route" for steam pressure. Moreover, it has been observed during test conducted at TNO Centre for Fire Research at the Netherlands that monofilament fibres behave better than coarser fibrillated fibres and monofilament fibres may prevent or at least limit explosive spalling to an acceptable level [23].

8.6.2 Structural Steel

Although structural steel is not a part of reinforced concrete, a brief description of protective measures for steel (being extensively used in buildings) from the fire effects is provided here. Fire safe steel buildings use active and passive fire protection systems. An active fire protection system is where sprinklers eliminate the heat source by fire suppression. Smoke or heat detectors generate an alert that initiates the extinguishing action.

Passive fire protection materials insulate steel structures from the effects of the high temperatures that may be generated in fire. The passive methods used to protect steel are:

- Fire-resisting boards
- Vermiculite cement sprays
- Cementitious and fibre sprays
- Dry linings
- Mineral wool
- Intumescent coatings
- Concrete encasement

Above methods can be categorised into four systems [24]:

- Directly applied systems: Under these systems, fire protection materials are used to insulate steel structures from the effects of the high temperatures, such as spray-applied fire-resistive materials and intumescent coatings.
- Membrane systems: These systems provide a thermal barrier against heat, such as gypsum wallboard.
- Water systems: These systems provide a cooling effect, such as water-filled hollow structural sections within steel structural frames to dissipate heat from fire.
- Concrete systems: Under these systems, structural steel is encased in concrete to slow down the conduction of heat.

Of these systems, intumescent paint has the capability of enhancing the appearance of exposed steelwork. Intumescent coatings are being used extensively in large public buildings simply because they allow the steel designer greater creative freedom without the disruptive element of sprayed fibre or board. Intumescent coatings provide the fastest, thinnest, and most aesthetically pleasing of all fireproofing methods. An additional advantage to intumescent fireproofing is that it can be applied off-site by the steel fabricator or in a specialised painting shop, reducing costs and minimising disruption for other trades.

Low-pressure spray-applied commercial and industrial structural steel fire protection products use exfoliated vermiculite to improve the application characteristics and to impart a high degree of fire resistance. Exfoliated vermiculite is very efficient at retaining moisture, and in the event of a fire this turns to steam which has a cooling effect on the steel substrate and thus delays its temperature rise. Vermiculite concentrate is also used in the production of fire-resistant gypsum plasterboard (drywall or wallboard). Vermiculite spray consists of factory-produced blend of exfoliated vermiculite, cementitious binders, and rheological and mix dispersing agents supplied as a dry mix to which clean water is added on site. It does not respond to any form of expansion, foaming, or chemical reaction to impart its fire protection properties.

Fig. 8.14 Spray-applied cementitious material fireproofing

Cementitious materials are also used commonly to protect structural steel from fire effects. Cementitious materials contain binders of Portland cement or gypsum in the range of 50–80% by weight which when mixed with water forms a slurry suitable for pumping and wet spray application. The high binder content and wet mixing assure a strong, durable, and homogeneous coating (Fig. 8.14) that is totally integrated throughout the applied thickness.

Bibliography

1. Road Research Laboratory, Ministry of Transport, Concrete Roads Design and Construction, published by Her Majesty's Stationary Office, printed by Lowe and Brydone Ltd, London, 1966.
2. Portland Cement Association, Building Movements and Joints, Illinois, USA, 1982.
3. BS 5337-76 (Amended-82), Code of Practice for the Structural Use of Concrete for Retaining Aqueous Liquids.
4. Technical Report presented by Peter Coombs and Brian Davies of FOSROC in Seminar on "Sealant Technology", Bahrain, November 1993.
5. British Standard Institution, BS 8110- Part 1: 1985, Structural Use of Concrete: Code of Practice for Design and Construction.
6. BS 5328, Concrete: Part-1, Guide to Specifying Concrete-1991 (Amended 1995); Part-2, Methods for Specifying Concrete Mixes- 199 1 (Amended 1995); Part-3, Specification for the Procedures to be Used in Producing and Transporting Concrete-1990 (Amended 1992); Part-4, Specification for the Procedures to be used in Sampling, Testing and Assessing Compliance of Concrete-1990 (Amended 1995).
7. ACI 318-95 (Revised 2014), Building Code Requirements for Structural Concrete.
8. Canadian Standards Association (CSA) A23.1-14 - Concrete materials and methods of concrete construction/Test methods and standard practices for concrete.
9. A.M. Neville and J.J. Brooks, Concrete Technology, E.L.B.S. Longman, Singapore, 1993.
10. ACI 201.2R-92 (Revised 2016), Guide to Durable Concrete.

11. American Concrete Institute (ACI), Publication SP-15(95), Field Reference Manual, ACI 301-96 (Revised 2016).
12. Mark Mitchell, Freeze/Thaw Protection Concrete, Concrete Society Journal CONCRETE Vol. 34, No. 6, June 2000.
13. Concrete Information- R&D Serial No. 2617- 2002; Portland Cement Association.
14. Peter H. Emmons, Concrete Repair and Maintenance Illustrated, R. S Means Company Inc. USA 1993.
15. Hubert Woods, Durability of Concrete Construction, ACI Monograph No. 4, USA 1984.
16. S. Chandra, Chalmers University of Technology, Sweden- "Durability problems in concrete". 25th Conference on **OUR WORLD IN CONCRETE & STRUCTURES:** 23 - 24 August 2000, Singapore.
17. David N. Bilow and Mahmoud E. Kamara (Portland Cement Association)- Fire and Concrete Structures- Structures 2008: Crossing Borders; ASCE (American Society of Civil Engineers).
18. ASTM E119 (Revised 2018), Standard Methods of Fire Tests of Building Construction and Materials.
19. CAN/ULC-S101-14 Standard Methods of Fire Endurance Tests of Building Construction and Materials- Fifth Edition
20. International Code Council – ICC, International Building Code, 2015.
21. 2012 Ontario Building Code Compendium- MMAH Supplementary Standards SB-2 Fire Protection Ratings- September 14, 2012.
22. ACI 221R-89 (Reapproved 2001), Guide for Use of Normal Weight Aggregate in Concrete.
23. Nick Varley, Fire Protection of Concrete Linings in Tunnels, Concrete Society Journal CONCRETE, Vol. 33, No. 5, May 1999.
24. George S. Frater, Ph.D. P.Eng, Canadian Steel Construction Council- **Fire and Structural Steel,** Canadian Consulting Engineer - Magazine for professional engineers in construction, May 1, 2006.
25. ASTM Special Technical Publication No: 169- A- (1975) USA. Significance of Tests and Properties of Concrete and Concrete- making Materials.
26. M.J. Tomlinson, *Foundation Design and Construction,* E.L.B.S. Longman, Singapore, 1992.

Chapter 9
Chemical Attack

9.1 Introduction

In general, concrete has a low resistance to chemical attack. There are several chemical agents, which react with concrete. The effects of some of the more common chemicals on the deterioration of concrete based on ACI 201.2R [1] are summarised in Table 9.1.

Deterioration is usually caused by the chemical reaction between the hardened cement constituents of concrete and the chemicals of a solution. The products that are formed after chemical reaction may be either water soluble, and may get removed from the internal structure of concrete by a diffusion process, or the reaction products if insoluble in water. These reaction products may get deposited on the surface of concrete as an amorphous mass, having no binding properties, with the result that it can be easily washed out from the concrete surface.

The factors which accelerate or aggravate chemical attack on concrete are usually high porosity, cracks, leaching, and liquid penetration due to flowing liquid or ponding or hydraulic pressure. Thus, where chemical attack of concrete is expected, special precautions should be taken with the choice of cement, aggregates, and use of admixtures. The resistance to chemical attack improves with increased impermeability and with the additional provision of membranes and protective-barrier systems.

The common forms of chemical attack on concrete and reinforcement can be classified under the following headings:

1. Chloride attack
2. Sulphate attack
3. Carbonation
4. Alkali-aggregate reaction
5. Acid attack

© Springer Nature Switzerland AG 2019
A. Surahyo, *Concrete Construction*,
https://doi.org/10.1007/978-3-030-10510-5_9

Table 9.1 Effect of commonly used chemicals on concrete, based on ACI-201.2R-92 [1]

Rate of attack at ambient temperature	Inorganic acids	Organic acids	Alkaline solutions	Salt solutions	Miscellaneous
Rapid	Hydrochloric Nitric Sulphuric	Acetic Formic Lactic	–	Aluminium chloride	–
Moderate	Phosphoric	Tannic	Sodium[a] hydroxide >20%	Ammonium nitrate Ammonium sulphate Magnesium sulphate Calcium sulphate	Bromine (gas) Sulphite liquor
Slow	Carbonic	–	Sodium[a] hydroxide 10–20% Sodium hypochlorite	Ammonium chloride Magnesium chloride Sodium cyanide	Chlorine (gas) Seawater Soft water
Negligible	–	Oxalic Tartaric	Sodium[a] hydroxide <10% Sodium hypochlorite Ammonium hydroxide	Calcium chloride Sodium chloride Zinc nitrate Sodium chromate	Ammonia (liquid)

[a]The effect of potassium hydroxide is similar to that of sodium hydroxide

9.2   Chloride Attack

Chlorides can be introduced into the concrete either during or after construction as follows:

1. Before construction: Chlorides can be admitted into concrete at the mixing stage through admixtures containing calcium chloride, through using mixing water contaminated with saltwater, or improperly washed aggregates.
2. After construction: Soluble chlorides in deicing salts or occurring naturally in soils, seawater, and groundwater can enter concrete by absorption, through its surface, by capillary attraction along interconnected voids, or by direct access through cracks in the concrete.

9.2.1   Effects of Chloride Attack

Chloride ions in concrete adjacent to the steel can readily destroy, even at high alkalinity, the passive oxide film, which normally protects the steel against corrosion. The presence of chloride ions not only effectively initiates the electrochemical process of corrosion, subject to the availability of moisture and oxygen, but it also increases the electrical conductivity of the concrete and thus results in an increased corrosion. Chlorides are not consumed in the corrosion process, but instead act as catalyst to the process and remain in the concrete.

As the rust layers build, tensile forces generated by expansion of the oxide cause the concrete to crack and delaminate. Spalling of delamination occurs, if the natural forces of gravity or traffic wheel load act on the loose concrete. When cracking and delamination progress, accelerated corrosion takes place because of easy access of corrosive salts, oxygen, and moisture. Corrosion then begins to affect rebars buried further within the concrete.

The concentration of chlorides necessary to promote corrosion, among other factors, is greatly affected by the concrete pH. It was demonstrated that a threshold level of 8000 ppm of chloride ions was required to initiate corrosion when the pH was 13.2. As the pH was lowered to 11.6, corrosion was initiated with only 71 ppm of chloride ions [2].

Discoloration of concrete flat work has been associated with the use of calcium chloride (Greening and Landgren 1966 [23]). Two major types of mottling discoloration can result from the interaction between cement alkalis and calcium chloride. The first type has light spots on a dark background and is characteristic of mixtures in which the ratio of cement alkalis to chlorides is relatively high.

Available evidence indicates that the magnitude and permanence of discoloration increase as the calcium chloride concentration increases from 0 to 2% by weight (mass) of cement. This second type of discoloration can be aggravated by high rates of evaporation during curing and improper placement of vapour barriers. Use of continuous fog spray or curing compounds can help alleviate this problem.

9.2.2   Protective Measures

Adequate protection against external sources of soluble chlorides can generally be achieved by the use of properly designed mixes producing dense impermeable concrete, which can be obtained by using low w/c ratio and special additives, such as microsilica. The use of slag cement or Portland-pozzolana (type P) cements is found to be helpful in restricting the mobility of chloride ions within the hydrated cement paste.

Good-quality concrete cover also plays a very important role to control the chloride-induced corrosion. Tables 7.4, 7.5, and 7.6 specify the requirements for durability of reinforced and prestressed concretes exposed to various conditions. Extra protection can be obtained by applying surface-applied penetrating sealers (silane or siloxane based) and coatings/membranes (epoxies, urethanes, chlorinated rubber, and methacrylate).

For using calcium chloride or chloride-based admixtures, most of the codes specify very low limits that effectively ban the use of these admixtures in concrete containing embedded metal. The limits for total chloride contents recommended by

Table 9.2 Limits of chloride content of concrete, specified by BS-8110 [7]

Type or use of concrete	Maximum chloride content expressed as percentage of chloride ion in weight of cement (inclusive of PFA or slag)
Prestressed concrete, heat-cured concrete containing embedded metal	0.1
Concrete made with sulphate-resisting Portland (type V) cement or supersulphate cement	0.2
Concrete containing embedded metal and made with the following cements: 1. Ordinary Portland (type I) 2. Rapid-hardening Portland (type III) 3. Portland blast-furnace (type IS) 4. Low-heat Portland (type IV) 5. Low-heat Portland blast furnace (type I and IP) or combination with slag or PFA	0.4

Table 9.3 Maximum allowable chloride ion content of concrete specified by ACI-301-96 [3]

Type	Maximum water-soluble chloride ion in concrete by weight of cement
Prestressed concrete	0.06
Reinforced concrete exposed to chloride in service	0.15
Dry reinforced concrete or concrete which will be protected from moisture in service	1.00
Other reinforced concrete construction	0.30

BS-8110 are given in Table 9.2 and those of ACI-301 [3] in Table 9.3. CSA [4] also recommends that the water-soluble chloride ion content by mass of the cementing materials in the concrete before exposure shall not exceed the values proposed by ACI-301 [3] in Table 9.3.

The use of calcium chloride as an accelerator will aggravate the effects of poor-quality concrete construction, particularly when the concrete is exposed to chlorides during service. Adherence to the limits just mentioned does not guarantee absence of corrosion if good construction practices are not followed. The use of calcium chloride also increases shrinkage and creep, and lowers the resistance of air-entrained concrete to freezing and thawing at later ages [5].

The use of saline water from sea or other sources for mixing or curing purposes should be avoided as it contains chlorides. The use of such water is already discussed in Chap. 6. The chlorides may also be present in aggregates; hence washed fine and coarse aggregates should be used in concrete to avoid the intrusion of chlorides by this source.

In order to control chloride penetration to the reinforcement through cracks, the use of an elastomeric joint sealant placed in a cut recess has been proved very effective. Such crack repair method is known as "rout and seal" [6]. The method involves enlarging the crack along its exposed face and filling and sealing it with a suitable joint sealant. The sealing materials, which remain elastomeric, can be selected from silicones, polysulphides, asphaltic materials, urethanes, or epoxies.

9.3 Sulphate Attack

A concrete structure that is in contact with sulphates can be subjected to varying degrees of attack. Sodium, calcium, and magnesium sulphates are most active and occur widely in soils (particularly clays), groundwater, and seawater. Ammonium sulphate is also aggressive, which is mostly present in agricultural land, where it has been used as a fertiliser.

9.3.1 Effects of Sulphate Attack

Salts attack concrete only when present in solutions and not in solid form. Usually, the damage to concrete attacked by sulphates starts at the edges and corners, followed by cracking and spalling of the concrete. All sulphates are potentially harmful to concrete. Most sulphate solutions react with calcium hydroxide $Ca(OH)_2$, and calcium aluminate C_3A, of hydrated cement to form calcium sulphate and calcium sulpho-aluminate. Crystallisation of the new compounds is accompanied by an increase in molecular volume, which causes expansion and disintegration of the concrete at the surface. The attack is greater in concrete exposed to wet/dry cycling. When water evaporates, sulphates can accumulate at the concrete surface, increasing in concentration and their potential for causing deterioration.

The intensity and rate of sulphate attack depend on a number of factors such as type of sulphate (magnesium sulphate is the most vigorous), its concentration, and the continuity of its supply to concrete. The concentration of sulphates in solution is expressed in parts of SO_4 or SO_3 per million (ppm) by weight. When results are expressed as SO_3 they may be converted to SO_4 by multiplying with a factor of 1.2. A concentration of 1000 ppm (1.0 g/L) is considered to be moderately severe, and 2000 ppm very severe, especially if magnesium sulphate is the predominant constituent. Permeability and presence of cracks also affect the severity of the attack.

9.3.2 Protective Measures

The type of cement is a very important factor. Since it is C_3A that is attacked by sulphates, the vulnerability of concrete to sulphate attack can be reduced by the use of cement low in C_3A, viz. sulphate-resisting (type V) cement. Improved resistance is also obtained by the use of Portland blast-furnace cement (type IS) and Portland-pozzolana cement (type P). However, the essential feature of resisting sulphate attack is to produce a dense and an impermeable concrete.

Typical requirements given in ACI-318 [9] and BS-5328 [8] for concrete exposed to sulphate attack in soil, groundwater, or seawater are given in Tables 9.4 and 9.5 (based on cement group as per Table 9.6). CSA-A23.3 [10] recommends that w/c ratios for submerged portions of a structure should not exceed 0.45 by mass and for portions in splash zone and above should not exceed 0.40 by mass. W/C ratio as

Table 9.4 Requirements for concrete exposed to sulphate attack based on ACI-318 [9]

Sulphate exposure	Water-soluble sulphate (SO₄) in soil % by weight	Sulphate (SO₄) in water ppm	Types of cement	Normal-weight aggregate concrete Maximum free w/c ratio	Lightweight aggregate concrete Minimum compressive strength in MPa (psi)
Negligible	0.00–0.10	0–150	–	–	–
Moderate (seawater)	0.10–0.20	150–1500	Modified (type II), Portland pozzolana (type IP/MS), Portland blast-furnace (type IS/MS)	0.50	28 (4000)
Severe	0.2–2.0	1500–10,000	Sulphate-resisting Portland (type V)	0.45	31 (4500)
Very severe	Over 2.0	Over 10,000	Sulphate-resisting Portland (type V) plus pozzolana	0.45	31 (4500)

high as 0.50 by mass may be used for submerged areas provided that the C_3A content of the Portland cement does not exceed 8%. Proper proportioning, silica fume, fly ash, and ground slag generally improve the resistance of concrete to sulphate attack. Calcium chloride shall not be used as an admixture in reinforced concrete to be exposed to seawater. The resistance of concrete to sulphate attack can be tested by storing the specimen in a solution of sodium or magnesium sulphate or in a mixture of the two.

For concrete structures housed in sulphate-bearing soils, protective coatings may also be applied, which will reduce the penetration of salt into the concrete, and in severe exposure it may be wise to give this kind of protection to new concrete to increase its durability. When concrete already shows signs of deterioration and tests indicate that enough salt is present at the reinforcement to allow it to rust, using a waterproof coating will not help in reducing the possibility of further salt penetration as it is too late to get benefit from such coatings.

Due to waterproof coatings, water vapour pressure is built up behind them especially if water can get into the concrete from another face. This buildup of vapour pressure can cause the coating to blister and peel off if adhesion to the concrete is inadequate. This problem can be overcome by using a coat of primer, which penetrates into the surface of the concrete and gives a foothold for the surface coating [11].

BS-5328 [8] suggests that for very high sulphate concentrations, as mentioned at serial No. 6 of Table 9.5, some form of protection such as sheet polyethylene or polychloroprene or surface coating based on asphalt, chlorinated rubber, epoxy, or polyurethane materials should be used to prevent access by the sulphate solution. Moreover, the general information on type of coatings, material, and efficiency is summarised in Table 9.7.

Table 9.5 Requirements for concrete exposed to sulphate attack based on BS-5328 [8]

Concentration of sulphate and magnesium						Requirements		
	In soil or fill					Dense fully compacted concrete made with 20 mm nominal maximum size aggregate		
In groundwater (g/L)		By acid extraction	By 2:1 water:soil extract		Cement group from Table 9.6	Minimum cement content	Maximum free w/c/ ratio	
SO$_4$ (g/L)	Mg (g/L)	SO$_4$%	SO$_4$ (g/L)	Mg (g/L)	–	kg/m^3	–	
(1) <0.4	–	<0.24	<1.2	–	1, 2, 3	–	–	
(2) 0.4– 1.4	–	Classify on the basis of a 2:1 water:soil extract	1.2– 2.3	–	1 2 3	330 300 280	0.50 0.55 0.55	
(3) 1.5– 3.0	–		2.4– 3.7	–	2 3	340 320	0.50 0.50	
(4) 3.1– 6.0	≤1.0		3.8– 6.7	≤1.2	2 3	380 360	0.45 0.45	
(5) 3.1– 6.0	>1.0		3.8– 6.7	>1.2	3	360	0.45	
(6) >6.0	≤1.0		>6.7	≤1.2	Same as 4 plus surface protection			
(7) >6.0	>1.0		>6.7	>1.2	Same as 5 plus surface protection			

Note: Adjustments to minimum cement contents should be made for aggregates of nominal size other than 20 mm as mentioned in note 2 of Table 4.1

Table 9.6 Types of cement based on BS-5328 [8]

Group	Description
1	(a) Portland cement conforming (b) Portland blast-furnace cements (c) High slag blast-furnace cement (d) Portland pulverised-fuel ash cements (e) Pozzolanic pulverised-fuel ash cement (f) Combinations of Portland cement conforming with ggbs (g) Combinations of Portland cement conforming with pulverised-fuel ash
2	(a) Portland pulverised-fuel ash cements containing not less than 26% of pfa by weight of the nucleus or combination of Portland cement with pfa with not more than 40% pfa and not less than 25% pfa by weight of the combination (b) High-slag blast-furnace cement containing not less than 74% slag by weight of nucleus or combination of Portland cement with ggbs not less than 70% and not more than 85% by weight of the combination
3	Sulphate-resisting Portland cement

Note: The nucleus is the total weight of the cement constituents excluding calcium sulphate and any additives such as grinding aids. *Pfa* Pulverised fly ash, *ggbfs* ground granulated blast-furnace slag

Table 9.7 Waterproof treatment to reduce the penetration of saltwater into new concrete (general information) [11]

Type of coating	Protective material	Efficiency	Life
1. Hot-applied mastic asphalt tanking (black)	Bitumen	Very high if the coating is complete. Able to span cracks	Over 50 years if covered by backfill. Probably 20 years or more if exposed, depending on climate
2. Tanking with heavy-performed sheet materials	1. Bitumen on inorganic fabric base alone or in combination with other materials (black) 2. PVC, polyurethane, butyl rubber (black), chloro-sulphonated polyethylene	Very high if properly bonded and if coating is complete. Able to span cracks	Over 20 years if covered by backfill. Up to 20 years if exposed, depending on exposure. Between 5 and 20 or more years depending on material and exposure
3. O-building paper	Bitumen on organic paper	Can be high but only if properly bonded to the concrete so that protection is complete. Able to span cracks	Short
4. Polythene sheet	Low-density polyethylene	–	10 years or more if not exposed to UV light
5. Liquid surface coatings	Pitch or coal tar epoxy with 250 μm dry coating thickness (black)	Very high if well applied and free from pin holes. Not able to span active cracks. Should be applied in more than one coat	Up to 20 years depending on exposure
	Epoxy resin, two pack polyurethane coatings 250 μm thick	Very good, but recoating is difficult because of poor inter-coat adhesion. Not able to span active cracks. Should be applied in two coats, the second coat of contrasting colour while first is still tacky	Depends on thickness and conditions. Thick (250 μm) coatings can have long life of more than 10 years
	Solvent-based acrylic, methacrylate, styrene-acrylic, one-pack polyurethane chlorinated rubber	Good, these paints combine protective and decorative qualities. Multi-coat system essential to reduce incidence of pin holes. Not able to span active cracks	Up to 10 years depending on exposure
	Emulsion-based acrylic or styrene-butadiene polymers and co-polymers with or without other materials	Fair, emulsion-based paints are not truly impermeable, but still give some resistance to water penetration. Some emulsion paints are as effective as solvent-based paints in protecting against CO_2 corrosion. Not normally able to span active cracks	Up to 10 years depending on exposure

(continued)

Table 9.7 (continued)

Type of coating	Protective material	Efficiency	Life
6. Water-repellent treatment	Silicone, silane, siloxane	These coatings make the surface water repellent without making it waterproof. These materials are easy to reapply after the years and some form good primers for other type of coatings. Concrete must be dry to ensure good penetration	Probably about 10 years, but can be longer

9.4 Carbonation

Carbon dioxide (CO_2) is present in air in small concentrations. In normal conditions, the concentration is about 0.03 percentage of the volume of air. CO_2 in the presence of moisture forms carbonic acid (H_2CO_3), which reacts with calcium hydroxide Ca $(OH)_2$ present in hydrated concrete to form calcium carbonate ($CaCO_3$). This process is known as the carbonation of concrete.

Carbonation is slow in saturated concrete because the pores are blocked with water, and also slow in very dry concrete because there is little pore fluid to react with. Carbonation begins at the concrete surface and slowly penetrates deeper, typically at a rate of about 0.039 in. per year in high-quality concrete with a low water-cementitious material ratio [12]. For carbonation to progress in severity to bi-carbonation, where carbonation reactions continue to penetrate the concrete at greater depths, additional carbon dioxide is needed at those levels. Additional carbon dioxide can travel further into concrete via cracks that allow ingress of gases, water, and other contaminants. A high w/cm also increases the likelihood of bi-carbonation, which ultimately results in porous, friable, and weak concrete.

9.4.1 Effects of Carbonation

Carbonation is one of the important factors that cause reduction of alkalinity in the concrete. Due to carbonation, the alkalinity of concrete is reduced to a pH value as low as 8.5 [13] and consequently concrete protection of the reinforcing steel is lost. As a result of this process, the concrete itself is not harmed, but the reinforcement can be seriously affected by corrosion process, which leads to cracking and subsequent spalling of the concrete.

The carbonation of concrete also causes shrinkage. Carbonation shrinkage may amount to as much as one-third of the total shrinkage [14]. The ambient humidity conditions and the sequence of CO_2 application are major factors affecting the extent of shrinkage. One of the causes of surface crazing of concrete, however, is the shrinkage that accompanies natural air carbonation of young concrete.

The concrete will have greater depth of carbonation with a high w/c ratio and inadequate curing. Carbonation penetrates more deeply into the concrete where it is poorly compacted (honeycombed). On hardened concrete, carbonation may reach a depth of an inch (25 mm) [15] under varying exposure conditions over a period of many years. The process requires constant change in moisture levels from dry to damp to dry. Carbonation will not occur when concrete is constantly under water. The extent of carbonation can be determined by treating a freshly broken surface with phenolphthalein; the free $Ca(OH)_2$ is coloured pink white while the carbonated portion remains uncoloured [5].

Carbonation or another kind can also occur to fresh unhardened concrete, and cause a soft chalky surface. This carbonation usually takes place during cold-weather concreting when there is an unusual amount of carbon dioxide in the air due to direct-fired heating or gasoline-powered equipment. It is not accompanied by significant movements or cracking, and after the concrete is 24 h old this danger no longer exists.

The environmental conditions that are also important are the relative humidity and temperature. The rate of carbonation increases with the temperature and humidity. Carbonation takes place most quickly at relative humidity around the level that is required for greatest comfort, about 65%. Hence it tends to be fast inside occupied buildings.

9.4.2 Protective Measures

In order to minimise the attack of carbonation and to prevent the ingress of CO_2, it is important to achieve relatively impermeable concrete with adequate cover to reinforcement like in case of chloride attack prevention. The properly designed concrete mix with controlled w/c ratio, adequate compaction, and curing is the main important factor in producing dense impermeable concrete. For various exposure conditions the importance of cover to reinforcement including maximum w/c ratio and minimum required cement content is discussed in Chap. 7. In good-quality concrete, the carbonation process is very slow. It has been estimated that the process will proceed at a rate up to 0.04 in. (1 mm) per year [16].

However, extra protection can be obtained by applying surface protection systems, e.g. a high-density, low-vapour transmission coatings like acrylate, membrane, or impregnating material like sodium silicate. Mortar surfacing of polymer applied by shotcrete has also proved helpful. These surface-applied barriers due to low-vapour transmission allow the un-carbonated concrete to re-alkalise (increase pH) the carbonated concrete, pushing the carbonated front (carbonation depth) back towards the surface.

Re-alkalisation can also be done by electrochemical treatment method. In this method, highly alkaline electrolyte is transported by applying an electrical field between the reinforcement steel in the concrete, as the cathode, and an externally mounted electrode mesh, the anode. The electrode mesh is embedded in an alkaline solution, usually sodium carbonate. During the treatment, the alkaline solution is transported into carbonated concrete, through a porous medium, the concrete, with applied electrical field as the driving force [17]. Cracks in concrete may allow carbonation to penetrate relatively quickly to the areas around reinforcing steel. Elastomeric crack sealants have proved very helpful in crack bridging as is discussed for control of chloride penetration through cracks.

9.5 Alkali-Aggregate (Silica) Reaction (ASR)

Deterioration of concrete in bridges, pavements, and other structures has occurred in numerous places due to a chemical reaction between the active silica constituents of the aggregate and the alkalis in the cement; this process is known as alkali-silica or alkali-aggregate reaction.

9.5.1 Effects of ASR

The reaction starts with the attack of siliceous minerals, in the aggregate by alkaline hydroxides, derived from the alkalis (Na_2O and K_2O) in the cement, and forms a gel around the reacting aggregates. The alkali-silicate gel formed attracts water by absorption or by osmosis, and thus tends to increase in volume. Since the gel is confined by the surrounding cement paste, internal pressure results and eventually leads to expansion, cracking, and disruption of cement paste, and to map-cracking or pattern-cracking of concrete.

The rock materials, which are potentially reactive with the alkali in cement, are [18] opal (amorphous), quartz, chalcedony (cryptocrystalline fibrous), cristobalite and tridymite, rhyolitic, dacitic, latitic, or andesitic glass. All these materials are highly siliceous. Some igneous and metamorphic siliceous rocks are also potentially reactive. The expansion may exceed 1.500% in mortar or 0.500% in concrete and can cause the concrete to fracture and break apart. The alkali-aggregate reaction may go unrecognized for some period of time, possibly years, before associated severe distress will develop.

9.5.2 Protective Measures

For ASR to occur, three conditions must be present:

- Reactive forms of silica in the aggregate
- High-alkali (pH) pore solution (water in the paste pores)
- Sufficient moisture

If any one of these conditions is absent, ASR gel cannot form and deleterious expansion from ASR cannot occur. Therefore, the best way to avoid ASR is through good mixture design and material selection.

BS-8110 [7] states that the reaction only occurs when cement with high alkali content is used with aggregate having reactive material. However, a high moisture level, within the concrete, is also one of the causes for this reaction.

To minimise the risk of alkali-aggregate reaction, the following precautions are recommended by BS-8110 and ACI-201 [1]:

1. Avoid the use of reactive aggregates, particularly where the concrete is to be exposed to seawater or other environments where alkali is available to enter the concrete in solution from an external source.
2. After curing, keep the concrete dry.
3. Take measures to reduce the degree of saturation of the concrete like provision of impermeable membranes.
4. Use Portland cements with an alkali content of not more than 0.6% (low-alkali cement) expressed as Na_2O.
5. As a mixing water for concrete containing reactive aggregates, the use of seawater, alkaline soil waters, or others containing significant amounts of alkalis should be avoided.
6. Limit the alkali content of the concrete mix to 3.0 kg/m^3 of Na_2O equivalent.
7. Use suitable pozzolanic material or blast-furnace slag that does not increase drying shrinkage. Use ground granulated blast-furnace slag (ggbfs) or pfa (fly ash) as composite cements or replacement materials in order that at least 50% ggbfs or 30% pfa by weight of the combined material is introduced in the mix.

CSA-A23.1-14 [4] also suggests that the risk for deleterious expansion and cracking of concrete due to alkali-silica reaction can be minimised by avoiding the use of reactive aggregates. Reducing the cement content or using a cement with lower alkali content, or both, may also be useful. Using appropriate amounts of supplementary cementing materials (e.g. fly ash, ground granulated blast-furnace slag, silica fume) and lithium-based admixtures can be effective in preventing or reducing expansion due to alkali-silica reactions.

CSA-A23.1-14 further specifies that fly ashes with low-to-moderate calcium contents that meet the requirements of CSA A3001 for type F and type C1 Fly Ash are generally effective in controlling expansion of concrete when used at moderate levels of replacement of 15–30%. Fly ashes with higher calcium contents that would be classed as type CH fly ash by CSA A3001 are generally not effective unless they are used at replacement levels in excess of 30% and in many cases up to 50%. To control ASR reaction, experiments are also underway on Lithium compounds (as concrete admixtures). Lithium ions minimise the reaction between the alkalis and aggregate damaging expansivity by modifying chemical composition of the reaction products. However, it has been found that the level of lithium required to control deleterious expansion due to alkali-silica reaction varies depending on the alkali content of the concrete and the nature and reactivity of the aggregate.

9.6 Acids

Concrete is susceptible to acid attack because of its alkaline nature. The components of the cement paste break down during contact with acids. In damp conditions, sulphur dioxide (SO_2) and carbon dioxide (CO_2) present in the atmosphere form acids, which attack concrete. Sewage also leads to formation of sulphuric acid. Industrial and mine waters may contain or form acids, which attack concrete. Clay and peat soils sometimes contain pyrite (iron sulphide), which produces sulphuric acid.

9.6.1 Effects of Acids on Concrete

Concrete is not acid resistant and acid attack may remove a part of the set concrete. The reaction between acid and calcium hydroxide $Ca(OH)_2$ of hydrated Portland cement produces water-soluble calcium compounds, which are leached away. When limestone aggregates are used, the acid may dissolve them. Strong solutions of sulphuric, sulphurous, hydrochloric, nitric, hydrobromic, or hydrofluoric acids are very harmful, and concrete will be destroyed by prolonged contact with any of these. Weaker solutions will attack more slowly, but sometimes significantly. Such attacks take place in various industrial conditions, such as chimneys, and in some agriculture conditions, such as floors of dairies, and in domestic sewers where sulphuric acid is produced. Animal wastes contain substances which may oxidise in air to form acids which attach concrete. In general, the attack occurs at values of pH below 6.5, whereas a pH of less than 4.5 leads to severe attack.

Acid rain, which comprises mainly of sulphuric acid and nitric acid with a pH value between 4.0 and 4.5, causes surface weathering of exposed concrete. When concrete is attacked by acid rain, neutralisation reaction occurs between acid rain and $Ca(OH)_2$ in the concrete. The pH value in the concrete drops and the calcium-silicate-hydrate (C-S-H) decomposes. Since the cement paste erodes from the surface, sand and aggregates are exposed to corrode reinforcement bars resulting in concrete cracking.

Concrete is also attacked by water containing free CO_2, such as mineral waters, which may also contain hydrogen sulphide. Flowing pure water, formed by melting ice and containing little CO_2, also dissolves $Ca(OH)_2$, thus causing surface erosion. Peaty water with CO_2 is particularly aggressive; it can have a pH value as low as 4.4. To avoid this, the use of calcareous, rather than siliceous aggregate is advantageous because both the aggregate and the cement paste are eroded [19].

Acid Attack in Sewers

Problems of acid attack on concrete above water level in sewers are of very serious nature and are usually assumed to be the result of sulphuric acid. This acid is formed from hydrogen sulphide gas, generated by bacteria from sulphur compounds in the sewage, which rises and combines with oxygen and with moisture condensed on upper surfaces of the sewer conduit/pipe. The cement of conduit is gradually dissolved and progressive deterioration of concrete takes place. The acid continuously eats into the concrete, and then the steel reinforcing, leading to failure of the sewer.

9.6.2   Protective Measures

For most sewers, the problems of acid corrosion in the crown portions can be solved by methods that prevent the formation of sulphuric acid. These include proper sewer design or operating modifications that result in the sewers running full or with ventilation at higher velocities. Lower temperature also tends to reduce the production of acid-forming hydrogen sulphide gas. When sewage is stagnant or moves slowly, bacterial creation of the sulphide is often too rapid to be oxidised by air dissolved in the sewage. Use of chemicals to prevent conversion of sulphur compounds to hydrogen sulphide, or use of toxic materials to decrease or eliminate the activity of aerobic bacteria, can also be helpful.

For all new concrete construction, protection inside of manholes and the inside crown of pipes above the waterline can be achieved with a sheet of acid-resistant PVC, mechanically anchored to the concrete. However, for the repair of old concrete construction, concrete surface needs to be refurbished and then it can be protected with an applied coating or lining. Current methods used to protect concrete include the application of roll-on plastic sheets which are cemented to the concrete. Extra protection can be obtained by applying surface treatments with coal-tar pitch, rubber or bituminous paints, epoxy resins, and other agents. For more details, the typical protection systems are listed in Table 9.7.

Additionally, as presented in 78th WIOA Victorian Water Industry Operations Conference in Australia (2015), one of the solutions to stop corrosion and deterioration of reinforced concrete sewers is by spraying magnesium hydroxide liquid (MHL) on the surface of concrete sewers. Compared to existing corrosion protection technologies like plastic lining, MHL spray coating is seven times cheaper over the life cycle of the asset and does not require flow diversion or man entry for application. This technology has been tested and used as a mainstream corrosion protection mechanism over the last 12 years in the USA and Australia. This technology can be easily implemented for assets like large sewer pipes, manholes, wet wells, and parts of sanitary treatment plants. The asset is cleaned with a high-pressure water jet and then MHL is sprayed using a gear- or airless pump. Coating thickness of 1.2 mm can be achieved easily. The MHL itself is very safe to use and any overspray into the sewer itself actually helps reduce odour and formation of H_2S [20].

9.6.3 Other Acid Attacks

Common among other acid attacks are lactic, acetic, and butyric acids in dairy and fruit products spilled on concrete floors of food-processing plants. Although attack from these products is comparatively mild, it is persistent and can result in softening a working floor so that it wears rapidly and becomes uneven and unsatisfactory for smooth operation of vehicles. Another common area of acetic acid attack (pH 3.4–3.9) is a concrete silo where dried food is stored for animals.

The American Concrete Pipe Association reports some results from tests in which pipe concrete with type I (ordinary Portland) cement was immersed for 9 months in acetic acid (pH percent 2.5) after 6 months of water curing. Clearly superior resistance was found in specimens in which 25% of the cement had been replaced with a calcined opaline shale pozzolana. This report suggests, however, that reliance should not be placed on serviceability of such Portland-pozzolana cement concrete if pH value of attacking acids is less than 5.0 [21].

Protective Measures in General

Concrete made with Portland cement is not recommended in acidic conditions with pH 5.5 or less without careful consideration of the soil conditions. Concrete made with supersulphated cement can have some acid-resistant properties, but the rate of erosion of concrete surfaces in acidic conditions is affected much less by the type of cement than by the quality of the concrete. Hence, the best way to resist chemical attack, inclusive of an acid attack, is to use dense concrete with low w/c ratio along with chemically resistant materials in the concrete mix.

Some pozzolanic materials, and particularly silica fume, increase the resistance of concrete to acids. Attack of $Ca(OH)_2$ can also be prevented by treatment with diluted water glass (sodium silicate) to form calcium silicates in the pores. Silicate aggregates being acid resistant are sometimes specified to improve the chemical resistance of concrete. To protect from acid attack, however, surface-applied barrier coatings, membranes, or surfacing systems are mostly recommended. For detailed information on concrete protection treatments in light of various acids ACI-515.2R-13 [22] can be referenced.

Bibliography

1. ACI 201.2R-92 (Revised 2016), Guide to Durable Concrete
2. D.A., Hausmann, Steel Corrosion in Concrete: Materials Protection, November 1967.
3. American Concrete Institute (ACI), Publication SP-15(95), Field Reference Manual, ACI 301-96 (Revised 2016).
4. Canadian Standards Association (CSA) A23.1-14 - Concrete materials and methods of concrete construction/Test methods and standard practices for concrete.
5. A.M. Neville and J.J. Brooks, Concrete Technology, E.L.B.S. Longman, Singapore, 1993.
6. ACI 224. 1R-93 (Revised 2007), Causes, Evaluation and Repair of Cracks in Concrete Structures, ACI Manual of Concrete Practice, Part-3, 1995.

7. British Standard Institution, BS 8110- Part 1: 1985, Structural Use of Concrete: Code of Practice for Design and Construction.
8. BS 5328, Concrete: Part-1, Guide to Specifying Concrete-1991 (Amended 1995); Part-2, Methods for Specifying Concrete Mixes-1991(Amended1995); Part-3, Specification for the Procedures to be used in Producing and Transporting Concrete-1990 (Amended 1992); Part-4, Specification for the Procedures to be used in Sampling, Testing and Assessing Compliance of Concrete-1990 (Amended 1995).
9. ACI 318-08 (Revised 2014), Building Code Requirements for Structural Concrete.
10. Cement Association of Canada- CSA- A23.3-04- Concrete Design Handbook.
11. Peter Pullar-Strecker, Corrosion Damaged Concrete, Assessment and Repair, an outcome of CIRIA (Construction Industry Research and Information Association) London, 1987. In preparation of this document, CIR/A refers following publications:
12. National Precast Concrete Association- IN, USA- Understanding Carbonation- Precast Inc. Magazine July 20, 2015.
13. Narayan Swamy, Arvind K. Suryavanshi and Shin Tanikawa, Protective Ability of an Acrylic-based Surface Coating System against Chloride and Carbonation Penetration into Concrete-ACI Materials Journal, Vol 95, No.2, April 1998.
14. Project Analysis and Control Systems Co. W. L.L Kuwait, Production, Problems and Protection of Concrete in the Arabian Gulf Region, 1989.
15. Portland Cement Association, Building Movements and Joints, Illinois, USA, 1982.
16. Peter H. Emmons, Concrete Repair and Maintenance Illustrated, R. S Means Company Inc. USA 1993.
17. Emmanuel E. Velivasakis, Sten K. Henriksen and David W. Whitmore, Halting Corrosion by Chloride Extraction and Realkalization, ACI Vol. 19, No. 12, 1997.
18. Hubert Woods, Durability of Concrete Construction, ACI Monograph No.4, USA 1984.
19. A.M. Neville, Properties of concrete- Pearson Education Limited Essex CM20 2JE England, 2011.
20. Nitin Apté, Calix Ltd- KNOW YOUR SEWER- CORROSION PROTECTION OF SEWER ASSETS- 78th WIOA Victorian Water Industry Operations Conference & Exhibition Page No. 22 Bendigo Exhibition Centre Australia, 1 to 3 September, 2015.
21. ASTM Special Technical Publication No: 169- A- (1975) USA. Significance of Tests and Properties of Concrete and Concrete- making Materials.
22. ACI 515.2R-13, Guide to Selecting Protective Treatments for Concrete- Farmington Hills, MI 48331 U.S.A.
23. N.R. Greening and R. Landgren - Surface Discolouration of Concrete Flat Work- Journal of the PCA Research and Development Laboratories Vol. 8, No. 3, 34-50 (September 1966)- Portland Cement Association, USA.

Chapter 10
Corrosion of Embedded Metals in Concrete

10.1 Introduction

Corrosion of reinforcing steel and other embedded metals is the leading cause of deterioration in concrete. According to NACE International (National Association of Corrosion Engineers), corrosion is "the destruction of a substance (usually a metal) or its properties because of a reaction with its environment".

The corrosion of structural steel is an electrochemical process that requires the simultaneous presence of moisture and oxygen. Essentially, the iron in the steel is oxidised to produce rust, which occupies a greater volume than the steel. This expansion creates tensile stresses in the concrete, which causes cracking, delamination, and spalling. The rate at which the corrosion process progresses depends on a number of factors, but principally the conditions immediately surrounding the structure. Corrosion also produces pits or holes in the surface of the reinforcing steel, which reduces its strength capacity due to the result of reduced cross-sectional area.

Metals have a tendency after production and shaping to revert back to their lower energy, more natural state of ore (e.g. iron ore). Corrosion is the reverse process of metallurgy. In other words, the energy used to transform ore into a metal is reversed as the metal is exposed to oxygen and water. As the metal is exposed to these elements, the corrosion process begins—oxides are formed on the steel surface and in some cases combine with sulphides and carbonates. The process of corrosion is further explained in the following pages.

10.2 Reinforcing Steel

Concrete normally provides reinforcing steel with excellent corrosion protection. While preparing concrete, calcium hydroxide $Ca(OH)_2$ is produced due to hydration of cement. The high alkalinity (pH) of this $Ca(OH)_2$ forms a thin protective film of

© Springer Nature Switzerland AG 2019
A. Surahyo, *Concrete Construction*,
https://doi.org/10.1007/978-3-030-10510-5_10

iron oxide on the steel reinforcement, which protects it from corrosion. This protection is known as passive protection. The pH of newly produced concrete is usually between 12 and 13.

As illustrated in Fig. 10.1 [1], the corrosion rate of iron is reduced as the pH increases. If concrete has a pH of between 12 and 13, it is usually an excellent medium for protecting steel from corrosion. In addition, concrete can be proportioned to have a low permeability, which minimises the penetration of corrosion-inducing substances. Whenever the passivating film is disrupted or pH value falls below, about 11, corrosion takes place. Figure 10.1 shows the relationship between pH and corrosion.

When concrete is permeable or when the cover is inadequate, carbon dioxide, CO_2, which is always present in the atmosphere, reduces this alkalinity to as low as pH 8.5 by carbonating the alkalis and thus increases the vulnerability of the steel to corrosion. The gases of sulphur and nitrogen oxides present in a polluted atmosphere are also known to accelerate the progress of the carbonation front, and reduce the alkalinity of concrete to as low as pH 2.5 [2].

Corrosion of reinforcement or prestressing steel can also occur if calcium chloride ($CaCl_2$) is used as an admixture, or $CaCl_2$ enters the concrete through contaminated aggregates, and mix water or salts penetrate into concrete due to its low quality. In fact, the passivity of steel is destroyed by aggressive chloride ions. Chloride ions are considered to be the major cause of premature corrosion of steel reinforcement. Chloride ions are common in nature and small amounts are usually unintentionally contained in the mix ingredients of concrete.

Fig. 10.1 Relationship between pH and corrosion rate (reprinted with permission from Concrete Repair and Maintenance)

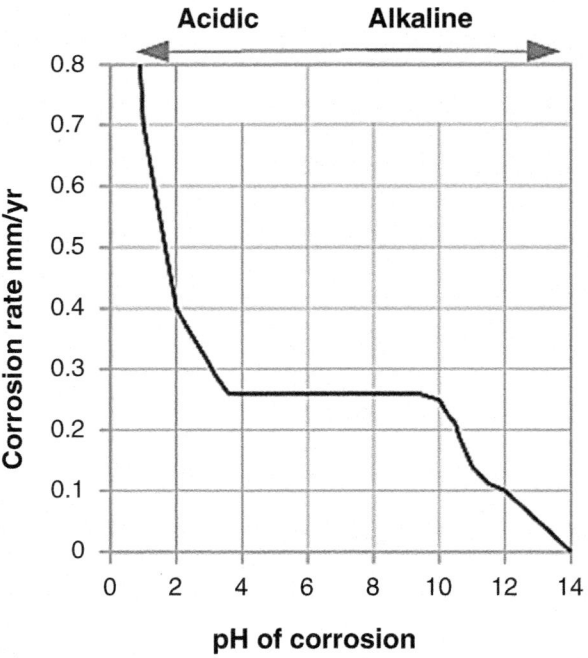

10.2.1 Effects of Corrosion

Corrosion of the steel produces iron oxides and hydroxides, which have a volume much greater than the volume of the steel reinforcement and eventually exert enough force on the concrete to cause cracks. The cracks provide easy access for oxygen, moisture, and chlorides due to which corrosion and cracking are accelerated resulting in spalling of the concrete. Figure 10.2 shows the example of corroded reinforcement causing cracking and spalling of concrete.

The structural capacity of a concrete member is affected by reinforcing bar corrosion and cracking of surrounding concrete. Corrosion also produces pits or holes in the surface of reinforcing steel, reducing strength capacity as a result of the reduced cross-sectional area. The reinforcing bars disintegrate, which reduces the load-carrying capacity of the concrete structure. Research conducted on flexural beams found that in steel with more than 1.5% corrosion, the ultimate load capacity began to fall, and at 4.5% corrosion the ultimate load was reduced by 12%, probably a result of reduced diameter of the corroded bar [3].

In compressive members, cracking and spalling of concrete reduce the effective cross section of the concrete, which results in the reduction of the ultimate compressive load capacity.

10.2.2 Protective Measures

To protect steel from corrosion for new reinforced concrete structures, several corrosion prevention methods have been developed. However, the best procedure is the use of concrete with low permeability and adequate cover. Generally, mix with low w/c ratio, properly compacted, finished, and properly cured will produce less permeable concrete, and hence provide greater protection against corrosion. Tables 7.4, 7.5, and 7.6 give the limiting values for adequate cover to

Fig. 10.2 The badly delaminated concrete soffit of a school corridor

reinforcement, proposed by ACI-318, BS-8110, and CSA, respectively, in various conditions, so that the passivity of embedded steel is maintained by the usually high alkalinity of concrete.

As per ACI-201 [4], numerous tests have been conducted, which indicate that concrete made with a w/c ratio of 0.40 and adequate cover over steel resist penetration by deicing salts better than concretes made with w/c ratios of 0.50 and 0.60. Concrete with 40 mm (1.5 in.) cover and 0.4 w/c ratio was found sufficient to protect embedded steel against corrosion for 800 applications of salt. Concretes with low w/c ratio can be produced by [5]:

- Increasing the cement content.
- Reducing the water content, by using water reducers and super-plasticisers.
- By using amounts of fly ash, blast-furnace slag, or other cementitious materials: These materials reduce the permeability of the concrete to the penetration of chloride ions.

Additionally, the use of concrete ingredients containing chlorides should be limited. Tables 9.2 and 9.3 specify the limits on use of chloride ion contents in concrete provided by BS-8110 [6] and ACI-301 [7] including recommendations by CSA-23.1-14 as mentioned in Chap. 9. Moreover, for providing good-quality concrete, air entraining also plays an important role. Air entraining protects the concrete from freezing and thawing and also reduces bleeding and permeability.

In very severe exposure conditions the following additional protective measures may be adopted:

1. Using corrosion-resistant reinforcing bars
2. Using concrete surface treatments
3. Using corrosion-inhibiting admixture
4. Using electrochemical technique (cathodic protection)

10.2.2.1   Using Corrosion-Resistant Reinforcing Bars

In recent years, efforts have been made to develop corrosion-resistant rebars to prevent reinforcement corrosion. Some of the common types of corrosion-resistant rebars currently being used are discussed in the following sections.

Stainless Steel Bars: Stainless steel reinforcing bars can provide long-term corrosion resistance when concrete is exposed to chloride-containing environments, e.g. road salt and seawater. They have been used as rebar in highway bridges, ramps and barrier walls, parking garages, tunnels, sea walls, marine facilities, and buildings. They are produced in various types/grades; however, the most common grades of stainless steel for reinforcement are type 316LN and type 2205 Duplex, both of which have excellent corrosion resistance. OPSS.PROV-1440 [8] also recommends that stainless steel reinforcing bars, stainless steel spirals, stainless steel spiral spacers, and mechanical connectors shall be of stainless steel types 316 LN and 2205 Duplex. It further specifies that fabrication of stainless steel reinforcing bars shall be so that the bar surfaces are not contaminated with deposits of iron or non-stainless steels, and the bar is not damaged by straightening from coil.

Stainless steel reinforcing bars and spirals shall be according to ASTM A276 [9] and ASTM A955M [10], minimum grade 420 as specified by OPSS.PROV-1440 [8]. Depending on the stainless steel grade selected, yield strengths >75 ksi (520 MPa) and tensile strengths >100 ksi (690 MPa) can be achieved. OPSS.PROV-905 [11] provides that tie wire used to tie stainless steel reinforcing bars to stainless steel reinforcing bars, and shear studs shall be type 316 LN or type 316L, stainless steel wire, 1.2 or 1.6 mm in diameter. It further specifies that bar chairs for supporting stainless steel reinforcing bars shall be non-metallic. Concrete chairs shall not be used to support stainless steel reinforcing bars.

Nickel used in the composition of stainless steel improves the mechanical properties of stainless steels and helps to boost their corrosion resistance. Stainless steel rebars also have superior toughness, ductility, and fatigue resistance. These properties are beneficial in earthquake-prone regions and in low-temperature applications [12]. Figure 10.3 shows the stainless steel bars.

Galvanised steel reinforcing bars: Zinc-coated (galvanised) steel bars for concrete reinforcement are produced according to ASTM-A767 [13]. Galvanised steel reinforcement has been used in reinforced concrete structures since 1930s. The primary advantage of hot-dipped galvanising is that the zinc not only forms a barrier coating but also acts as a sacrificial anode. Thus, any scratches or other flaws in the coating are not critical and do not lead to active corrosion of the underlying steel.

Zinc has the advantage over the traditional black steel that it can tolerate more chlorides. Galvanised rebar can withstand chloride concentration at least four to five times higher than black steel, and remains passivated at lower pH levels, slowing the rate of corrosion. Figure 10.4 shows the galvanised steel bars.

Under hot-dip galvanising process a metallic zinc coating to steel rebar is continuously applied by immersing the bars in a bath of molten zinc at about 450 °C. A reaction takes place between the steel and the zinc producing a coating with an outer

Fig. 10.3 Stainless steel reinforcing bars

Fig. 10.4 Galvanised steel reinforcing bars

layer of pure zinc adhering to the outer surface as the rebar is removed from the zinc bath. Further, using a small amount of aluminium in a zinc bath, the process produces an almost pure zinc coating that adheres well and is resistant to corrosion in concrete.

The galvanised coating provides greater adhesion than fusion-bonded epoxy coatings. In addition, the iron-zinc alloy layers of the coating are actually harder than the underlying steel producing an extremely tough and abrasion-resistant coating. Galvanised rebar can generally be treated in the same manner as uncoated rebar and does not require special handling precautions to protect the coating during handling and transporting to the job site. Further, the bond strength of galvanised rebar to concrete is no less than that of uncoated bar, and in many cases is somewhat better [14].

Epoxy-coated reinforcing bars: Epoxy-coated rebar, also referred to as green rebar, is used in concrete subjected to corrosive conditions. These may include exposure to deicing salts or marine environments. The primary function of an epoxy coating is to act as a physical barrier to prevent corrosion-causing chloride ions and oxygen from reaching steel surface. According to Concrete Reinforcing Steel Institute (CRSI), epoxy-coated steel reinforcing bars have been used in many projects for corrosion protection, since 1973.

Epoxy coatings are 100% solid, dry powders. These dry epoxy powders are electrostatically spread with automatic spray guns over blast-cleaned and preheated (to about 250 °C) steel reinforcing bars. When the powder hits the hot bars, it melts and becomes fluid to form a smooth coating. The coatings achieve their toughness and adhesion to the steel bars as a result of a chemical reaction initiated by heat. These epoxy powders are thermosetting materials and their physical properties do not change readily with changes in temperature.

Epoxy-coated steel reinforcing bar protects the steel using several mechanisms. If undamaged, the coating prohibits the passage of chloride ions, thus protecting the steel from corrosion damage. If minor holes or damage are present in the coating and sufficient chloride ions are available in the concrete, then localised corrosion may occur. Hence, handling procedure for epoxy-coated reinforcing bars is of great importance. When handling epoxy-coated rebar, plants must take special care not to damage the coating. Handling requirements are covered by ASTM D3963 [15]. Using nylon strapping and multiple lift points along the bar will help to ensure that the coating is not marred or damaged by cables or chains and is not allowed to sag and rub during offloading.

Bars that have damaged coatings need to be repaired using an approved repair material and process prior to being placed in formwork. OPSS 1442 [16] recommends that the repairs shall be completed before rusting and shipment occur, with the following exceptions:

1. Damage shall not exceed a surface area of 10 mm^2 in any linear metre of coated bar, not including sheared ends.
2. There shall be no more than four defects per coated bar length.
3. Bars containing hairline cracks that are associated with bond loss shall be rejected.
4. Coated reinforcing steel bars with damage exceeding the above shall be rejected.

Epoxy-coated rebar also needs to be protected from direct sunlight as ultraviolet light degrades the epoxy coating over time. CRSI recommends that bar be exposed to ultraviolet light no longer than 30 days unprotected. Specification for epoxy-coated steel reinforcing bars is also provided by ASTM A775/A775M-17 [17]. Figure 10.5 shows the epoxy coated bars.

Fig. 10.5 Epoxy-coated steel reinforcing bars

Glass Fibre-Reinforced Polymer Bars (GFRP): The GFRP rebar is a structural ribbed reinforcing bar made of high-strength and corrosion-resistant glass fibres that are impregnated and bound by an extremely durable polymeric epoxy resin. OPSS.MUNI-950 [18] describes GFRP as a fibre-reinforced composite with a polymeric matrix and continuous fibre reinforcement of glass. It further specifies that bars used shall be grade III and binding material for bars shall be composed of thermoset vinyl ester resin that is homogeneous throughout the cross section of the bar. Fibre reinforcement in the bars shall be continuous E-glass or E-CR glass fibres according to ASTM D 578 [19]. According to the CSA S807-10 FRP [20] specification, glass fibre-reinforced polymer (GFRP) is produced in three grades: 40 GPa Grade I (LM), 50 GPa Grade II (Std), and 60 GPa Grade III (HM).

GFRP is permanently resistant to chemical acids and alkaline bases; therefore extra concrete cover, anti-shrink additives, and even cathodic protection are not required. GFRP significantly improves the longevity of engineering structures where corrosion is a major factor.

Many researches have shown that GFRP reinforcement exhibits high strengths, is lightweight, and is corrosion resistant. Being a corrosion resistant, GFRP reinforcement is a promising material for structures that operate in marine and sensitive environments.

Some of the advantages and disadvantages of GFRP rebar can be summarised as follows:

Advantages:

- GFRP rebar is almost ¼ the weight of traditional reinforcement material.
- Tensile strength of GFPR is greater than normal steel.
- GFRP rebar is non-conductive to electricity and heat, making it an ideal choice for facilities like power generation plants and scientific installations.
- It is impervious to chloride ions and other chemical elements and does not corrode.
- It can be manufactured in custom lengths, bends, and shapes.
- It is transparent to magnetic fields and radar frequencies.

Disadvantages:

- FRP has lower modulus of elasticity, for example: Normal steel with 60 ksi (400 MPa) tensile strength has 29,000 ksi (200,000 MPa) modulus of elasticity, whereas GFPR steel with 100 ksi (690 MPa) tensile strength has only 6000 ksi (41,000 MPa) modulus of elasticity.
- Due to its inelastic behaviour design codes significantly reduce the allowable stress capacity.
- Higher initial cost.
- Storage and handling requirements for FRP reinforcing on the construction site can be more restrictive due to FRP's susceptibility to damage by overexposure to UV light, improper cutting, or aggressive handling.
- Tests have shown bond strength of GFRP bars to be inferior to steel bars [21].

- FRP reinforced concrete typically has lower shear strength than steel reinforced concrete of equal flexural strength [21].

While using GFRP reinforcing bars, it is advisable that all bends must be made at the factory. Field bending of FRP bars is not possible. This is because the bent bars must be formed in the factory while the thermoset resin is uncured. In general, the field handling and placement of FRP bars are similar to coated steel rebar (epoxy or galvanised), but with the benefit of weighing one-fourth the weight of steel. The finishing of concrete using GRFP bars must be properly done as the quality of poorly finished composite would affect durability of the produced concrete. Figure 10.6 (and Fig. 1.1) from one of the sites in Toronto shows installation of GFRP rebars in combination with traditional rebars.

10.2.2.2   Using Concrete Surface Treatments

Covering the concrete with coatings, waterproof membranes, or penetrating sealers slows down the rate of concrete deterioration and ultimately improves protection to steel in the following different ways:

- By increasing the resistance of concrete to chloride and carbon dioxide penetration
- By reducing the moisture content of the concrete

Fig. 10.6 GFRP and traditional rebar installation in progress

Coatings: Organic coatings include epoxy coatings, polyvinyl chloride, polypropylene, phenolic nitrite, polyurethane, etc. All these coatings act as a barrier to the aggressive ions, moisture, and oxygen and remain cathodic with respect to the steel. Of these, epoxy coatings have been most popular.

To protect concrete in potable water tanks, polymer-modified cementitious coatings have been used with much success. A polymer-modified cementitious coating can provide an extremely dense protective layer on the surface of concrete while at the same time providing a degree of flexibility which allows it to seal the hairline cracks. A good polymer-modified cementitious coating can bridge these small cracks and be flexible enough to withstand a small amount of crack movement from thermal expansion and contraction of the concrete. However, the quality of the polymer component of the coating is an important ingredient that dictates just how flexible and dense the coating will be. Acrylic- and styrene-acrylic-based polymers provide the desired properties [22].

Using alkaline slurry coating also enhances the alkaline environment around the bar. The cement slurry made for this purpose should be selected from a well-designed cementitious repair mix. Paints containing active ingredients such as red lead, zinc chromate, or zinc phosphate make the steel passive by forming a fine protective oxide layer.

A concrete surface coating with low gas permeability can resist carbon dioxide penetration much better than even high-quality concrete cover. Polymer-based coatings on concrete surfaces have proved satisfactory in this protection system [2].

Sealers: Applying a sealer to the concrete can be an effective and initially inexpensive method of tackling corrosion problem, thus increasing the service life of a reinforced concrete structure. However, not all sealers have an equal service life; some will require more maintenance costs and more frequent reapplications. The purpose of a sealer is to reduce corrosion of reinforcement in concrete by preventing capillary action at the surface, therefore preventing water and chloride ions from penetrating the concrete.

Sealers can either be pore blockers or penetrate into the concrete and act as hydrophobic agents. Most pore blockers are not appropriate for use on bridge decks because they do not offer good skid resistance and do not hold up under traffic wear. The penetrating sealer system includes boiled linseed oil emulsions, silane sprays, siloxanes, migrating corrosion inhibitors, certain epoxies, and high-molecular-weight methacrylate. Epoxies have been used as both penetrating sealers and coatings. Epoxies were chosen for use as sealers because they have good adhesion to concrete.

Waterproof Membranes: Surface-applied membrane or sheet membrane forms a barrier against water penetration on the outside of the concrete. Another option is a fluid-applied membrane. In the same manner as a sheet membrane, the fluid-applied membrane forms a barrier on the surface of the concrete to stop water penetration. The membrane system includes urethanes, acrylics, epoxies, neoprene, cement polymer concrete, certain methyl methacrylate, and asphalt products.

Another method of reducing concrete permeability is to use integral crystalline waterproofing (ICW) admixture. An ICW admixture is included with the concrete mix at batching plant or directly to the ready-mix truck. Through the use of crystalline technology, the ICW admixture reduced the penetration of water and water-born chemicals through following three primary mechanisms:

- Crystallisation and lowering the concrete permeability, thereby enhancing concrete durability
- Reducing the size and quantity of cracks in the concrete
- Self-sealing cracks and microcracks that form later in the structure's life

The effects of ICWs have been seen not only in numerous projects worldwide, but also in a unique long-term study that has been performed by the University of Hawaii [23]. ICWs can control the corrosion in reinforced concrete by impending the development of corrosive conditions caused by the moisture flow.

10.2.2.3 Using Corrosion-Inhibiting Admixture

Corrosion inhibitors are meant to supplement the concrete's natural ability to protect the embedded reinforcing by forming a passivating oxide layer on the steel. Corrosion inhibitors are chemical admixtures usually added to concrete in very small concentrations during batching as a corrosion protection measure. Inhibitors are often used in combination with low-permeable concrete and usually they have the effect of increasing the threshold chloride concentration required to initiate corrosion. Calcium nitrite is the most commonly used corrosion-inhibiting admixture.

Corrosion inhibitors are either inorganic or organic and are usually classified on their protection mechanism. An active type of inhibitor (anodic) helps to form an oxide film on the surface of the steel reinforcing bars. However, passive systems protect by reducing the rate of chloride ion migration. Calcium nitrite (commercial product name: Darex Corrosion Inhibitor) is an inorganic inhibitor and protects the steel reinforcing bars through oxidation-reduction reactions at the steel surface. Organic inhibitors consist primarily of amines and esters (commercial product names: Rheocrete 222, Armatec 2000, Ferrogard 901, MCI 2000, Catexol 1000CI) [24]. They form a protective film on the surface of steel reinforcing bars and sometimes delay the arrival of chloride ions at the steel reinforcing bars.

The mixture of $NaNO_2$ with a sodium phosphate derivative and an acyl derivative of tetra amine was found to be very effective for protection from corrosion and corrosion cracking [25]. To avoid cracking in concrete, correct amount of steel also helps to keep cracks tight. ACI-224R [26] provides guidance for minimising the formation of cracks that could be detrimental to embedded steel. In general, the maximum allowable crack widths are 0.007 in. in deicing salt environments and 0.006 in. in marine environment.

10.2.2.4   Use of an Electrochemical Technique/Cathodic Protection (CP)

Cathodic protection systems for reinforced concrete structures were originally developed for existing concrete structures that were experiencing reinforcing steel corrosion. However, this method is now being used successfully in new reinforced concrete projects also. It is widely used in the oil and gas industries.

There are generally two types of cathodic protection systems—galvanic (sacrificial) and impressed current.

Impressed Current Cathodic Protection: Impressed current systems use an external power source to provide the electrochemical protection. An impressed current cathodic protection (CP) system for concrete structures may require the following basic components: DC power supply (rectifier), inert anode material such as catalysed titanium anode mesh, wiring, and conduit instrumentation such as embedded silver/silver chloride reference electrodes.

Depending on the case of application there are different kinds of anode systems that can be used. A very cost-effective solution is a conductive coating which reaches a lifetime of up to 20 years. However, titanium anode meshes or titanium anode ribbon meshes guarantee a lifetime of at least 40 years.

In fact, corrosion of a metal is an electrochemical process that mostly takes place when two dissimilar metals, i.e. having different electrode potentials, are in electrical contact in the presence of moisture and oxygen. The current flows through the system causing attack on the more anodic metal, i.e. the metal with more negative electrode potential. The metal with the more positive electrode potential or cathodic metal remains un-attacked. However, the same process takes place in steel alone. This happens when chloride enters concrete from different sources (discussed in Chap. 9) and surrounds the reinforcement, reacting at anodic sites to form hydrochloric acid, which destroys the passive protective film on the steel. The surface of the steel then becomes activated locally to form the anode, whereas the rest of the steel which is undestroyed acts as a cathode. Thus, electrons flow from the anode to the cathode leading to corrosion at the anode.

The idea of cathodic protection is to stop rusting by supplying all the required reinforcement with electron flow from an outside source using a direct current (DC) supply. This process artificially makes every part of the required reinforcement slightly cathodic to an externally applied anode at or near the concrete surface. The electrons from the outside source supplied to all parts of the reinforcement will flow from it in anodic areas that are corroding. This results in the transportation/migration of the negatively charged chloride ions out of steel towards the positively charged external anode system as shown in Fig. 10.7. Once the chloride content has been reduced to acceptable levels as described in Tables 9.2 and 9.3, the temporarily installed anode system is removed from the concrete structure.

Along with the removal of chlorides, hydroxyl ions and hydrogen are produced at the cathode, i.e. at the reinforcing bars due to the following reaction [27]:

$$2H_2O + 2e^- \rightarrow 2(OH)^- + H_2$$

Fig. 10.7 Current and
ionic charge flow in
cathodic protection

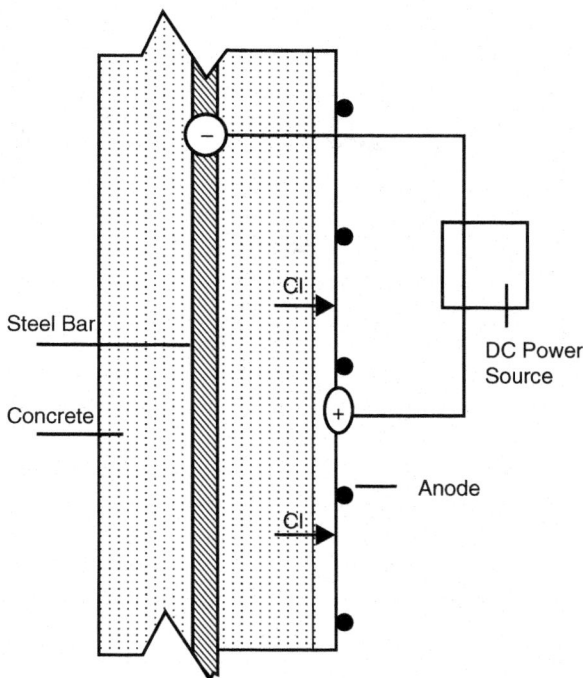

This reaction raises the pH level and re-establishes the corrosion protection properties of the concrete. The anodes used as mentioned above for cathodic protection
are usually made of zinc or magnesium or alloys containing one or the other of these
metals in the form of conductive paints, mesh, plates, or wires.

In case cathodic protection is adopted for a new concrete structure, the major
advantages can be summarised as below [28]:

1. The effective period is not dependent upon the exposure conditions.
2. Corrosion of steel reinforcement can be prevented even if the concrete is cracked.
3. Chloride contamination levels accumulated in the concrete during the structure
   life do not affect performance.
4. Conventional concrete can be used without special design considerations for
   corrosion.
5. In case of inadequate concrete cover, there will be less probability of shrinkage
   cracking.

Thus, the structural performance is independent of the reinforcing steel corrosion
during the entire structure's life when the cathodic protection method is used.
Cathodic protection, however, must not be used on prestressed concrete because
there is a danger that the steel will become embrittled by the presence of hydrogen
produced by the reaction mentioned above.

Sacrificial Anode Cathodic Protection: A sacrificial anode is a metal anode elec-
trically linked to the structure to be protected that is more reactive to the surrounding

corrosive environment. The sacrificial anode corrodes, protecting the metal of the structure being protected.

A sacrificial or galvanic anode system for reinforced concrete uses a more reactive metal (anode) such as zinc or aluminium-zinc-indium (Al-Zn-In) [29], to create a current flow. The sacrificial anodes commonly used are aluminium, zinc, or magnesium. Sacrificial anode systems are based on the principle of dissimilar metal corrosion and the relative position of different metals in the galvanic series. In a galvanic cathodic protection system, the anodes connected to the protected structure have a natural electrochemical potential that is more negative than the steel reinforcement of the structure. When connected, current flows from the anode (more negative potential) to the structure (less negative potential) in a DC circuit. The sacrificial anode will corrode during the process and is consumed. Galvanic anodes may be installed as cast anodes in soil or thermally sprayed onto atmospherically exposed concrete to form a sacrificial coating.

Galvanic CP systems have the benefit of no auxiliary power supply and the advantage of being used for prestressed or post-tensioned concrete without the risk of elevated potential levels, which can lead to hydrogen embrittlement of the steel. The anode life, however, may be relatively short as compared to the inert anodes, which are used with impressed current systems. Also, the current that is produced by a galvanic anode is a function of its environment (i.e. moisture and temperature conditions) and the output cannot be easily adjusted or controlled as with the impressed current method [29].

10.3   Other Metals

Nonferrous metals such as copper, zinc, aluminium, lead, and their alloys may be subjected to corrosion, when embedded in concrete or in surface contact with it.

Copper and its alloys are generally less reactive to fresh concrete and mortar but may cause corrosion, if soluble chlorides are present. When copper is adjacent to steel reinforcement, and an electrolyte such as chloride is present, steel corrosion is likely to occur due to galvanic action. Under these circumstances, copper in concrete can be protected with a suitable coating. As per Portland Cement Association copper should also be protected when it comes in contact with concrete mixtures that contain components high in sulphur, such as cinders and fly ash, which can create an acid that is highly corrosive to most metals including copper.

Moreover, copper pipes are used successfully in concrete except under the unusual circumstance where ammonia is present. Very small amounts of ammonia and nitrates can cause stress corrosion cracking [30].

Zinc is sometimes used in concrete as a coating for steel. Galvanised corrugated steel sheets have also been used as a formwork for concrete. Zinc has good resistance to atmospheric corrosion owing to the formation of an insoluble basic carbonate on its surface; being anodic to iron, zinc gives electrochemical protection even when the coating is incomplete. It is widely used for galvanised products where it is

applied by hot dipping and electroplating, but it may also be applied by metal spraying. Galvanising usually furnishes sacrificial protection to steel, and these coatings
are so thin that their expensive pressures generally do not cause any damage to the
surrounding concrete. However, calcium chloride admixtures corrode galvanised
steel and may lead to severe cracking and spalling of the surrounding concrete.

Lead is reactive to fresh or moist concrete. Although lead is deemed to have
superior corrosion properties to those of steel and aluminium under atmospheric
conditions, it reacts with calcium hydroxide in concrete to form soluble lead oxides
and hydroxides, which do not protect the metal from further corrosion. When a lead
pipe is partially embedded in concrete and partially exposed to air, its destruction
may occur within a few years. The embedded lead has a different electrical potential
than that in the air and an electrical current is created that will cause corrosion and
gradual disintegration of the embedded lead. Generally, no damage will be observed
in the concrete because of the softness of the lead, which will absorb expansive
pressures caused by the formation of corrosion products. Preferably, a protective
coating or covering should always be used when lead pipe or cable sheaths are to be
embedded in concrete. Bitumen or synthetic plastic coatings or sleeves have been
used successfully.

As mentioned above in cathodic protection section, corrosion also takes place in
concrete when two different metals are cast into a concrete structure, along with an
adequate electrolyte. A moist concrete matrix forms a good electrolyte. This type of
corrosion is known as galvanic. Each metal has a unique tendency to promote electrochemical activity. For example, copper (electrode potential = +0.34 V) is more
active while zinc (electrode potential = −0.75 V) is almost inactive.

When two such metals are in contact via an active electrolyte, the less active
metal, or we can say the metal with more negative electrode potential, is corroded.
The list of metals in order of electrochemical activity proposed by Portland Cement
Association [31] is shown in Table 10.1.

One of the most common examples found is the use of *aluminium* cast into reinforced concrete in the shape of electrical conduit or handrails. Aluminium has less
activity than steel, because aluminium has an electrode potential of −1.66 V, whereas
steel or iron has an electrode potential of −0.44 V. When the metals are in contact
in an active electrolyte, the less active metal (lower number) in the series corrodes.

Table 10.1 List of metals in order of electrochemical activity

Name of metal	Electrochemical activity number	Name of metal	Electrochemical activity number
Zinc	1	Lead	7
Aluminium	2	Brass	8
Steel	3	Copper	9
Iron	4	Bronze	10
Nickle	5	Stainless steel	11
Tin	6	Gold	12

The order of activity is from less active (1) to more active (12)

Therefore, aluminium is the metal that corrodes. The steel will remain un-attacked, but the aluminium surfaces will grow a white oxide, which will cause tensile forces to crack the surrounding concrete. Aluminium should not be used in concrete containing admixtures with chlorides, nor in or near seawater. However, corrosion inhibitors, such as calcium nitrate, have been found to improve the corrosion resistance of aluminium in concrete.

Chromium and *nickel* alloyed metals generally have good resistance to corrosion in concrete and are being used in the manufacture of corrosion-resistant stainless steel reinforcing bars. However, the corrosion resistance of these metals may be adversely affected by the presence of soluble chlorides from seawater or deicing salts. If pinholes are present in the coating, they may lead to corrosion in the presence of chlorides.

Cadmium coatings will satisfactorily protect steel embedded in concrete, even in the presence of moisture and normal chloride concentrations. However, when cadmium is applied on other metals (e.g. iron, steel) it behaves as a sacrificial coating by corroding before the substrate metal to which it is adhered. Cadmium is a soft white metal that can be electroplated as a versatile coating.

Bibliography

1. Peter H. Emmons, Concrete Repair and Maintenance Illustrated, R. S Means Company Inc. USA 1993.
2. Narayan Swamy, Arvind K, Suryavanshi and Shin Tanikawa, Protective Ability of an Acrylic-Based Surface Coating System Against Chloride and Carbonation Penetration into Concrete- ACI Materials Journal, Vol. 95, No.2, April 1998.
3. AL-Sulaimani, Kaleemullah, Basunbal and Rasheed, Influence of Corrosion and Cracking on Bond Behavior and Strength of Reinforced Concrete Members, ACI Structural Journal, March-April 1990.
4. ACI 201.2R-92 (Revised 2016), Guide to Durable Concrete.
5. NRCMA (National Ready Mixed Concrete Association, USA), Concrete in Practice- What, Why & How- Corrosion of Steel in Concrete- www.nrcma.org.
6. British Standard Institution, BS 8110- Part l: 1985, Structural Use of Concrete: Code of Practice for Design and Construction.
7. American Concrete Institute (ACI), Publication SP-15(95), Field Reference Manual, ACI 301-96 (Revised 2016).
8. Ontario Provincial Standard Association- OPSS.PROV1440, Material Specification for Steel Reinforcing in Concrete- November 2014.
9. ASTM A276/A276M-17, Standard Specification for Stainless Steel Bars and Shapes.
10. ASTM A955/A955M - 17a, Standard Specification for Deformed and Plain Stainless-Steel Bars for Concrete Reinforcement.
11. Ontario Provincial Standard Association- OPSS.PROV 905, Construction Specification for Steel Reinforcing in Concrete- November 2014.
12. Nickel Institute- Stainless Steel Reinforcement- www.nickelinstitute.org.
13. ASTM A767/A767M – 16, Standard Specification for Zinc-Coated (Galvanized) Steel Bars for Concrete Reinforcement.
14. International Zinc Association Belgium- Hot dip Galvanized Reinforcing Steel- A Concrete Investment, www.zincworld.org. This Article is based on information provided in the book

entitled *Galvanized Steel Reinforcement in Concrete* edited by Stephen R. Yeomans, University of New South Wales, Canberra, Australia, published by Elsevier Science November 2004.

15. ASTM D3963/D3963M – 15, Standard Specification for Fabrication and Jobsite Handling of Epoxy-Coated Steel Reinforcing Bars.

16. Ontario Provincial Standard Association- OPSS-1442, Material Specification for Epoxy Coated Reinforcing Steel Bars for Concrete- November 2007.

17. ASTM A775/A775M – 17. Standards Specification for Epoxy-Coated Steel Reinforcing Bars.

18. Ontario Provincial Standard Association- OPSS.MUNI 950, Construction Specification for Glass Fiber Reinforced Polymer Reinforcing Bar- November 2017.

19. ASTM D578/D578M - 05(2011) Standard Specification for Glass Fiber Strands.

20. Canadian Standards Association CSA-S807-10 (R2015) - Specification for fiber-reinforced polymers.

21. Carin L. Roberts- Wollmann- FRP Bars as Internal Reinforcement for Concrete Overview- FRP Showcase 2006- Virginia Department of Transportation.

22. Greg Illig- Protecting Concrete Tanks in Water and Wastewater Treatment Plants- WWD (Water & Wasts Digest) - December 28, 2000.

23. Alireza Biparva, Research and Development manager/Concrete Specialist at Kryton International Inc- "Reinforced Concrete Corrosion: A Silent Killer"- Construction Canada, February 2015.

24. U.S. Department of Transportation, Federal Highway Administration- "Materials and Methods for Corrosion Control of Reinforced and Pre-stressed Concrete Structures in New Construction- Publication No: 00.081" August 2000.

25. Vinod Kumar- Protection of Steel Reinforcement for Concrete: A Review. R & D Centre for Iron & Steel, Steel Authority of India Limited- RANCHI - 834 002 (BIHAR) INDIA.

26. ACI 224R-01 (Reapproved 2008), American Concrete Institute, Farmington Hills, MI, USA.

27. Emmanuel E. Velivasakis, Sten K. Henriksen and David W. Whitmore, Halting Corrosion by Chloride Extraction and Re-alkalization, ACI Vol. 19, No. 12, 1997.

28. Miki Funahashi, Cathodic Protection Systems For New RC Structures, ACI Journal, Concrete International- Vol 17, No. 7, July 1995.

29. Steven F. Daily, Corrpro Companies, Inc. Understanding Corrosion and Cathodic Protection of Reinforced Concrete Structures.

30. Hubert Woods, Durability of Concrete Construction, ACI Monograph No. $, USA 1984.

31. PCA- Portland Cement Association, Corrosion of Embedded Metals, Skokie, Illinois USA, 2007.

Chapter 11
Hot and Cold Weather Concreting

11.1 Introduction

Weather conditions greatly affect concrete quality. The temperature of the air, humidity level, wind speed, and temperatures of the surface where concrete is to be placed, all play an important role in the quality of concrete and must be taken into consideration. High temperatures affect concrete at all stages of the production and placing process and most of the effects have consequences for long-term strength or durability.

In general, due to high temperature, the evaporation of water from the mix takes place that results in loss of workability and higher plastic shrinkage. The early setting due to high temperature results in greater difficulty with handling, compacting, and finishing of concrete. On the other hand, the fresh concrete with low temperatures starts freezing due to which the mixing water converts to ice resulting in an increase in the overall volume of the concrete. As no water is then left for chemical reaction, the setting and hardening of concrete are delayed, and concrete produced will be of low strength. Hence proper protections are vital, when concrete is placed in abnormal weather conditions either hot or cold.

11.2 Concreting in Hot Weather

ACI-305 [1] defines hot weather as the job-site conditions that accelerate the rate of moisture loss or rate of cement hydration of freshly mixed concrete, including an ambient temperature of 27 °C (80 °F) or higher, and an evaporation rate that exceeds 1 kg/m^2/h, or as revised by the architect/engineer.

However CSA [2] specifies that precautions should be taken to protect concrete in place from the effects of hot and/or drying weather conditions when the ambient

© Springer Nature Switzerland AG 2019
A. Surahyo, *Concrete Construction*,
https://doi.org/10.1007/978-3-030-10510-5_11

air temperature is at or above 27 °C or when there is probability of the air temperature rising above 27 °C during the placing period.

In hot weather, concrete in plastic state must be kept workable during the entire period of placing so as to achieve satisfactory compaction and finishing. High temperatures may lead to problems in mixing, placing, and curing of concrete that can adversely affect the properties and serviceability of concrete. The Middle East in general and the Gulf region in particular experience very severe climatic conditions. In Bahrain, the temperature during the hottest summer months is usually between 44 °C and 48 °C. This temperature is usually coupled with values of relative humidity between 20 and 90%.

The harmful effects of such aggressive, climatic conditions, however, may be minimised by selection and proper control of concrete-making materials and mixture proportioning, and control on factors like initial concrete temperatures, wind velocity, ambient temperature, and humidity conditions during placing and curing periods. Generally, if concrete strengths are satisfactory and curing practices are sufficient to avoid undesirable drying of surfaces, durability of hot weather concrete should not be greatly different from similar concrete placed at normal temperature.

11.2.1 Problems and Effects of Hot Weather

The problems and effects of hot weather on concrete in plastic as well as in hardened state can be summarised as below:

1. Due to high temperature, the evaporation of water from the mix takes place that results in loss of workability and higher plastic shrinkage. ACI-305 [3] specifies that concrete at a temperature of 70 °F (21 °C) placed at an air temperature of 70 °F (21 °C), with a relative humidity of 50% and a moderate wind speed of 10 mph (16 km/h), will have six times the evaporation rate of the same concrete placed when there is no wind.
2. The early setting due to high temperature results in greater difficulty with handling, compacting, and finishing of concrete.
3. The risk of cold joints increases due to increased rate of setting of concrete.
4. A high temperature increases the slump loss that results in the tendency to add water on construction sites, resulting in lower strength concretes.
5. While placing concrete, if its temperature is higher than the normal, rapid hydration takes place that leads to quick setting and reduction in long-term strength of hardened concrete.
6. At high temperature in dry air, curing water evaporates rapidly that results in slowing down the process of hydration. This also leads to lower strength and shrinkage cracking of hardened concrete.
7. Concrete placing for large masses in hot weather will have an increased rate of hydration and heat evolution that will increase differences in temperature between the interior and the exterior concrete resulting in thermal cracking.

8. Durability of concrete is decreased due to shrinkage and thermal cracks resulted by hot weather concreting.
9. Concretes mixed, placed, and cured at higher temperatures normally develop higher early strengths in comparison to concretes produced and cured at lower temperatures, but strengths are generally lower at 28 days and later ages [3].
10. Permeability and porosity of the cement paste matrix play a major role influencing the resistance of concrete to adverse conditions. The tests conducted on concrete samples by Associate Professor Q.A. Kayyali of the University of Kuwait indicate that hot environment exposure causes an increase in the total pore volume of hardened Portland cement paste. It also causes development in its pore size distribution characteristics resulting in the reaction of coarser pores whose diameter exceeds 100 (nanometre) [4]. At this diameter of pores, the chance of chloride ions and detrimental gases is enhanced which increases the chance of steel corrosion.

11.2.2 Remedial Measures and Precautions

In order to minimise the problems of concreting in hot weather, the major precaution is to control the temperature of concrete at the initial stage. However, further control during placing and curing along with proper selection of materials to produce a suitable concrete mix is also of great importance.

11.2.2.1 Temperature Control of Concrete

The temperature of concrete delivered or made at site should be kept low, preferably about 16 °C (60 °F) with a maximum limit of 32 °C (90 °F) [5]. However, ACI-305 [1] limits the maximum allowable fresh concrete temperature to 35 °C (95 °F), unless otherwise specified, or unless a higher allowable temperature is accepted by architect/engineer, based upon past field experience or preconstruction testing using a concrete mixture similar to one known to have been successfully used at a higher concrete temperature.

The temperature of the concrete can be reduced by cooling the water and aggregate and by selection of suitable concrete mix.

Water: Water greatly influences many significant properties of concrete, both in the freshly mixed and hardened state. High water temperature causes higher concrete temperatures, and as the concrete temperature increases more water is needed to obtain the same slump. Figure 11.1 illustrates the possible effect of concrete temperature on water requirements. The extra water increases the water cement or water-cementitious material ratio and accordingly will decrease the strength, durability, water tightness, and other related properties of concrete. Hence, special attention should be paid to control the use of additional water in hot weather concreting, which could be done by controlling its temperature.

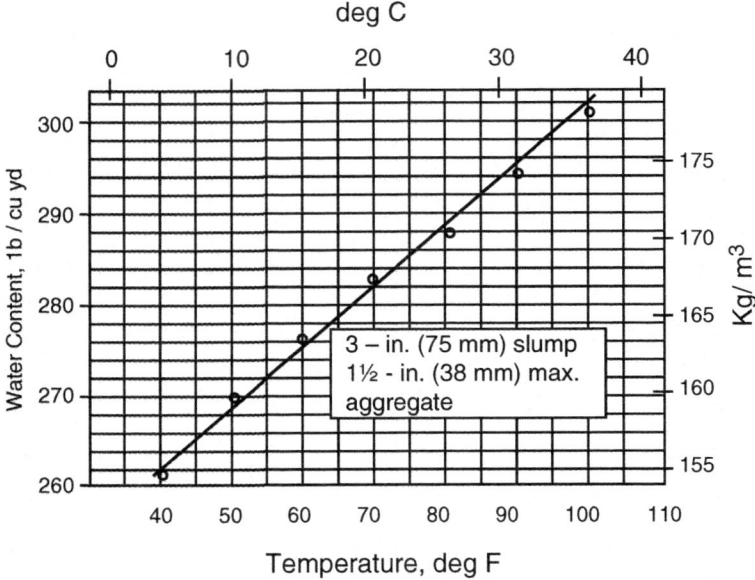

Fig. 11.1 Effect of temperature increase on the water requirement on concrete (U.S. Bureau of Reclamation 1975) [3] (Copyright ASTM—Reprinted with Permission)

As such the water has a specific heat of about four to five times that of cement or aggregates, the temperature of the mixing water has the greatest effect per unit weight on the temperature of concrete. The temperature of water is easy to control in comparison to other constituent materials. Water can reduce the concrete placing temperature, maximum up to 8 °F. In general, lowering the temperature of the batch water by 3.5–4 °F will reduce the concrete temperature approximately 1 °F [3]. Efforts should, therefore, be made to obtain cold water. In order to keep water cold, the equipment or vehicles like tanks, pipes, or trucks used for storing and transporting water should be insulated or painted white or both. By using water chillers, ice, heat pump technology, or liquid nitrogen, water can be cooled to as low as 1 °C (33 °F).

When ambient temperatures exceed 86 °F (30 °C), chilled water alone is insufficient and must be supplemented with ice. Use of ice as part of the mixing water is the common practice in hot climates for reducing concrete temperature. However, care should be taken that all the ice is melted completely before the completion of mixing. To be more effective, the ice should be used in the form of crushed or chipped ice or 2 mm thick ice flakes and should be placed directly into the mixer as part of the added mixing water. Owing to its larger surface area, flake ice absorbs heat in seconds and will chill 140 times faster than water at 0 °C [6]. For most concrete, the maximum temperature reduction with ice is achieved at approximately 20 °F [3]. Moreover, the use of ice blocks should not be allowed, as there is a danger that they may remain partially solidified and so cause voids in the concrete.

When greater temperature reductions are required, injection of liquid nitrogen into mixer holding freshly mixed concrete is an effective method for reduction of concrete temperature. Liquid-injected nitrogen does not affect the mixing water requirement except by reducing concrete temperature.

Aggregate: Aggregates are the major constituent of concrete, since they form 60 to 75% of the volume of normal-weight concrete used in most structures. The changes made in its temperature have, therefore, a considerable effect on concrete temperatures. For example, a moderate 1.5–2 °F temperature reduction will lower the concrete temperature to 1 °F [3].

In order to reduce the temperature of aggregates, its stockpiles may be protected from direct sunrays by erecting temporary sheds. Other measures to assist temperature control include totally enclosed batching and mixing plant, and reflective-paint mixer and cooling drums with chilled water before loading. Spray of cool water on aggregate also helps in reducing the temperature. Cooling of coarse aggregate can also be accomplished by blowing air through the moist aggregate. The airflow will enhance evaporative cooling and can bring the coarse aggregate temperature reasonably down.

The aggregates selected should result in lower water demand during mixing and impart favourable thermal properties to the concrete. Gradation, particle shape, and absence of undersized material are very important in minimising water demand. A low coefficient of thermal expansion of an aggregate causes less expansion of concrete in a hot environment, less shrinkage upon subsequent cooling, and thus a lower risk of thermal cracking. Materials having a high rate of thermal diffusivity dissipate internal heat of hydration of the concrete more rapidly and help minimise peak temperatures and thermal expansion.

Cement and Mix Design Selection: In addition to above, selection of suitable concrete mix also reduces the effects of a high air temperature. The mix selected should have cement content as low as possible so as to have lower heat of hydration, but sufficient to meet strength and durability requirements. In fact, the temperature increase due to hydration of cement in a concrete mix is proportional to its cement content. Hence, only enough cement should be used to provide the required strength and durability and not more. Concrete mixtures usually consist of approximately 10–15% cement.

Inclusion of supplementary cementitious materials, such as fly ash or slag, should be considered to delay setting and to mitigate the temperature rise from heat of hydration. In general, use of normally slower setting type II (modified) Portland cement or type IP or IS blended cement (slag cement) may improve the handling characteristics of concrete in hot weather. Use of slower hydration cements results in slower rate of heat development; hence there will be less thermal expansion and risk of thermal cracking upon cooling of concrete on setting.

Preferably, the concrete should be proportioned for a slump of not less than 3 in. (75 mm) to allow quick placement and proper consolidation in the form. Use of water-reducing or set-retarding admixtures is also helpful in increasing workability and setting time, which are usually affected in such environment. The use of super-plasticisers has proved beneficial under hot weather conditions when used to produce flowing concrete. The improved handling characteristics of flowing concrete provide rapid placement and consolidation and hence the period between mixing and placing is reduced. The rate of slump loss of flowing concrete may also be less at higher temperatures than in concrete using conventional retarders. Concrete strengths are mostly found to be little higher than those of comparable concrete without admixture and with the same cement content [1].

11.2.2.2 Control on Placing and Curing of Concrete

Particular attention should be given to placing, compacting, and protecting the concrete as soon as possible after mixing. The concrete must be transported, placed, compacted, and finished at the fastest possible rate. ASTM C 94/C 94M-04 [7] requires that the concrete to be discharged within 1–1/2 h or before the truck-mixer drum has revolved 300 revolutions, whichever comes first, unless otherwise specified. Similarly, ACI-305 [1] also specifies that the concrete mixture shall be held in the mixer for 90 min, unless otherwise specified. During the entire 90-min period, keep the mixer agitated at 1–6 rpm. At the end of 90 min, the concrete mixture should be mixed at full mixing speed designated by the manufacturer (6–18 rpm) for 2 min. During mixing and agitation periods, the addition of water, chemical admixture, or both to adjust slump is permitted provided that the specified concrete mixture *w/cm* is not exceeded.

However, as mentioned in Chap. 7, OPSS-1350 [8] limits the concrete delivery time based on air and concrete temperature. It requires that, when concrete is transported to the site by means of agitating or mixing equipment, discharge of the concrete shall be completed within 1.5 h after introduction of the mixing water to the cement and aggregates; except when the air temperature exceeds 28 °C and the concrete temperature exceeds 25 °C, the concrete shall be discharged within 1 h after the introduction of the mixing water. When concrete is transported by means of non-agitating equipment, discharge shall be completed within 30 min after introduction of the mixing water to the cement and aggregates. Use of retarders does not change the specified concrete discharge time.

Maintaining a continuous flow of concrete to the construction site is important to avoid the possible development of a cold joint. Thermometer should be kept ready to monitor the concrete temperature of each load reached at job site. Concrete should not be placed against surfaces of absorbent materials that are dry and against surfaces that have free water. Concrete should be placed in layers shallow enough to assure vibration well into the layer below. The placing time between successive layers should be minimised to avoid cold joints.

After placing and finishing, evaporation of water from the concrete surface must be prevented or minimised by protecting the concrete from high temperature, direct sunlight, low humidity, and drying winds. Soon after concrete is placed, the horizontal surfaces should be covered by white polythene sheets or burlaps and vertical surfaces by hanging burlaps, like curtains, to avoid direct sunrays. As soon as the concrete begins to harden, it should be protected from quick drying by moistening the burlaps or with light spray of water to concrete under the cover of polythene sheet. Generally, the surface should not be cooled to dissipate heat from the concrete, as this will induce thermal gradients.

Once the concrete is set, continuous curing is important because the volume changes due to alternate wetting and drying which results in surface cracking. The surface should be cured by continuous spray or ponding of water under the cover of polythene sheet and these sheets may be securely fastened down at all corners to prevent wind from blowing underneath to avoid moisture loss. Curing must be continued for at least the first 7 days.

Curing compounds of light colour to reflect sun radiation and hence to minimise the rise in the concrete temperature can also provide an effective curing. Other means of curing such as covering the concrete with constantly wetted sand, earth, and straw or sawdust can be successfully used after the concrete is hard enough to resist damage. Curing means and methods are further addressed in detail within Chap. 7.

11.3 Concreting in Cold Weather

OPSS-904 [9] defines cold weather as those conditions when the ambient air temperature is at or below 5 °C. It is also considered to exist when the ambient air temperature is at or is likely to fall below 5 °C within 96 h after completion of concrete placement. Temperature refers to shade temperature. However, ACI-306R-16 [10] defines cold weather as a period when the average daily temperature has fallen to or is expected to fall below 4 °C (40 °F) during the protection period. However, CSA-A23.1-14 [2] considers requirement for cold weather precautions when there is a probability of the air temperature falling below 5 °C within 24 h of placing.

Since cold weather affects concrete properties and concrete gains strength and sets very slowly, it must be adequately protected from freezing and thawing or special precautions may be made to minimise cold weather effects.

11.3.1 Problems of Cold Weather

Because of low temperatures, the fresh concrete starts freezing due to which the mixing water converts to ice resulting in an increase in the overall volume of the concrete. As no water is then left for chemical reaction, the setting and hardening of concrete are delayed, and concrete produced will be of low strength. Rapid cooling of concrete surfaces or large temperature differences between exterior and interior of concrete, particularly before the concrete has developed sufficient strength to withstand induced thermal stresses, will cause cracking, which will affect adversely the strength and durability of the concrete structures.

Concrete that is allowed to freeze while in its plastic state can have its potential strength reduced by more than 50% and its durability properties will be dramatically reduced [11].

11.3.2 Remedial Measures and Precautions

Newly placed concrete must be protected from drying so that adequate hydration can occur. When concrete that is warmer than 60 °F (16 °C) is exposed to air at 50 °F (10 °C) or higher, it is essential that measures be taken to prevent drying. In order to minimise the problems of concreting in cold weather, the major precaution is to control the temperature of concrete at the initial stage. However, further control during placing and curing along with proper selection of materials to produce a suitable concrete mix is also of great importance.

11.3.2.1 Control of Concrete Placement Temperature

In order to avoid problems like frost in cold weather concreting, it should be ensured that the placing temperature is high enough to prevent freezing of the mix water and that the concrete is thermally protected for a sufficient time to develop an adequate compressive strength which is usually considered at least 500 psi (3.5 MPa). Concrete that is protected from freezing until it has attained a compressive strength of at least 500 psi (3.5 MPa) will not be damaged by exposure to a single freezing cycle (Powers 1962). Based on the recommendations of ACI-306 [12] Table 11.1 provides minimum concrete placing temperatures for various air temperatures and

Table 11.1 Recommended concrete temperature for cold weather concreting, based on ACI-306R-88 [12]

Air temperature	Minimum dimensions of section			
	Smaller than 300 mm (12 in.)	300–900 mm (12–36 in.)	900–1800 mm (36–72 in.)	Above 1800 mm (72 in.)
Minimum concrete temperature as placed and maintained				
–	13 °C (55 °F)	10 °C (50 °F)	7 °C (45 °F)	5 °C (40 °F)
Minimum concrete temperature as mixed for indicated temperature				
Above −1 °C (30 °F)	16 °C (60 °F)	13 °C (55 °F)	10 °C (50 °F)	7 °C (45 °F)
−18 °C to −1 °C (0 °F to 30 °F)	18 °C (65 °F)	16 °C (60 °F)	13 °C (55 °F)	10 °C (50 °F)
Below −18 °C (0 °F)	21 °C (70 °F)	18 °C (65 °F)	16 °C (60 °F)	13 °C (55 °F)
Maximum concrete temperature drop permitted in first 24 h after end of protection				
–	10 °C (50 °F)	5 °C (40 °F)	−1 °C (30 °F)	−6.6 °C (20 °F)

sizes of sections when concreting in cold weather. The temperature of concrete at the time of placement should always be near the minimum temperatures given in the table. Concrete that is placed at low temperatures, i.e. 5–13 °C (40–55 °F), is protected against freezing and receives long-time curing, thus developing a higher ultimate strength and greater durability.

The table also suggests that after placing the concrete, the maximum allowable temperature drop during initial 24 h must not exceed the values given in the table. This can be accomplished by gradually reducing sources of heat, or by allowing insulation to remain until the concrete attends the mean ambient temperature.

The concrete mixing temperature should also be controlled, so that when the concrete is placed, its temperature is not below the recommended values shown in Table 11.1. As the ambient air temperature decreases the concrete temperature during mixing should be increased to allow for heat loss during transportation and placing.

Before placing of concrete, it should be ensured that the temperature of surfaces in contact with fresh concrete should not cause early freezing or delay setting of the concrete. Hence, all snow, ice, and frost must be removed from the surfaces, including reinforcement against which the concrete is to be placed. ACI-306 [12] recommends that the temperature of these contact surfaces, including subgrade materials, should not be higher than a few degrees above freezing, say 2 °C (35 °F), and preferably not more than 5 °C (40 °F) higher than the minimum placement temperatures given in Table 11.1.

CSA-21.3-14 [2] provides that concrete shall not be placed on or against any surface that will lower the temperature of the concrete in place below the minimum value of 10 °C for slabs less than 1 m thick or 5 °C for slabs more than 1 m thick, except when non-chloride, non-corrosive accelerators are used. Some non-chloride, non-corrosive accelerators conforming to ASTM C494, types C and E, have been found to accelerate setting and strength gain at ambient temperatures of 5 °C and below. When adequate information pertaining to past performance records is available, concrete containing non-chloride, non-corrosive accelerators may be placed at ambient temperatures as low as −5 °C.

11.3.2.2 Temperature Control of Concrete by Heating of Ingredients

The temperature of concrete at the time of placing can be raised by heating the ingredients of the concrete mix.

Water: The mixing water can be heated easily, but care should be taken that its temperature does not exceed 65 °C (150 °F). For concrete containing silica fume, the water shall not be heated to more than 40 °C (104 °F). It is also important to prevent the cement coming in direct contact with hot water as this will result in formation of cement balls. It is therefore suitable to arrange the feeding of mix ingredients in such a way that the hot water should first come in contact with aggregates instead of cement. However, ACI-306R-88 [12] suggests that water with a temperature as high as the boiling point may be used provided that resulting concrete temperatures are within the limits as shown in Table 11.1 and no flash setting occurs. If loss of effectiveness of the air-entraining admixture is noted due to an initial contact with hot water, the admixture must be added to the batch after the water temperature has been reduced by contact with the cooler solid materials.

Aggregates: When aggregates are free of ice and frozen lumps, the desired temperature of the concrete during mixing can usually be obtained by heating only the mixing water, but when air temperatures are consistently below −4 °C (25 °F) it is usually necessary to heat the aggregates as well.

The heating of aggregates is arranged by passing steam through pipes embedded in aggregate storage bins. Aggregate should be heated sufficiently to eliminate ice, snow, and frozen lumps of aggregate. Often 3 in. (76 mm) frozen lumps will survive mixing and remain in the concrete after placing. Overheating should be avoided so that spot temperatures do not exceed 100 °C (212 °F) and the average temperature does not exceed 65 °C (150 °F) when the aggregates are added to the batch [12]. For concrete containing silica fume, the aggregate shall not be heated to more than 40 °C. Materials should be heated uniformly since considerable variation in their temperature will significantly vary the water requirements, air entrainment, rate of setting, and slump of the concrete.

11.3.2.3 Monitoring Concrete Temperature

The actual concrete temperature determines the effectiveness of protection, regardless of air temperature. Therefore, it is desirable to monitor and record the concrete temperatures up to the end of its curing period. OPSS.PROV-904 [9] calls for use of thermocouples and data loggers for monitoring concrete temperature. Temperature data loggers are used for recording temperatures whereas thermocouple sensors are used for measuring temperatures (see Fig. 11.2). There are many types of

Fig. 11.2 Picture showing temperature data logger and thermocouple sensor (source: Vacker LLC) [13]

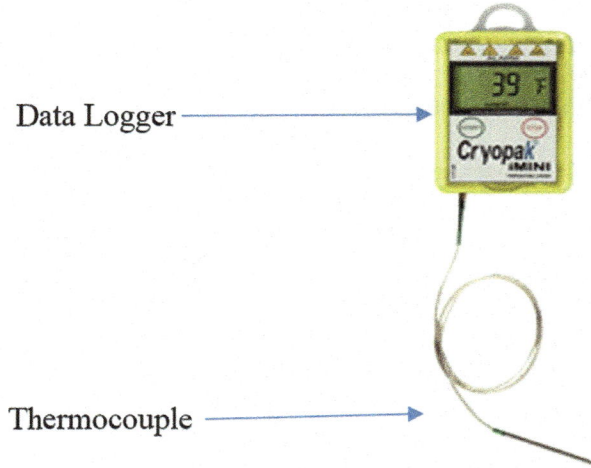

Data Logger

Thermocouple

Fig. 11.3 Picture showing thermocouple with exposed wire sensors fastened to top and middle layers of rebar

thermocouples, such as some have metal probes at the end as shown in Fig. 11.2 and some are without any probe but have exposed wire sensor as shown in Fig. 11.3. A thermocouple is a temperature sensor formed by connecting wires of two dissimilar metals together. These thermocouples are embedded within the concrete prior to casting by binding them to the reinforcement bars. Installed at several locations within the same element, these thermocouples can provide an idea on the heat of hydration within the concrete and provide information on the temperature gradients of the monitored concrete. The current temperature is displayed on the LCD screen of data logger and the recording also takes place. When the recording is over, thermocouples are removed from the data logger and are left in the concrete since they cannot be removed now after the concrete is cured. The data logger can be reused using a new thermocouple. The data logger can be connected to a computer and the data can be downloaded.

OPSS.PROV-904 [9] specifies that the contractor shall supply and install thermocouple wires and associated instrumentation with a combined accuracy of ±1 °C capable of recording and displaying temperature. The instrumentation shall include data loggers capable of recording at hourly intervals or less and shall allow direct reading of temperature.

The thermocouples for concrete temperature measurement shall be installed according to Table 11.2 prior to placing concrete. Thermocouples for monitoring ambient air temperature shall be installed in the shade close to the surface of the concrete at a minimum frequency of 1 thermocouple per stage.

Recording of concrete temperatures shall begin at the start of placement. The temperature shall be recorded automatically at intervals no greater than 1 h until the

Table 11.2 Minimum number of thermocouple sets for concrete temperature measurement as per OPSS.PROV-904 [9]

	Concrete elements requiring temperature monitoring	Number of thermocouple sets in each element	Number of thermocouples in each set	Thermocouple set locations
Cold weather protection	Each concrete element	Minimum of three per element or stages thereof	2	In locations where the concrete is expected to reach the highest temperature and at the surface of concrete
Bridge decks	All	(1) Minimum of 3 per stage, or per deck if deck is not placed in stages (2) When diaphragm is cast together with a deck a minimum of 4 per stage	3	The beginning, middle, and final portions of the deck placement and in the diaphragm. In locations where the concrete is expected to reach the highest temperature and at the surfaces of concrete (Note 1)
HPC	Substructure elements: abutments, pier columns, and pier caps	Minimum of 3 per element or stages thereof	2	In locations where the concrete is expected to reach the highest temperature and at the surface of concrete
Large concrete components where the smallest dimension is 1.5 m	Elements with smallest dimension of 1.5 m or more	Minimum of 3 per element or stages thereof	2	In locations where the concrete is expected to reach the highest temperature and at the surface of concrete

Notes: 1. For bridge decks, thermocouples shall be installed in sets of three consisting of one mid-depth thermocouple and two surface thermocouples. The surface thermocouples shall be placed immediately above or the shortest distance from the corresponding mid-depth thermocouple. The surface thermocouples shall be installed beneath the burlap in contact with the surface concrete or imbedded in the concrete within 5 mm of the surface and, for bridge decks, the second surface thermocouple shall be placed inside the bottom form

end of the monitoring period. The monitoring period shall be 7 days, or longer when necessary in order to meet the requirements of the Withdrawal of Protection clause. The digital temperature indicators shall be left in place until the end of the monitoring period.

OPSS.PROV-904 [9] further requires that the contractor shall also physically monitor and verify concrete and ambient air temperature readings every 6 h, or more frequently, for the first 3 days and every 12 h for the remainder of the monitoring period. The contractor shall take necessary action to maintain the temperature within the specified limits.

ACI-306R-88 [12] provides that the inspection personnel should keep a record of the date, time, outside air temperature, temperature of concrete as placed, and weather conditions (calm, windy, clear, cloudy, etc.). Temperatures of concrete and the outdoor air should be recorded at regular time intervals but not less than twice per 24-h period. The record should include temperatures at several points within the enclosure and on the concrete surface, corners, and edges. There should be a sufficient number of temperature measurement locations to show the range of concrete temperatures. Temperature-measuring devices embedded in the concrete surface are ideal, but satisfactory accuracy and greater flexibility of observation can be obtained by placing thermometers against the concrete under temporary covers of heavy insulating material until constant temperatures are indicated.

11.3.2.4   Use of Special Cements and Admixtures

During frost conditions, it is also helpful to use high alumina cement as it generates higher heat of hydration during initial hours while setting. During this period concrete develops sufficient strength, which makes it safe against frost action. In addition to this, use of rapid-hardening Portland (type III) cement or accelerating admixtures and air-entraining agents is generally helpful in cold weather. Air entrainment increases the resistance of hardened concrete to freezing and thawing and improves the workability of fresh concrete. The resistance to freeze is also improved by use of lowest practical w/c ratio, total water content, and durable aggregates. These items are already explained under the topic "Freezing and Thawing", Chap. 8.

The use of calcium chloride to accelerate the setting and hardening should be avoided, as it increases the chance of reinforcement corrosion. However, noncorrosive accelerating admixtures are also available and can be used to accelerate the setting of concrete as mentioned above.

11.3.2.5   Temperature Control of Concrete by Other Methods

After placing concrete, the temperature could also be controlled by insulating it from the atmosphere. This could be achieved by constructing temporary enclosures around the structure and heating the air inside by steam heating or central heating with circulating water. Hoarding enclosures shall be constructed to withstand wind and snow loads and shall be reasonably airtight. The housing shall provide sufficient space between the concrete and the enclosure to permit free circulation of warm air. The heating apparatus shall be so positioned that there is no direct discharge of heat on the concrete surfaces or formwork-containing concrete. The relative humidity within the enclosure should be maintained at no less than 65%.

During the protection period, care should be taken that the maximum decrease in temperature measured at the surface of the concrete in a 24-h period shall not exceed 11 °C (52 °F) as mentioned in Table 11.1. When the surface temperature of the concrete is within 11 °C (52 °F) of the ambient or surrounding temperature, all protection may be removed.

In some cases, heating from external sources may be avoided to prevent freezing of the concrete if the generated heat is retained. Since most of the heat of hydration of the cement is generated during the first 3 days, it could be retained to maintain the temperature at the recommended level by covering the concrete with insulating materials such as polystyrene foam sheets, urethane foam, and foamed vinyl blankets.

The ill effects of concreting in cold weather can also be minimised by using precast members that are usually manufactured in controlled environment of factories.

11.3.2.6   Curing Requirements

Concrete at a low temperature has a slower rate of set and strength gain. A rule of thumb is that a drop in concrete temperature of 10 °C will approximately double the set time. Newly placed concrete should not be allowed to freeze during the first 24 h. The reaction between cement and water called hydration generates heat. Protecting that heat from escaping the system using insulating blankets may be enough for a required good-quality concrete. More severe temperatures may require heating enclosure.

Live steam exhausted into an enclosure around the concrete is an excellent method of curing because it provides both heat and moisture. Liquid membrane-forming compounds can also be used within heated enclosures for early curing of concrete surfaces. Use of unvented heaters should be avoided, as carbon dioxide from the heaters can cause soft, dusting floors.

Additionally, "concrete curing blankets" are also available in market nowadays. After the final finish is completed, concrete should be covered with a concrete curing blanket. The heated concrete blanket will prevent freezing and keep the concrete at an optimal curing temperature. After about 3 days, heated concrete blankets may be removed to allow the concrete to air-dry. Manufacturers of Powerblanket, USA, claims that if Powerblanket is used to pour in cold weather, it will cure concrete 2.8 times faster than with conventional insulated blankets.

Rapid cooling of the concrete upon termination of the heating period should be avoided. Temperature differences between the surface and the interior of the concrete should be controlled. Thermal cracking may occur when the difference exceeds 20 °C. Insulation and protection should be gradually removed to avoid thermal shock.

Like in hot weather, it is also good practice in cold weather to leave forms in place as long as possible. Even within heated enclosures, forms serve to distribute heat more evenly and help prevent drying and local overheating. Also consider using insulated concrete forms (ICF) for walls. These will prevent freezing and reduce or even eliminate the need for a heated enclosure. Covering the top of these

Table 11.3 Minimum cold weather protective measures as per OPSS.PROV-904 [9]

Footings and slabs on the ground				
Anticipated minimum ambient air temperature °C	Thickness			
	>1.0 m	1.0–0.5 m	<0.5–0.25 m	<0.25 m
+ 5 to 0	PM1	PM1	PM1	PM2
−1 to −10	PM2	PM2	PM2	PM3
−11 to −20	PM3	PM3	PM4	PM5
Less than −20	PM3	PM4	PM5	PM5
All other components				
+ 5 to 0	PM1	PM1	PM1	PM2
−1 to −10	PM2	PM2	PM3	PM4
−11 to −20	PM3	PM3	PM4	PM5
Less than −20	PM4	PM5	PM5	PM5

Notes:

A. Protective Measures

PM1: Cover components with a moisture vapour barrier (white polyethylene film according to ASTM C 171, at least 100 μm thick)

PM2: Cover components as for PM1, and then cover the moisture vapour barrier with insulation having an *R*-value of 0.67

PM3: Cover components as for PM1, and then cover the moisture vapour barrier with insulation having an *R*-value of 1.33

PM4: Cover components as for PM1, and then cover the moisture vapour barrier with insulation having an *R*-value of 2.00

PM5: Housing and heating

B. All R-values are metric

C. The conversion factor from metric to imperial units is

Metric *R*-value × 5.678 = imperial *R*-value

forms is essential to retain all the heat of hydration. These forms take advantage of all the heat produced and cool gradually preventing thermal shock.

OPSS-904 [9] recommends that for cold weather conditions (+5 °C and less), minimum protection of concrete shall be according to Table 11.3 and shall be maintained for the duration of curing period. According to this requirement, curing compound shall not be used in exposed cold weather conditions.

Bibliography

1. ACI 305.1-06, Specification for Hot Weather Concreting. First Print March 2007.
2. Canadian Standards Association (CSA) A23.1-14 - Concrete materials and methods of concrete construction/Test methods and standard practices for concrete.
3. ACI 305R-99 (Revised 2010), Hot Weather Concreting.
4. Project Analysis and Control Systems Co.W.L.L Kuwait, Production, Problems and Protection of Concrete in the Arabian Gulf Region, 1989.
5. A.M. Neville and J.J. Brooks, Concrete Technology, E.L.B.S. Longman, Singapore, 1993.
6. Prentis Polhill, Concrete Production in Hot Environments, Concrete Society Journal, CONCRETE, Vol. 34, No. 7, July/August 2000.

7. ASTM C 94/C 94M-04 (Revised 2017), Standard Specification for Ready-Mixed Concrete.
8. Ontario Provincial Standard Specification-OPSS. PROV-1350, Material Specifications for Concrete – Materials and Production- November 2013.
9. Ontario Provincial Standard Specification (OPSS.PROV)-904- Construction Specification for Concrete Structures- 2010
10. ACI 306R-16, Guide to Cold Weather Concreting.
11. Ontario Cast-in-place Concrete Development Council (OCCDC)- Reinforced Concrete Reference Guide- Concrete Technical Information, "Cold Weather Concreting" 2009.
12. ACI 306R-88, Cold Weather Concreting.
13. Vacker- Authorized Partner of Cryopak, Measuring Temperature of Concrete Using Data Logger or Thermocouple- Email:sales.en@vackerglobal.com.

Chapter 12
Errors in Design and Detailing

12.1 Introduction

Errors in design and detailing are one of the common causes of failure and cracking in concrete structures. Design error is basically a deviation from a drawing or specification including omissions and ambiguities. A large percentage of defects in buildings arise through decisions or actions taken during the design stages. Lack of communication, insufficient documentation, missing input information, and lack of coordination between disciplines are the main problems in design management [12]. However, major design quality problems occur during construction when errors, omissions, and ambiguities in plans and specifications become evident [13].

A study by Building Research Advisory Service, UK, was carried out for analysing the reasons for causes of failures in buildings, which indicates that 58% of all failures are attributed to faulty design; 35% defects are attributed to faulty execution, and the rest to material failures or poor performance [1]. This assessment indicates that more failures occur through inadequacies in design or execution than through faulty materials.

Some of the contributing factors to design error can be summarised as follows:

- Lack of proper subsoil investigations including assessment of dewatering volumes
- Lack of coordination between parties and within their in-house teams
- Low budget and time pressure on the designer
- Insufficient oversight and late design changes
- Designer rushes out tender drawings before proper review
- Lack of construction experience by the designer including insufficient knowledge
- Unclearly defined responsibilities within design team
- Inexperienced drafting and design staff

© Springer Nature Switzerland AG 2019
A. Surahyo, *Concrete Construction*,
https://doi.org/10.1007/978-3-030-10510-5_12

Errors in design and detailing may include improper design of foundations resulting in differential movements and foundation settlements, lack of adequate movement joints and proper detailing of construction joints, improper detailing of reinforcement, restraint of members subjected to thermal volume changes, improper grades of slab surfaces, and overlooking the considerations of groundwater pressure and earth pressure while designing basements, water-retained structures, etc. Some of the examples of improper design and common problems along with possible remedial measures are discussed in this chapter.

12.2   Foundation Settlements

Settlement can be defined as the permanent downward displacement of the foundation. Its effect upon the structure depends on its magnitude, its uniformity, the length of the time over which it takes place, and the nature of the structure itself. There are many reasons for foundation settlement; however broadly they can be grouped under the following two types:

1. Settlement directly due to weight of the structure: For example, the weight of a building may cause compression of an underlying sand deposit or consolidation of an underlying clay layer.
2. Settlement due to secondary influences: The second basic type of settlement of a building is caused by secondary influence, which may develop at a time long after the completion of the structure. This type of settlement is not directly caused by the weight of the structure. For example, the foundation may settle as water infiltrates the ground and causes unstable soils to collapse. The foundation may also settle due to yielding of adjacent excavations or the collapse of limestone cavities or underground mines and tunnels. Other causes of settlement that would be included in this category are natural disasters, such as settlement caused by earthquakes or undermining of the foundation from floods. Another cause is the downward displacement of the foundation due to the drying of underlying wet clays, which may occur due to adjacent trees.

Some of the examples discussing the above reasons of settlement are briefed hereunder.

12.2.1   Soil Conditions

Mostly, the construction of single- and double-storey building structures is started without proper site investigations, which sometimes results in foundation failures/settlements. In fact, settlement of foundations is not necessarily confined to very large and heavy structures. In soft and compressible silts and clays, appreciable settlement can occur under light loadings. In Scotland, settlement of concrete foundation occurred in two-storey houses founded on soft silty clay. In less than 3 years from the time of completion, differential settlement and cracking of the blocks of houses were so severe that a number of the houses had to be evacuated [2]. Actually, more extensive investigations for light structures are needed where there are problems of deep-seated swelling and shrinkage of clays or where layers of peat or loose fill materials are encountered.

Similarly, for multistorey buildings, if it is to be constructed on rock the idea of site investigation should not be overlooked. It is always advisable to excavate down to expose the rock in a few places to ensure that there are no zones of deep weathering or heavily shattered or faulted rock.

In order to avoid any future complications, the designer must satisfy himself/herself about the soil conditions, either asking for shallow trial pits or hand auger boring or mechanical boring, depending on the nature/volume of the structure and site conditions. Additionally, materials used for construction below foundation should be durable including properly compacted so as to avoid danger of the disintegration of the foundations.

12.2.2 Load Variations

Variations in foundation loading mostly result in foundation settlement. This happens when a building is constructed with a high tower at the centre having low projecting wings. If some portion of the building is higher, the soil below the higher portion will be subjected to higher pressure than the soil below the lower portion of the building. Differential settlement between the tower and wings would be expected unless special methods of foundation design are considered to prevent it.

Provision of an expansion joint between various blocks of the building so that each block can move as a separate unit will be of great help in this case. This case is already explained in Chap. 8. Additionally, in designing the foundation, the pressure on soil should be reduced to such value that the difference in settlement between the two portions of the structure is reduced to minimum. In no case the pressure on the soil should exceed the safe bearing capacity of soil.

Differential settlement must be considered inevitable for every foundation, unless the foundation is supported by solid rock. The effect of the differential settlement on the building depends to a large extent on the type of construction. Karl Terzaghi (1938) [14], the founder of soil mechanics, summarised his studies on several buildings in Europe where he found that walls 60 ft. (18 m) and 75 ft. (23 m) long with differential settlements over 1 in. (2.5 cm) were all cracked, but four buildings with walls 40 ft. (12 m) to 100 ft. (30 m) long were undamaged when the differential settlement was 3/4 in. (2 cm) or less. This is probably the basis for the general design guide that building foundations should be designed so that the differential settlement is 3/4 in. (2 cm) or less.

12.2.3   Tree Roots

The problem of settlement in existing structures constructed close to growing trees or by planting trees and shrubs close to them have generally been overlooked. The roots of trees and shrubs can extract large quantities of water from the soil and if the soil is of clay this will lead to a drying shrinkage. The removal of water from the soil can take place both vertically and horizontally. Hence, the precautions must be taken against settlement occurring due to such cause. As a rule, for construction near trees, clay soils are more problematic than porous/sandy soils because of their increased water retention and their potential to swell in heavy rain. Conversely, healthy trees take large amounts of water out of soil, often forcing clay soils to shrink. Each such action exerts a significant pressure on the foundation, causing cracking and subsidence.

Some trees have a seasonal impact on ground moisture content, with winter rainfall rehydrating soil that has dried over summer, a time when dormant root action takes less water out of the ground. Given that mature elms, oaks, horse chestnuts, planes, and ashes can draw up to 50,000 L of water a year from the surrounding soil, the consequent soil-water retention, or frost, can lead to significant heave. Worst-case examples have resulted in concrete slabs being pushed into humps as the soil expansion exerts an upward pressure on the floor slab. Similarly, trench foundations can crack, with consequent movement affecting the structure above.

All trees have radial "zones of influence" on buildings that diminish the further away from the tree the construction takes place. As a rule, it is recommended that properties be built at least a distance equivalent to the tree's height away from that tree. Attempts to insert a root barrier around a construction (e.g. a polypropylene or similar geomembrane) to dissuade root growth near a foundation often only cause roots to grow under and around it. However, while using root barriers to block an existing tree's root system will tend not to work, installing these barriers to control a new tree's growth is more likely to be successful [3].

The root systems of isolated trees generally spread to a radius greater than the height of the tree. Fast-growing trees are especially dangerous and within 5–6 years the roots extend to a distance of 15–18 m and abnormally dry out the clay below the foundations of the nearest part of the house. Trees planted along side roads have caused marked depressions along the edges of the roads and underneath cement concrete paths.

On sites where the soil is of firm shrinkable clay, it is advisable that fast-growing and water-seeking trees should not be planted within 18 m (7.6 m minimum) of buildings. In addition to this, the foundation should be taken down to a depth of minimum 1 m. Usually, strip footing of concrete is suggested with minimum width of 450 mm and 150 mm thick for load-bearing structures.

It is also most important to understand that when trees are cleared, clay soil will gradually swell as water returns to the ground. A clay site cleared of trees needs to be allowed to recover before construction begins, which may need many years, or foundations need to be specially designed to prevent damage caused by this swelling.

The recommended special foundations consist of using bored piles and ground beams.

12.3 Inadequate Design of Pavement Joints

In concrete pavements, cracking of slab is a common defect. To control the stresses resulting from the combined effects of temperature and moisture changes and wheel loadings and to minimise cracking in rigid pavements, transverse and longitudinal joints are used in reinforced and unreinforced slabs. Transverse joints may be provided in the form as contraction joints or construction joints or expansion joints. These joints must be capable of opening and closing and transferring load between adjacent slabs.

The cracks in concrete pavements mostly happen because the joints are not correctly spaced and are inadequately designed. Cracking occurs most frequently in the transverse direction; this is usually due to the slabs being too long so that warping caused by temperature gradients is usually restrained. Transverse cracking in plain concrete pavement may be prevented by a closer spacing of joints, i.e. about 4 m (13 ft.) apart. Distance between transverse contraction joints may be greater when the pavement is reinforced. Joint spacing depends on many factors including the quantity of steel and the thickness of slab. The Portland Cement Association recommends joint spacing of not more than 12 m (40 ft.) for reinforced pavements [4].

A longitudinal joint in a concrete pavement is a joint running continuously the length of the pavement. Longitudinal joints are used on highways to control cracking along the pavement centre line. Longitudinal cracking is usually due to the omission of a longitudinal joint and can be prevented by providing such joints or by increasing the amount of transverse reinforcement at the centre. Concrete joints are further described in Chap. 8.

12.4 Improper Steel Reinforcement

Reinforcement is provided mainly to resist internal tensile forces calculated from analysis. Also, reinforcement is provided in compression zones to increase the compression capacity, enhance ductility, reduce long-term deflections, or increase the flexural capacity for beams. It is important to provide the adequate area of reinforcement required to resist internal tensile or compression forces required to attain the design section strength. In addition to providing the sufficient areas of reinforcement and the required development lengths, good detailing should be done considering the overall structural integrity.

The use of an inadequate amount of reinforcing may result in excessive cracking. A typical mistake is to reinforce a member by considering it as a non-structural member. However, the member, such as a wall, may be tied to the rest of the structure in such a manner that it is required to carry a major portion of the load once the structure begins to deform. The element which was treated as non-structural earlier then begins to carry loads. As this member was not detailed to act structurally, cracking may result.

In any reinforced concrete structure, the reinforcing steel has two parts: primary reinforcement and secondary reinforcement (distribution steel). Primary reinforcement is the steel in the concrete that helps carry the loads placed on a structure. Without this steel, the structure would certainly collapse. Secondary reinforcement is the steel placed in a structure that enhances the durability and holds the primary reinforcement and the structure together. It provides the resistance to cracking, shrinkage, temperature changes, and impacts necessary for a long service life of the structure. Both types of reinforcement must be placed in their specified locations including spacing as designed to avoid any disaster.

A 66 ft. section of the De la Concorde overpass in Montreal (Canada) suburb, built in 1970, collapsed in 2006. Five people were killed, six were seriously injured, and two vehicles were crushed under concrete. Some of the factors to this collapse were concluded as poor initial design and incorrectly placed steel reinforcement.

It should therefore be ensured that reinforcing steel placed in a structure is:

- Of the correct grade and type
- Of the correct size, shape, and length
- Placed in its specified location and spaced as designed
- Placed in the correct quantities

12.5   Thermal Volume Changes

Concrete changes slightly in volume for various reasons; however, this section addresses volume changes of hardened concrete due to temperature variation.

12.5.1   Thin Concrete Elements

Like most of the materials, concrete changes volume when subjected to temperature variations. Volume changes create stresses when the concrete is restrained. The resulting stresses can be of any type: tension, compression, shear, etc. Designers should give special consideration to structures in which upper surfaces are exposed to temperature changes, while the bottom surfaces of the structure are either partially or completely protected, like deck slab, grade slab, road, or runway concrete pavements. A drop in temperature may result in cracking in the exposed element, while increase in temperature may cause cracking in the protected bottom portion of the structure. The curling of slabs may also occur as discussed in Chap. 8.

Allowing for movement by using properly designed expansion and contraction joints and proper reinforcement will help minimise these problems. Additionally, thickened edges, shorter joint spacing, permanent vapour-impermeable sealers, and large amounts of reinforcing steel placed 50 mm (2 in.) below the surface will all help reduce curling [1].

12.5.2   Thick Concrete Elements

For thin items such as pavements (mentioned above), heat dissipates almost as quickly as it is generated. For thicker concrete sections (mass concrete), heat dissipates more slowly than it is generated. To prevent damage and minimise cracking in concrete structures, it is necessary to control and manage these temperatures. Usually, members with a minimum dimension of 4 ft. (1.3 m) are considered as mass concrete, as suggested by ACI-301. However, ACI-116R [5] defines mass concrete as "any volume of concrete with dimensions large enough to require that measures be taken to cope with generation of heat from hydration of the cement and attendant volume change to minimize cracking".

In mass concrete, the internal temperature rises and drops slowly, whereas the exterior surface cools rapidly to ambient temperature. The contraction of the exterior surface due to cooling is restrained by the interior hot concrete as it does not contract as rapidly as the exterior surface of the concrete. Thus, the tensile stresses are developed due to this restraint, resulting cracking of the exterior surface of the mass concrete. Thermal cracking in mass concrete can be minimised by:

- Using suitable concrete mix that produces low heat of hydration, such as using supplementary cementitious materials like fly ash or slag as cement replacement within the concrete mix design. Class F fly ash generates about half as much heat as the cement that it replaces and is often used at a replacement rate of 15–25%. Ground granulated blast-furnace slag is often used at a replacement rate of 65–80% to reduce heat [6].
- Controlling concrete temperature during placing and transportation.
- Pre-cooling of concrete: Methods to pre-cool concrete include shading and sprinkling of aggregate piles, and use of chilled mix water and ice as discussed in Chap. 11.
- Using concrete surface insulation or insulated formwork: The concept of exterior surface insulation of freshly poured mass concrete is not to increase the maximum concrete temperature, but to decrease the abrupt rate of cooling.
- Using an aggregate with a low thermal expansion: Concretes containing low-thermal-expansion aggregates such as granite and limestone generally permit higher maximum allowable temperature differences in comparison to using high-thermal-expansion aggregates.

12.5.3   Roof Surfaces

Temperature movements are most severe at the roof. Black roofing under a hot sun can rise 90 °F (50 °C) above the ambient air temperature during the day. This will raise the temperature in the concrete roof slab, creating a tendency to curl, expand horizontally, and move masonry parapets. Partitions and facing materials in the top storey of a building must be detailed to accommodate these movements.

The use of insulation properly placed in the building can play a large part in minimising the amount of movement in concrete members due to temperature variations. The heating effects of direct sunshine on roofs can be reduced by light colour-reflective treatments.

12.6   Improper Grades of Slab Surfaces

Insufficient slope to the roof slabs requiring drainage for proper run-off needs special attention. Drains should be at low, not high, points. Proper slope-to-pitch for quick run-off is important to prevent deterioration and leakage within the structure which ultimately result in rusting of steel and spalling of concrete. Standing water provides concrete with the potential for saturation. The quicker the water runs off the structure, the less leakage can occur through joints and cracks.

Usually designers propose a fall of 1 in 100 (1%), which practically does not work properly. For positive drainage, ACI-302.1R-96 [7] recommends a minimum slope of 1 in 50 (2%), i.e. 1/4 in./ft. (20 mm/m), to prevent ponding, which preferably should be achieved by sloping the structure rather than doing it in screeds.

Misdirection of roof water run-off is also a common defect. Generally it has been observed that the roof water run-off is either directed toward the building or directed to an area of grade sloped negatively toward the building, creating ponding water that saturates the soils directly under the slab. All drainage pipes/downspouts should be directed at least 5 ft. away from the foundation. Roof gutters and downspouts should be regularly maintained and cleaned of all debris.

12.7   Basements

Water pressure has been known to buckle basement floors. This happens when an ordinary slab on ground has not been designed for hydraulic uplift. With or without a membrane under the floor, a rapid rise in groundwater levels may not be relieved by seepage through temperature and shrinkage cracks. Groundwater is water that is naturally located below the ground's surface. The groundwater level can be, at times, above the level of the basement floor. In some locations, groundwater can be above the level of the floor at all times. Some of the reasons for basement failures are the following [8]:

Seepage: If the water table rises, water can enter the basement via cracks, holes, and other unintended flow paths. This is generally considered to be part of the aging process of the building and the materials used to build it. Regardless of the condition of the drainage materials and pipe work around the foundation, water can also enter the foundation floor or walls via cracks and holes or other defects during heavy rains, ground-thaw, or snow-melt periods, when there's lots of water in the ground. Settlement of the lot grading around the building and downspouts discharging run-off water too close to the home can also increase the quantity of water around the foundation and increase the risk of water entering via cracks.

Sump pump failure: If the basement is equipped with a sump and sump pump(s) due to high water table, which are working properly and adequately maintained, it can safely pump excess water above the foundation and away from it. If the pumps cannot keep up, or fails to operate (perhaps due to a power failure, or malfunction), the groundwater level around the foundation can rise to the point that it flows up and out of the sump onto the basement floor.

Weeping tile failure: Over time, the foundation drainage system can deteriorate. As a result, the weeping tile system around the basement of the building can fail. This may be due to a partially or fully collapsed pipe, or due to sediments plugging the pipes. If the weeping tile fails, the drainage of water around the foundation is either impeded or blocked altogether. As a result, the groundwater level around the foundation gets too high and it may spill into the basement via the sump, if one exists, or via leaks in the foundation. In situations where there are leaks in the sewer lateral or plumbing beneath the foundation, groundwater can inundate the sanitary lateral and restrict the flow of sanitary wastewater. This could result in both ground-water and wastewater entering the basement by way of the floor drain or lowest sanitary fixture.

Gravity does its best to move water from high to low. If either the groundwater level or sewer level around the building is above the basement floor, gravity will try to move that water into the basement. A crack in the foundation floor, for example, provides gravity with a perfect path for water to be pushed into the basement.

Better drainage under the floor and its perimeter will help avoid water pressure if the drains themselves do not become overloaded. Sump pumps, if required, should be designed having proper size and should be maintained regularly. The outlet pipe from the sump pump should not be connected to sanitary sewer system. The best option is to direct the sump pump to the lawn, where it can infiltrate naturally into the ground. Cracks in floor or walls and surrounding the structure should be repaired on regular basis.

Weeping tile systems, as with the sewer lateral, require maintenance with time. In general, over time, the effectiveness of the weeping tile system will deteriorate, as the pipe ages, or gets plugged with fine sediments, or collapses. Maintenance of the weeping tile system is critical, but unfortunately it is not easy to get at. Sometimes built-up sediment can be flushed out without totally excavating the weeping tile. However, if the pipe is collapsing or has deficiencies all around, full replacement may be required. Ideally during replacement, additional access points shall be designed to permit future flushing and inspection of the system without excavation [8].

12.8   Inadequate Design Loads

All structures are designed to support loads without deforming excessively. These loads include live load, which is the weight of people and objects, rain, wind, and snow, and dead load of the building itself. The major causes of structural failure are defective designs that have not determined the actual loading conditions on the structural elements. Design defects don't only mean errors of computation, but a failure to account for loads the structure will be expected to carry, reliance on inaccurate data, ignorance of the effects of repeated or impulsive stresses, improper choice of materials, and impact of heavy rains, earthquakes, hurricanes, and a defective site, with very unusual ground conditions. Structure may fail even if the design is satisfactory, but the materials are not able to withstand the loads. A durable and stable structure should be able to meet all these challenges as all or most of the above information/data is incorporated into the design.

Correct structural design is significant for all structures, because even a slight probability of failure is not acceptable since the results can be disastrous for human life and property. Therefore, design engineers are required to be exceptionally careful in ensuring an appropriate building design that can sustain the applied loads. Some examples of design load failures are briefed hereunder.

Lian Yak Building, Singapore: This building, also known as Hotel New World, was completed in 1971 consisting of six storeys and a basement garage. In March 1986, the building crumbled to the ground in under 60 s, trapping 50 people beneath rubble, 33 of whom were killed. Inquiry revealed that the building's design engineer had made a serious error. He calculated building's live load; however he did not properly assessed the dead load of the structure itself. Hence the building could not support its own weight and collapsed.

Hyatt Regency Walkway Collapses: Hyatt Regency Hotel in Kansas City, Missouri, was built in July 1980. One year after the opening, the walkways on the second and fourth storeys collapsed under the weight of partygoers, killing 114 people and injuring 200 people. Inquiry revealed that a change in the details of structural connections was left unchecked during the shop drawing submittal process, which doubled the load on the fourth-floor walkway connections. This extra load on these connections led to the collapse of the fourth-floor suspended walkway onto the second-floor walkway and then onto the ground floor below.

12.9   Water-Retained Structures

Concrete tanks are commonly used for water and other liquid storage purposes. Depending on requirements and site considerations, they may be elevated, underground and above ground, and covered or uncovered, and frequently have both full and empty liquid levels throughout operation. A tank's design must be compatible with materials stored within it, so as to meet the specific design requirements for tanks storing water, acids, bases, peroxides, flammable materials, sludge tanks, etc.

Underground tanks: In water-retained underground structures, the design should consider the cases where the structure is full of liquid and also when it is empty. The structure when empty must have the strength to withstand the active pressure of any retained earth. Since the passive resistance of the earth is never certain to be acting, it is generally ignored when designing for the full structure.

Buried concrete tanks can be damaged by groundwater-generated buoyancy force. A buried tank can fail due to the buoyancy force, when the groundwater exerts more pressure upward on the underside of the base slab than can be counteracted downward by any of the following [9]:

• The self-weight of the tank
• The weight of soil supported by the tank
• The weight of any liquid in the tank at the time of failure
• Any additional groundwater mitigation system, or GMS

There are two commonly encountered failure mechanisms for buried tanks due to the groundwater-generated buoyancy force: structural failure of the base slab and complete tank flotation. For one of these to occur, either the effects of groundwater are not properly quantified or the part of the groundwater GMS fails. It is important for designers, contractors, owners, and operators to understand the buoyancy force. Otherwise, costly and inconvenient repairs may be required during the construction, or design life, of any buried tank.

Elevated tanks: The manner in which elevated tanks are supported should be designed and reviewed carefully. Steel structures used to support a tank should be treated with retardant material and corrosion protection to prevent structural failure and buckling of steel structure from adjacent fires. Preference should be given to reinforced concrete support structures, which should be regularly inspected for cracks and other signs of structural failure.

Concrete surfaces are especially vulnerable to acid spills. If the tank stores corrosive materials or sludge and such materials as in wastewater treatment plants, the concrete surfaces should be coated with a chemically resistant coating. Winter deicing procedures should also be reviewed to ensure that corrosive salts are not needlessly damaging structural supports and containment areas. An Ontario provincial government study [10] has been undertaken on the 53 concrete water tanks built in the province between 1956 and 1981. The concrete tank deterioration observed was primarily caused by two separate factors: the first, freezing of water in the walls and the second, internal ice formations in the tank. Both resulted from the tanks being in a temperature environment below the freezing point of water. The more serious problems were associated with water freezing within the wall and may have resulted in delamination of coatings and the wall itself, jack rod spalls, freeze-and-thaw spalling, and even sudden progressive collapse of the tank. Other problems, such as widening of vertical cracks, were caused by hoop tensile forces acting on the tank because of the expansion of internal ice formations.

Tanks are designed to hold liquids or gases at a given temperature and pressure. Temperature differential effects are important and must be added to other loads. Changes in the internal and external tank pressure can result in catastrophic tank failure. Sunlight shining on a tank can increase internal tank pressure, which must be released to prevent tank failure. Concrete tanks subjected to freezing conditions should design for the additional temperature differential effects and forces from ice formations and ensure complete watertightness.

Many tanks use vents to regulate internal pressure. If these vents become blocked, excessive high tank pressure may result, causing tank failure. Tank vents may become blocked due to sludge buildup, accumulation of condensed product, ice, rust, bird nests, or improper maintenance. All tank vents should be periodically inspected to ensure that all vents, flame arresters, etc. are properly maintained. It is also important to check that vents are of the proper size when filling the tank with a different material. High-vapour-pressure materials can easily exceed the design requirements of a tank designed to hold a low-vapour-pressure liquid [11].

During construction of underground water-retained structures, usually base slab is cast along with some upstand of wall, as it is difficult to cast the wall and base slab in one go; hence construction joint is required. Most of the designers overlook to provide these details due to which the joint is not sealed properly and leads to leakage of water. For construction joint, it is suggested to form grooves (key joint) in the first pour so that the next pour is keyed into it. The other method to seal such joints is to incorporate centre bulb-type water bar in wall as shown in Fig. 8.7, which is more successful in water-retained structures.

12.10   Insufficient Drawing Details

An adequate design does not guarantee a satisfactory function without including design detailing. Detailing is an important component of a design. Poor detailing may or may not directly lead to a structural failure but it may contribute to the deterioration of the concrete.

It has been observed that sometimes the details available in architectural drawings do not tally with the details provided in the related structural drawings, and such mistakes may lead to a big problem. Before issuing both sets of drawings to the site, the designer must confirm that changes made in the architectural drawings are incorporated in the related structural drawings, and vice versa.

Thus, steps are needed to significantly improve the quality control of designs and specifications. Drawings and specifications should be complete and the information on or in them easily found by the construction engineer. A principal factor is to ensure that on all drawings for any one contract the same conventions are adopted and uniformity of appearance and size is achieved, thereby making the drawings easier to read. The scale employed should be commensurate with the amount of detail to be shown. The element detailing must comply with any of the standard code requirements. An independent check should be made of all design calculation to ensure that the section sizes, slab thickness, reinforcement sizes, and spacing specified are adequate to carry the worst combination of design loads. The check should include overall stability, robustness, serviceability, and foundation design.

The structural design must be able to withstand the loads acting on them during a reasonable lifetime. During its life the structure should have adequate structural integrity to be able to adopt to unforeseen influences and localised damages without major failure. In addition, the structures must behave satisfactorily in service life. Structural design is largely controlled by regulations or codes but even within such bounds the designer must exercise judgement in his/her interpretation of the requirements. There is also a need for designers to be aware, in a far more specific way, of the likelihood and causes of failures. They should also be trained on design principles and solutions, to avoid failures. Only a thorough understanding of structural behaviour can minimise the problems.

Bibliography

1. Ytterberg, Robert F., "Shrinkage and Curling of Slabs on Grade, Part I—Drying Shrinkage, Part II—Warping and Curling, and Part III— Additional Suggestions," Concrete International, American Concrete Institute, Farmington Hills, Michigan, USA April 1987.
2. M.J. Tomlinson, Foundation Design and Construction, E.L.B.S. Longman, Singapore, 1992.
3. Austin Williams- The Distance at Which Trees Can Affect a Building is Quite Significant- Architects Journal, 7 December 2006. London UK.
4. Joint Design for Concrete Highway and Street Pavements. Portland Cement Association, Skokie, III (1980).
5. ACI Committee 116, "Cement and Concrete Terminology (ACI 116R)," American Concrete Institute, Farmington Hills, Mich., 2000, USA.
6. John Gajda and Martha Vangeem (Construction Technology Laboratories Skokie)- Controlling Temperatures in Mass Concreting- Concrete International, 2002.
7. ACI 302. 1R- 96 (Revised 2015), Guide for Concrete Floor and Slab Construction.
8. Utilities Kingston Ontario Canada- Causes of Basement Flooding.
9. Mark Bruder, Storage Tanks Containment & Spills- Beware of buried tank buoyancy, *Environmental Science & Engineering Magazine 56, May 2013.*
10. W. M. SLATER, Canadian Water Tanks in Ontario- Canadian Journal of Civil Engineering, Canadian Science Publishing – Volume 12, June 1985.
11. David Patzer, CSRMA Risk Control Advisor- Preventing Above Ground Storage Tank Failures.
12. Rifat Akbiyikli and David Eaton- Design and Design Management in Building Projects: A Review- e-Journal of New World Sciences Academy- 2012, Volume: 7, Number: 1, Article Number: 1A0304
13. Kent Davis, W.B, Ledbetter- Measuring Design and Construction Quality Cost- Report Published by University of Texus, 1987.
14. Terzaghi, K (1938)- Settlement of Structures in Europe and Methods of Observation- ASCE, Vol. 103.

Chapter 13
Achieving Quality in Concrete Construction

13.1 Introduction

Quality of concrete construction includes steps for maintaining the required strength of concrete within the deviation permitted and construction of a durable structure. Quality of concrete construction depends on many factors. From the selection of construction materials to the curing of the structural member, each and every step must be carried out to maintain the quality requirement of the project. Any deviation from the required quality may result in failure of the structure or may impose financial implications on the contractor for not maintaining the quality.

As per ISO 9000 [1], quality can be defined as "The totality of features and characteristics of a product or service that bears on its ability to satisfy stated or implied needs". In other words, as mentioned by Phil Crosby [2], the famous author of several books, quality could be defined as "Conformance to requirements". In the design and construction industry, the architect or engineer develops a concept, which the owner approves; then a detailed scope of work is used to define the product. Drawings and specifications are prepared and the work is then awarded for construction. The point is that conformance to codes and specifications becomes the items which define the requirements to follow.

Now in order to further evaluate quality we must differentiate between its two types: quality control (QC) and quality assurance (QA). QC and QA are mostly used interchangeably or together to mean the same thing. The fact is that these two terms represent separate processes that may or may not coincide. Either way, they both have a direct impact on the final outcome of a work item with respect to construction of a structure.

The contractors are responsible for constructing the work in accordance with the plans and specifications. Each contractor is also responsible for controlling the quality of its work to meet contract plans, specifications, and related requirements. Quality control personnel typically take actions for the producer or general contractor to control the materials, processes, and quality of work so as to control

© Springer Nature Switzerland AG 2019
A. Surahyo, *Concrete Construction*,
https://doi.org/10.1007/978-3-030-10510-5_13

the level of quality being produced in the end product. The contractor's QC is the systematic implementation of a program of inspections, tests, and production controls to attain the required standards of quality and to preclude problems resulting from noncompliance. Furthermore, each contractor will define methods for ensuring that activities affecting quality will be accomplished under controlled conditions.

Quality assurance personnel typically take actions on behalf of the owner, or the owner's representative, to provide confidence and document assurance that what is being done and what is being provided are in accordance with the applicable project specifications and standards of good practice for the work. Independently of the contractors, the consultant's resident engineer (RE) will provide QA through daily monitoring and scheduled inspections to verify the effectiveness of the contractor's QC program and assure that the quality and contract requirements are met by the contractors. The RE assures that the contractor's QC is working effectively and that the resultant construction complies with the quality requirements established by the contract. QA/QC is further explained in the following pages.

13.2 Quality Control (QC)

As per ACl-121 [3] quality control is a production tool and is defined as "Those actions related to the physical characteristics of the materials, processes, and services which provide a means to measure and control the characteristics to predetermined quantitative criteria". Being a production tool, it is a part of the contractor's programme to assure compliance with the quality requirements mentioned in the documents that constitute the contract between the contractor and the owner. The contractor's responsibility includes ensuring that adequate QC services are provided for work accomplished on- and off-site by the contractor's own organisation, suppliers, manufacturers, subcontractors, technical laboratories, and consultants. The work activities include safety, submittal management, and all other functions relating to the requirements.

As per ISO 9000 [1], QC could be defined as "The operational techniques and activities that are used to fulfil requirements for quality". The requirement for a comprehensive quality control programme should be detailed in the contract documents. To be effective, quality control must be ongoing and proactive, not intermittent or reactive. An effective quality control programme is essential to the contractor and his/her responsibilities. Quality control does not cost; it pays. The savings from avoided replacements, repairs, reworking, claims, and litigation will exceed many times over the cost of an effective quality control programme.

To meet QC requirements, the contractor is required to submit a project-specific QCP and states how the process control of materials, equipment, and operations shall be maintained.

13.2.1   Quality Control Plan (QCP)

In QCP, it should be identified which quality standards are to be used in the project and how they will be implemented. The QC plan is intended to be a roadmap for the project, describing how, when, and where the contractor will administer the necessary control of their processes throughout all phases of construction in order to deliver a quality end product that will meet specification requirements. As a minimum, the QCP shall include the following information for each project:

- The name, telephone number, and duties of all quality control personnel necessary to implement the QCP.
- The name and location of the testing facility to be used for conducting concrete testing on-site and in laboratory. A list of the testing equipment proposed for process control testing, and the test methods and frequency of calibration and verification of the equipment.
- Name and details of concrete supplier to be used for the project.
- The transportation and handling procedures for delivering concrete to site: The description or plan drawing of the traffic control for delivery of the concrete mix to the site of work shall be included.
- Procedures for all material checks, including frequencies of testing, shall be identified. The methods to monitor ingredients used, and the record of each batch, shall be included.
- The location, procedures, and frequency for sampling and testing the concrete mix for slump, air content, and compressive strength.
- The procedures for placement of the concrete to include as a minimum the placing sequence, identification of the placing equipment, and a description of the pumping procedures.
- The methods for vibrating, finishing, and curing concrete. The description of the equipment shall be included.
- The procedure for determining when the forms and falsework may be removed.
- The procedures identifying what actions will be taken by the contractor to reject the concrete and/or make the appropriate adjustments on the delivery of concrete at site, when test results identify concrete that is not in compliance with the specifications.
- The procedures identifying what actions will be taken related to project site conditions for air temperature, wind speed, and relative humidity that may adversely affect concrete quality.

The purpose of QC in concrete projects is to measure and control the variations of the mix ingredients, and to measure and control the variation of operations like batching, mixing, placing, curing, and testing which affect the strength or the uniformity of the concrete. In view of these different processes in the manufacture of concrete, the problems of quality control are diversified. Hence in order to achieve this part of quality in concrete construction, it is necessary to analyse the different factors causing variations in the quality and the manner in which they can be controlled.

13.2.2 Factors Causing Variations in the Quality of Concrete

The main factors causing variation in concrete quality are as follows:

1. Personnel
2. Material, equipment, and workmanship
3. Field testing

13.2.2.1 Personnel

Inadequate knowledge of factors influencing the behaviour of concrete has harmful consequences in the operations of manual and technical staff. The basic requirement for the success of any quality control plan is the availability of experienced, knowledgeable, and trained personnel at all levels. The construction engineer must be able to understand the specification stipulations and design details, and should have the knowledge of construction operations as well. In order to ensure lasting and durable concrete structures, he should also be aware of the effects of climate, temperature, and exposure conditions. It is always helpful to provide training that guarantees that personnel understand what is expected. ISO (International Organization of Standardization) has discovered that the most frequent problems are those involving training, document control, and records.

Everything in quality control cannot be specified and much depends upon the attitude and orientation of the people involved. In fact, quality must be a discipline imbibed in the mind and there should be strong motivation to do everything right the first time.

13.2.2.2 Material, Equipment, and Workmanship

Material

The contractor usually submits the concrete mix design to the engineer for approval in accordance with the contract provisions. The concrete mix design for the required type of concrete generally specifies the following:

1. Required strength of concrete
2. Cementitious content in kilograms per cubic meter or equivalent units
3. Aggregate size, and gradation
4. Aggregate source location(s)
5. Weights of aggregates in kilograms per cubic meter or equivalent units
6. Maximum allowable water content in kilograms per cubic meter or equivalent units and the design water/cementitious ratio
7. The limits for slump
8. The limits for air content
9. Quantity and name of admixture if to be used

Any change in any one of the constituent materials of the concrete shall require a new concrete mix design. If, during the progress of the work, the mix design is found to be unsatisfactory for any reason, including poor workability, the contractor shall revise the mix design and submit the proposed changes to the engineer for review.

To avoid any variation in produced concrete, its constituent materials need to be handled properly and with care.

Cement: For uniform quality of concrete, the cement should preferably be used from a single source. When cement from different sources is used, various tests made by different agencies show that the crushing strength at 28 days of concrete varies up to 50% or more [4]. This variation in the strength is related to the composition of raw materials as well as variations in the manufacturing process.

The cement should be tested initially once from each source of supply. Adequate storage under cover is necessary for protection from moisture. Bags of cement should be stacked on pallets, or similar platforms, to permit proper circulation of air. For a storage period of less than 60 days, ACI-304 [5] recommends that the bags be stacked no higher than 14 layers and for longer periods no higher than 7 layers. As an additional precaution, it is recommended that the oldest cement be used first. However, set cement with hard lumps is to be rejected. Storage facilities for bulk cement should include separate compartments for each type of cement used. The interior of the cement silo should be smooth with a reasonable bottom having proper ventilation arrangement. Storage silos should be drawn down frequently, preferably once per month, to prevent cement caking.

While transporting cement the top 25–50 mm (1–2 in.) layer of cement in a bulk truck or railroad car is aerated by the air above it. Aeration can cause slump loss or increased water requirement and false set in cement. Cement in paper bags is subject to aeration; hence false set in such type of bags is common. This can occur long before there is evidence of hard lumps. Aeration of cement should therefore be avoided. To minimise the effects, cements should be transported in closed containers.

Contamination of cements by different materials should be avoided. Very small amounts of sugar and starch can cause significant retardation. Similarly, small amounts of lead, zinc, and copper compounds also have the same effects on cement. With ammonium sulphate contamination, the sulphate content of the concrete may be increased, which may result in production of unsound concrete.

Aggregate: In order to produce uniform quality of concrete, fine and coarse aggregates must be of good quality, uncontaminated, and uniform in grading and moisture content. Grading, shape and type, and moisture content of the aggregate are the major sources of variability.

The grading of the aggregate affects the workability, and hence the water/cement ratio of the concrete. It has been observed that the aggregate with the coarsest grading would produce concrete with a crushing strength some 20% greater than that with the finest grading. The variation in the grading of the aggregate can be reduced by using grading requirements provided under Tables 2.8–2.13. Graded aggregates should not be allowed to segregate. For effective control of gradation, it is essential that handling operations should not significantly increase the undersize materials in aggregates prior to their use in concrete, and should be uniform and within specified limits.

The use of flat or elongated aggregate particles should be avoided as they have detrimental effect on the workability of concrete, resulting in the necessity of more highly sanded mixes, and the consequent use of more cement and water. For the same degree of workability, an angular aggregate may produce concrete having a crushing strength some 50% lower than relatively rounded aggregate.

The aggregates should be obtained from one source, so that the effect of shape and type of aggregate may be controlled at this stage. Research work using 24 different crushed rock aggregates [6] has shown a variation of water requirements from 170 to 230 kg/m^3 resulting in a 28-day compressive strength variation between 29 and 53 N/mm^2. Similarly, using fine aggregate from different sources has proved different water requirements. In comparable concrete mixes, one sand needed 80 lb/y^3 (48 kg/m^3) more mixing water [7]. In fact, rough and angular sand required the greater amount of mixing water and also needed more Portland cement to maintain the water/cement ratio in comparison to smooth and rounded one.

The aggregate should be free from impurities and deleterious materials, since for every 1% of clay in sand there could be as much as 5% reduction in the strength of the concrete.

The moisture contents of aggregates should be taken into account while arriving at the quantity of mixing water. The use of aggregates having varying amounts of free water is one of the most frequent causes for loss in control of concrete consistency. In some cases, it may be necessary to wet the coarse aggregate to compensate for high absorption, or to reduce the temperature in hot seasons. In such cases, the coarse aggregates must be dewatered before shifting it to the mixer to prevent transfer of excessive free water.

As far as moisture variations in fine aggregate (sand) are concerned, bulking of sand is important in several ways. A given weight of moist sand generally occupies a considerably larger volume than the same weight of dry sand and gives erroneous results when volume batching is adopted. It increases the water/cement ratio, which in turn enhances the workability but reduces the strength. Errors due to bulking of fine aggregate can be entirely eliminated by using weigh batching instead of volume batching.

The aggregates are required to be tested once initially for the approval of each source of supply. Subsequently, periodic tests should be conducted at the site for grading and moisture content.

Water: The water used for mixing concrete should be free from silt, organic matter, alkali, and suspended impurities. Sulphates and chlorides in water should not exceed the permissible limits. As mentioned in Chap. 6, the water containing excessive amounts of dissolved salts reduces compressive strength by 10–30% of that obtained using freshwater. Generally, water fit for drinking may be used for mixing concrete.

The other factor which greatly causes variation in strength of concrete is excess water quantity. An increase of about 10% water than the required quantity in concrete mix can reduce the strength by 15%. Chapter 7 addresses in detail the effects of excess water quantity on concrete strength.

Supplementary Cementitious Materials (SCMs): Cementitious material includes Portland cement, blended cements, ground granulated blast-furnace slag, fly ash, silica fume, and other materials having cementitious properties.

SCMs shall conform to the requirements of contract documents and shall be free from lumps. These materials shall be suitably protected from the weather and dampness during storage. Bulk or bagged SCMs shall be stored separately to prevent contamination and to permit identification of each cementing material at all times. Where there is doubt as to the quality of the cementing material, it shall not be used in the work until testing has been completed ensuring that the test mixtures containing SCMs are achieving the desired results, have correct dosage, and are without any unintended effects. Changing an SCM source, type, or even dosage will require new trial batches.

Normal Portland cement, type GU or GUb (general-use Portland blended cement), or sulphate resistant, type HS or HSb (high sulphate-resistant Portland blended cement), shall be used based on requirements unless otherwise specified on the drawings or in the special provisions. Should the contractor choose to include a silica fume in the concrete mix design, the substitution of silica fume shall not exceed 8% by mass of normal Portland cement. If it is intended to use fly ash in the concrete mix design, the fly ash shall be Class CI (C Intermediate—refer Chap. 2) and the substitution shall not exceed 20% by mass of normal Portland cement.

However as mentioned in Chap. 2, the following proportions of SCMs specified by OPSS.PROV 1350 shall be preferred:

1. Slag up to 25%
2. Fly ash up to 10%, except for silica fume overlays and high-performance concrete (HPC), where up to 25% is permitted
3. A mixture of slag and fly ash up to 25%, except that the amount of fly ash shall not exceed 10% by mass of the total cementing materials in concrete other than silica fume overlays and HPC

Admixtures: Air-entraining and chemical admixtures shall conform to the requirements spelled out within the contract documents. All admixtures shall be compatible with all other constituent materials. The addition of calcium chloride, accelerators, and air-reducing agents shall not be done, unless otherwise approved by the engineer. Appropriate low-range water-reducing and/or super-plasticising admixtures shall be used in all concrete containing silica fume.

While selecting admixtures (as discussed in Chap. 6), their effect on water demand and required dosage rate must be examined. If admixtures are used in large quantities, side effects may occur. For example excessive amounts of super-plasticiser can cause shrinkage cracks in concrete. Similarly too much air entrainment may reduce the strength of concrete. Some accelerators begin to show retarding properties if they are added to concrete in excess.

Similarly, some of the water-reducing and -retarding admixtures are generally used in small doses which normally reduces mixing water requirements 5–8%. However, super-plasticisers are used in large quantities and reduce water requirements 12–25% or more. Hence, while using different water-reducing admixtures, their effects on water content must be taken into account when calculating the quantity for mix water.

Equipment

The equipment used for batching, mixing, and vibration should be of right capacity. Weigh batchers should be regularly overhauled and calibrated at frequent intervals. In general, batching by weight is more reliable than batching by volume. The normal method of batching by volume, i.e. using shallow gauge boxes, may produce variation in the concrete strength by 20–70%. Hence batching by volume should be avoided.

Mixer's performance should be checked for conformity to the requirements of the relevant standards. Inadequate mixing results in production of non-uniform concrete, and large variations in strength and workability of concrete will occur. All concrete should be thoroughly mixed until it is uniform in appearance and all ingradients are uniformly distributed. Mixers should not be loaded above their rated capacity and should be operated at approximately the speeds for which they are designed. Concrete should be mixed for the required time, and both under-mixing and over-mixing should be avoided.

The vibrators should have the required frequency and amplitude of vibration as per Table 7.3. Vibrators too large for the application result in the high concentration of cement paste, which leads to crazing of concrete surfaces. Insufficient or under-vibration results in concrete having entrapped air pockets and pores. These voids reduce density, strength, and durability.

Workmanship

The concrete in its plastic state should be handled, transported, and placed in such a manner that it does not get segregated. The time interval between mixing and placing the concrete should be reduced to the minimum possible. The concrete should be covered during transport, particularly on hot or wet days or when there is a drying wind. No water should be added to the concrete after it has been taken out from the mixer or as specified in the contract documents. The concrete should be placed in uniform layers and during placing the formation of cold joints in successive pours must be avoided. Hand shovelling and placing in large heaps and in sloping layers should also be avoided. Dropping concrete from excessive heights should be discouraged. The concrete from upper to lower elevations should always be dropped vertically through chutes or circular pipes as shown in Figs. 7.6 and 7.7.

In order to achieve the required strength, impermeability, and durability of concrete, proper attention must be given to the adequate compaction of concrete. Inadequate vibration or over-vibration leads to segregation. Inadequate compaction can cause porous and non-homogeneous conditions. Internal vibrators should not be used to move concrete laterally. They should be inserted and withdrawn vertically at close intervals using proper methods of vibration to ensure that all concrete has been adequately consolidated. Only 5% of the air voids left in concrete due to incomplete compaction can lower the compressive strength by nearly 30%. Inadequate concrete cover to reinforcement should be avoided, as this will help in promotion of corrosion. For adequate cover of concrete to reinforcement, Tables 7.4, 7.5, and 7.6 are recommended. Adequate curing is very essential for the development of strength of concrete. This topic is discussed in detail in Chap. 7.

13.2.2.3   Field Testing

Contractor quality testing refers to tests required by the specifications to be performed and recorded by the contractor. Concrete testing plays a vital role in the overall quality control plan. Apart from the tests on concrete materials, concrete is tested both in the fresh and hardened stages. The test on fresh concrete offers some opportunity for necessary corrective actions to be taken before it is too late. For fresh concrete, slump test, temperature, and air test are the most common and easy to perform on-site.

Tests on hardened concrete are usually performed on 150 mm cubes or 100 mm cylinders prepared from samples of the concrete used on-site. The importance of correct sampling and preparation, curing and testing of the cubes cannot be overemphasised. Usually three control specimens are made from each sample, out of which one is tested at 7 days, and the 28-day test result shall be the average of the strengths of the remaining two specimens as per CSA (Canadian Standards Association). Additional cylinders may be cast, at the discretion of the engineer or contractor. The early test results enable estimates to be made of the probable 28-day concrete strengths. Any necessary remedial action can then be taken at a much earlier stage than would otherwise be possible. It is, in fact, only an acceptance test which helps the engineer/architect decide whether to accept or reject the concrete.

It must be noted that non-compliance by a single test specimen, or even by a group, does not necessarily mean that the concrete from which the test specimen has been made is inferior to that specified; the engineer's reaction should be to investigate the concrete further. This could be done by performing non-destructive tests on the concrete in the structure or by taking test cores for assessing the strength.

Since, in spite of all pre-planning, the eventual success of a project depends on the actual work in the field, and the final products depend on each member of the team being completely prepared before construction starts. In order to accomplish a successful concrete project it is essential to take the following steps before any work is started:

- Have the specific objectives clearly defined.
- Ensure that the teams have the necessary resources and organisational support to carry out their designated functions.
- Select the entire team early and then use their expertise.
- Have a total understanding of the quality control that will be enforced, as well as the complete testing programme to be used throughout the construction period.
- Make it clear that management is committed to quality.
- Raise the quality awareness and personnel concern of all employees.
- Drive out fear so that everyone may work efficiently for the company.
- Encourage employees to communicate to management the obstacles they face in attaining their improvement goals.

13.3 Quality Assurance (QA)

As per ACI-121 [3], QA is a management tool and is defined as "All those planned and systematic actions necessary to assure that the final product will perform its intended function". Being a management tool it is a part of the owner's responsibility. Just as an effective quality control programme pays for the contractor, an effective quality assurance programme pays for the owner. The lack of an effective QA programme can have significant long-term costs for the owner. The owner is usually responsible for quality planning and overall management. Where the owner does not have skills required to fulfil these responsibilities, he/she can appoint an organisation or individual to perform this function. As per ISO 9000 [1], QA is defined as "All those planned and systematic actions necessary to assure that the final product will satisfy given requirements for quality".

The purpose of quality assurance is that the product achieved should serve its intended purpose satisfying requirements in conformation to codes, specifications, and drawings. The supervision exercised on-site must ensure that the materials used for making concrete are of the required quality and that the stated mix proportion is adhered to as closely as possible. It is also required to control testing of the properties of the concrete in both its fresh and hardened state to ensure that it conforms to the design requirements. In order to achieve this part of quality in concrete construction the following points should be considered before starting the project:

- Appoint early the project management/quality assurance team with clearly defined responsibilities.
- Develop a QA system for the project. Indicate which section of the QA team shall develop the QA plan and QA programmes at various stages of planning, design, selection of materials, construction supervision, and material testing. Table 13.1, based on ACI-121 [3], indicates the various stages of a project and how the overall QA system is developed.
- Establish who does what within the organisation.

Table 13.1 Development of a QA system based on ACI-121 [3]

Project stage	QA system stage	Source of QA requirements and reference	Responsible review organisation and action
Planning	Owner[a] develops project QA plan	Owner, manager, consultant, and engineer	Owner review and approval are required if project QA plan was developed by another organisation
Design	Engineer develops design QA programme and submits to the owner for review prior to start of design	Owner's project QA plan	Owner, manager, or quality consultant reviews and approves engineer's QA programme
Selection of material and services	Supplier develops and submits supplier QA programme	Any combination of owner's project QA plan, contract documents	Owner, manager, consultant, or engineer reviews supplier's QA programme
Construction	Construction contractors develop and submit contractor QA programme	Any combination of owner's project QA plan and contract documents	Owner, manager, consultant, or engineer reviews contractor's QA programme
Material testing	Material testing laboratory develops and submits a material testing laboratory QA programme	Any combination of owner's project QA plan and contract documents	Owner, manager, consultant, or engineer reviews material testing laboratory's QA programme

[a]Indicates owner or engineer or quality consultant

- Select the people who are qualified for the particular tasks, and train the personnel in their responsibilities.
- Conduct pre-construction meetings to ensure that everyone understand his/her contract responsibilities and all necessary activities. In fact, a pre-construction meeting is important for both the QC and QA programmes for the smooth running of the project. It establishes cooperative and non-adversarial relations between both the parties. During the meetings, both the programmes should be reviewed and the lines of authority and communication established and recognised.
- Quality assurance should not be viewed as an "Us vs. Them" scenario; rather all parties should be working together toward a common goal.
- Both the contractor and consultant should perform visual "inspections" of key items that influence performance, yet these are not easily measured.
- Monitor and review the progress of the teams regularly; this will facilitate alterations to the strategy at an early stage, if needed.

For implementing quality assurance programme, the following important elements need to be controlled:

1. Design
2. Material
3. Inspection
4. Testing and evaluation

13.3.1   Control of Design

The design calculations and procedures must be documented, checked, and approved. The drawings and specifications should be complete containing information to be easily found by the construction engineer. The element detailing must comply with any of the standard code requirements, and it should be identified. The provisions specify the cover to reinforcement, minimum thickness for fire resistance, maximum and minimum steel areas, bar anchorages, lap lengths, etc. Responsibility for overall coordination and verification of all required approvals and reviews should be identified in a QA programme. The responsibilities of design organisation may include:

- Organising soil investigation reports
- Clarifying design queries (RFI, i.e. Request for Information) from field staff
- Review and approval of field changes
- Issue of revised design drawings
- Review of contractor's shop drawings
- Resolution of non-conforming items
- Maintain record of as-built drawings

13.3.2   Material Control

The controls for the various construction materials are specific to the type and application. The material types are sensitive to permeability, segregation, and contamination and the requirements are intended to ensure that performance of the concrete produced in its final position is consistent and predictable and meets expectations over the lifetime of the installation.

The procedures for assurance that the concrete constituent and related materials meet the requirements of the contract documents prior to use should be controlled and documented. Material approval document should specify the description of material, manufacturer or supplier name and applicable code, specifications, drawings, and standards. The constituent material should be tested prior to use and such test report and relevant documents should be attached for requesting approval of material. The capability of the material supplier should be evaluated. The procurement of material to site should also be recorded to verify that the material conforms to the approved one. Material specifications and drawings, including those prepared by material suppliers, and used to place their products in the construction, should be retained in the file.

13.3.3   Inspection

Inspection should be performed by qualified, experienced, and trained personnel. A programme for inspection should be established and implemented to assure that concrete construction and materials comply with the contract specifications and drawings. The inspection programme should include written checklists for items of construction and required criteria for acceptance in accordance with the project documents. The checklist should identify the items to be verified on-site, like:

- Setting out.
- Concrete bed.
- Formwork.
- Steel reinforcement including chairs and specified tie wires.
- Embedment, e.g. conduits, pipes, water stops, and anchor bolts: Embedded items need to be securely fixed in place before concrete is placed and be held in position until concrete placements are finished. Typically, embedded items should not affect the positioning of reinforcement, unless specifically allowed in the specifications.
- Cleanliness.
- Joints.
- Access.
- Compaction arrangements.
- Curing arrangements.
- Concrete grade.
- Concrete temperature.
- Concrete slump and air content.
- Concrete arrival and completion timings.

The inspection call should contain items like:

- Date and time of inspection.
- Location and item of inspection.
- Acceptance or non-acceptance.
- Remarks.
- Contract, consultant, and contractor's name.
- Signature column for submittal and approval.
- Approval date.

Rectification works for materials or processes, which do not conform to the project specifications, should be carried out immediately. Procedures identifying the defects and steps taken for rectification, repair, and rejection or redone should be recorded.

The following record containing construction information may be kept in project file by inspection/construction supervision team:

- Correspondence between members of the construction team, design team, and owner or consultant
- Daily and monthly progress record
- Test results on fresh and hardened concrete
- Job progress photographs
- As-built and shop drawings
- The record of tender drawings, construction drawings, and superseded drawings
- Information concerning the foundation and soil bearing capacity
- Diaries and field inspection reports
- Record of monthly meetings
- Project evaluation report

The attitude of QA team should be proactive, identifying contractor's problems early and offering solutions, thinking ahead, and trying to control any future complications. The best results for all parties can be obtained if inspection for QC by the contractor and inspection for QA by the owner or consultant are handled in a co-operative non-adversarial manner by fully qualified personnel.

13.3.4 Testing and Evaluation

All tests on concrete and its constituent materials should be performed as per contract documents, and the results should be documented. All testing should be performed by qualified personnel and through approved laboratory or on-site, if proper testing equipment is available. ACI [8] recommends that acceptance testing be performed by the owner or engineer as a part of the overall QA programme. It further specifies that "The owner or engineer should avoid the undesirable practice of arranging payment for acceptance inspection and testing services through the contractor. Such practice is not in the owner's interest".

The record of test results on fresh and hardened concrete should be maintained properly. The test result should identify:

- Name of contract and contractor
- Name of supervising authority
- Name of lab testing authority
- Signature columns for testing and supervising authorities
- Test date
- Location where material is used
- Test method and result
- Statement of compliance or non-compliance
- Acceptance or non-acceptance
- Remarks

The test results should be evaluated by a qualified person of the construction management section in the light of the acceptance requirements mentioned in the contract documents. The non-conforming test results should be brought immediately to the attention of the engineer and contractor. The test result should be further evaluated by the engineer to determine its adequacy. Records for all non-confirming test results (NCRs) must be maintained. The QA/QC administrator should follow up on all NCRs to ensure that proper corrective actions have been taken and implement guidelines to prevent recurrence. All efforts should be made by both QA/QC team to be proactive in identifying potential NCRs with inspecting site conditions and ensuring that all materials are free from defects and meet the quality requirements of the contract documents.

For a compressive strength test result OPSS.MUNI-1350 [9] provides that it should be the average strength of two standard 100×200 mm or 150×300 mm (or 100×100 mm) concrete cylinders that are representative of concrete taken from one batch of concrete. It further specified that to conform to the specified nominal minimum 28-day compressive strength requirements:

1. The average of all groups of three consecutive strength tests shall be equal to or greater than the specified strength.
2. No individual strength test shall be more than 3.5 MPa below the specified strength.

OPSS.MUNI-1350 specifies that QA testing on concrete shall be conducted as shown in Table 13.2.

Successful concrete projects do not just happen; they require sound management of the whole construction process from inception through to completion. An additional prerequisite for good control is that the right decisions are taken at the right time. This is only possible when there is good data available on which to base the decisions, which in turn requires relevant information concerning the quality, material, and progress of the concrete projects. A well-functioning system for selecting

Table 13.2 Quality assurance tests

Required test	Test method
Slump and slump flow of concrete	CSA-A23.2-5C
Air content	CSA-A23.2-4C
Compressive strength	CSA-A23.2-3C and CSA-A23.2-9C
Accelerating the cure of concrete cylinders and determining their compressive strength (accelerated cured)	CSA-A23.2-10C
Yield	CSA-A23.2-6C
Chloride ion penetrability test	ASTM C 1202
Linear shrinkage test	ASTM C 157M (Note 1)

Note 1: Drying shall commence after 7 days of wet curing

and processing information is a necessity for controlling a project. Since many decisions will be required, up-to-date information must be continuously available.

When a job is completed, the results should be reviewed. The review shall be in the form of comparison between the goals defined and the results achieved. Major interest may be focused on the degree to which the functional and technical aspects of the goals were achieved. Complete documentation of the analysis and goal comparison are of great importance.

Environmental Considerations

Environmental protection during concreting process is an important requirement. The primary environmental safety risks are related to the potential for fresh concrete and concrete wash water to escape into watercourses and storm water drains. Cement, which represents about one-eighth of the volume of typical residential grade concrete, mixes with water to form a very high pH solution that is highly toxic to fish and other aquatic life. Once concrete has cured, it is no longer a threat to the environment.

The following considerations are necessary:

- Protect the surrounding environment and do not discard unused concrete samples on the job site.
- Do not discard equipment wash water in an area which could adversely affect the environment.
- Establish a protocol for dispose of unused concrete and wash water with the site superintendent.

Any spill should be contained immediately and removed as quickly as possible. If concrete or concrete wash water is spilled, it is required to report the spill to the Federal Department of Environment and the appropriate provincial authority.

Bibliography

1. International Organization for Standardization (ISO) 9000 to 9004-1987, Quality Management and Quality Assurance Standards: Guide Lines for Selection and Use.
2. Phil B. Crosby, Quality is Free, McGraw Hill, New York, 1979.
3. ACI 121R-85 (Revised 2004), Quality Assurance Systems for Concrete Construction.
4. Road Research Laboratory, Ministry of Transport, Concrete Roads Design and Construction, published by Her Majesty's Stationary Office, printed by Lowe and Brydone Ltd, London, 1966.
5. ACI 304R- 89 (Revised 2000), Guide for Measuring, Mixing, Transporting and Placing Concrete.
6. D.C. Teychenne, the Use of Crushed Rock Aggregates in Concrete, Building Research Establishment Report. Garston, CRC, 1978. Also Concrete Made with Crushed Rock Aggregate, BRE Current Paper CP63/78. Garston, BRE, 1978.
7. ACI 221R-89 (Reapproved 2001), Guide for Use of Normal Weight Aggregate in Concrete.
8. ACI 311.4R-05, "Guide for Concrete Inspection," American Concrete Institute.
9. Ontario Provincial Standard Specification OPSS. MUNI 1350- November 2014, Material Specification for Concrete- Materials and Production.

Annexure

Revisions to BS Standards Used in This Book

Used BS standard	Revised standard
BS 5337-76 (Amended-82)	Superseded by BS 8007:1987-Superseded by BS EN 1992 3:2006
BS 882-92	Superseded by BS EN 12620:2002 + A1:2008
BS 12-96	BS EN 197-1:2000
BS 4550-78 (Amended 1984), Part-3	BS EN 196-7:1992
BS 1881, Part 103-1993	BS EN 12350-4-2009
BS 5328 (Parts 1, 2, 3, 4)	Replaced by BS 8500 and withdrawn by BS EN 206:2013+A1:2016
BS 812, Part 112-90	BS EN 1097-2:2010
BS 812, Part 113	BS EN 1097-8:2009
BS 3148-80	BS EN 1008:2002
BS 8110- Part l:1985	BS EN 1992-1-1:2004+A1:2014
BS 4449:2005+A3:2016	Current—not changed

Note: British standards replaced by European standards are denoted by BS EN

© Springer Nature Switzerland AG 2019
A. Surahyo, *Concrete Construction*,
https://doi.org/10.1007/978-3-030-10510-5

Conversion Factors

Length	
(Imperial to metric)	(Metric to imperial)
1 in. = 25.4 mm	1 mm = 0.03937 in.
1 ft. = 0.3048 m	1 m = 39.37 in.
1 yd = 0.9144 m	= 3.281 ft.
1 mile = 1.609 km	= 1.094 yd

Note: In nominal sizing 300 mm is taken as equivalent to 1 ft and 25 mm as equivalent to 1 in.

Area	
1 in.^{2} = 645.2 mm^2	1 mm^2 = 0.00155 in.^{2}
1 ft^2 = 0.0929 m^2	1 m^2 = 10.76 ft.^{2}
1 yd^2 = 0.8361 m^2	= 1.196 yd^2
1 acre = 0.4047 hectares	1 hectare = 2.471 acres
1 km^2 = 0.386 square mile	= 10,000 m^2
= 100 hectares	1 square mile = 2.5900 km^2
	= 640 acres

Volume	
1 in.^{3} = 16,390 mm^3	1 mm^3 = 0.00006102 in.^{3}
1 ft.^{3} = 0.02832 m^3	1 m^3 = 35.31 ft.^{3}
1 yd^3 = 0.7646 m^3	= 1.308 yd^3
1 gallon = 0.003785 m^3	

Force (SI to imperial to metric)	Force/unit length
1 N = 0.2248 Ibf = 0.1020 kgf	1 N/m = 0.06852 Ibf/ft. = 0.1019 kgf/m
1 KN = 0.1004 tonf = 102.0 kgf	1 KN/m = 68.5217 Ibf/ft. = 101.9716 kgf/m
	= 0.0306 tonf/ft.

© Springer Nature Switzerland AG 2019
A. Surahyo, *Concrete Construction*,
https://doi.org/10.1007/978-3-030-10510-5

Force/unit area

1 N/mm² (MPa) = 145.0 lbf/in.²	1 N/m² (Pa) = 0.02089 lbf/ft.²
= 10.20 kgf/cm²	= 0.102 kgf/m²

1 Gigapascal (GPa) = 1000 Megapascal (MPa) = 1,000,000 Kilopascal (KPa)

Force/unit volume

1 N/m³ = 0.006366 lbf/ft.³ = 0.102 kgf/m³

Moment

1 Nm = 8.851 lbf in. = 0.7376 lbf ft. = 0.1020 kgf/m

Mass

1 kg = 2.2046 lb	1 Mg (megagram) = 19.684 cwt
50 kg = 0.9842 cwt	1 tonne (Metric) = 0.9842 ton (Imperial)

Temperature

Fahrenheit (F) to Celsius	Celsius (C) to Fahrenheit
F = 1.8 C + 32	C = (F − 32) ÷ 1.8

Glossary

Abrasion damage The wearing away of a surface by rubbing and friction.

Alkalinity The condition of having hydroxyl (OH−) ions, containing alkaline substances. Alkalinity is thus a measure of the ability of water to neutralise acids.

Anode The positive electrode in a solution of electrolytes, which attracts anions (negative ions).

Autogenous volume Change in volume produced by continued hydration of cement exclusive of effects of external forces or change of water content or temperature.

Blemish Any superficial defect, which causes visible variation from a consistently smooth and uniformly coloured surface of hardened concrete.

Blistering The irregular raising of a thin layer, frequently 25–300 mm in diameter, at the surface of placed concrete during or soon after completion of the finishing operation. It is usually attributed to early closing of the surface and may be aggravated by cool temperatures.

Catalyst A substance that initiates chemical reaction and does not itself alter or enter into the reaction.

Cathode The negative electrode in a solution of electrolytes, which attracts cations (positive ions).

Cathodic protection It is a form of corrosion protection in which one metal is caused to corrode in preference to another, thereby protecting the latter from corrosion.

Cavitation damage Pitting of concrete caused by the collapse of vapour bubbles in flowing water, which form in areas of low pressure and collapse as they enter areas of higher pressure.

Cementitious material Material having cementing properties.

Chalking Formation of a loose powder that results due to disintegration of the surface of concrete or of applied coating.

Charging Feeding or loading materials into a mixer machine.

Cold joint A joint or discontinuity resulting from a delay in placement of sufficient time to preclude a union of the material in two successive lifts.

© Springer Nature Switzerland AG 2019
A. Surahyo, *Concrete Construction*,
https://doi.org/10.1007/978-3-030-10510-5

Corrosion Destruction of metal by chemical or electrochemical attack.

Craze cracks Fine random cracks or fissures in a surface of cement paste or concrete.

Curling Distortion of a straight or flat member into a curved, warped, or dished shape.

Deformation A change in dimension or shape of a member due to stress.

Delamination A separation along a plane parallel to a surface, like layers of a coating from each other.

Deterioration The disintegration or chemical decomposition of a material during test or service exposure.

Discolouration Change of colour from that which is normal or desired.

Disintegration Deterioration into small fragments and subsequently into particles.

Dusting The development of a powdered material at the surface of hardened concrete.

Early stiffening The early development of an abnormal reduction in the working characteristics of a cement paste mortar or concrete.

Efflorescence A deposit of salts, usually white, formed on a surface, the substance having emerged in solution from within concrete and subsequently been precipitated by evaporation.

Electrolyte An electrolyte is a substance that conducts electricity when in the molten state or in solution. The electricity is carried through the electrolyte by ions. The electric current enters and leaves the electrolyte through electrodes.

Entrained air Incorporating microscope air bubbles in mortar or concrete during mixing.

Erosion Progressive disintegration of a solid by the abrasive or cavitation action of gases, fluids, or solids in motion.

Expansive/expanding A cement, which produces a paste on mixing with water that tends to increase in volume, after setting, to a significantly greater degree than does Portland cement paste. It is used to compensate for volume decrease due to shrinkage.

Flexural strength The property of a material or a structural member that indicates its ability to resist failure in bending.

Fly ash The finely divided residue that results from the combustion of ground or powdered coal, also known as pulverised fuel ash (pfa).

Free water Mix water, excluding the water absorbed by the aggregates.

Hessian cloth A coarse fabric of jute, used as a water-retaining covering for curing concrete surfaces, also called burlap.

Honeycomb Voids left in concrete due to failure of the mortar to effectively fill the spaces among coarse aggregate particles.

Hydraulic cement A cement that sets and hardens by chemical interaction with water and is capable of doing so underwater.

Laitance A layer of weak and non-durable paste containing cement and fines from aggregates.

Latex A water emulsion of a high-molecular-weight polymer used especially in coating, adhesives, levelling compounds, and patching compounds.

Matrix In the case of mortar, the cement paste in which the fine aggregate particles are embedded; in the case of concrete, the mortar in which the coarse aggregate particles are embedded.

No-slump concrete Freshly mixed concrete that shows a slump of less than 6 mm (1/4 in.).

Pattern cracking Fine openings on concrete surface in the form of a pattern, resulting from a decrease in volume of the material near the surface, or increase in volume of the material below the surface, or both.

Peeling A process in which thin flakes of mortar are broken away from a concrete surface, such as by deterioration or by adherence of surface mortar to forms as formwork is removed.

pH It measures the hydrogen ion concentration of a solution. It is used for expressing the acidity or alkalinity of a solution. A substance with a pH of less than 7 is an acid, whereas one with a pH of greater than 7 is alkaline.

Pitting Development of relatively small cavities in a surface of concrete such as a pop-out.

Plastic cracking Cracking which occurs in the surface of fresh concrete soon after it is placed and while it is still plastic.

Pop-out The breaking away of small portions of a concrete surface that leaves a shallow depression.

Reshoring Fromwork, in which the original posting (vertical post) or shoring is removed and replaced in such a manner as to avoid deflection of the shored element.

Restraint Restriction of free movement of hardened concrete.

Scaling Local flaking or peeling away of the surface of hardened concrete.

Self-desiccation The removal of free water by chemical reaction within mortars and cement pastes so as to leave in sufficient water to cover the solid surfaces causing decrease in the required curing water.

Shrinkage cracking Cracking of a structure or member due to failure in tension caused by external or internal restraints.

Spalling A fragment, usually in the shape of a flake, detached from a larger mass by a blow, by the action of weather, by pressure, or by expansion within the large mass.

Swelling Volume increase caused by wetting or chemical changes or both.

Tooling The act of compacting and contouring a material in a joint.

Transit-mixed concrete Concrete, the mixing of which is wholly or partially accomplished in a truck mixer.

Index

© Springer Nature Switzerland AG 2019
A. Surahyo, *Concrete Construction*,
https://doi.org/10.1007/978-3-030-10510-5

Printed by Printforce, the Netherlands